POTENTIAL FOR INDUSTRIAL ENERGY-EFFICIENCY IMPROVEMENT IN THE LONG TERM

ECO-EFFICIENCY IN INDUSTRY AND SCIENCE

VOLUME 5

Potential for Industrial Energy-Efficiency Improvement in the Long Term

by

Jeroen de Beer
ECOFYS,
Utrecht, The Netherlands

SPRINGER-SCIENCE+BUSINESS MEDIA, B.V.

A C.I.P. Catalogue record for this book is available from the Library of Congress.

ISBN 978-90-481-5444-9 ISBN 978-94-017-2728-0 (eBook)
DOI 10.1007/978-94-017-2728-0

Printed on acid-free paper

CONTENTS

PREFACE ix

1. INTRODUCTION 1
   1.1 Impetus for improving the efficiency of energy use 1
   1.2 Energy services and energy-efficiency improvement 3
   1.3 Energy for industry 6
   1.4 Scope and objective of this book 9
   1.5 Outline of this book 11

2. TOWARDS A METHOD FOR ASSESSING LONG-TERM OPPORTUNITIES FOR ENERGY-EFFICIENCY IMPROVEMENT 13
   2.1 Introduction 13
   2.2 Requirements of a method for assessing future opportunities for the improvement in energy efficiency 14
   2.3 Existing methods for assessing future opportunities for improvement in energy efficiency 15
      2.3.1 Thermodynamical methods 16
      2.3.2 Technology exploration 17
      2.3.3 Trend extrapolation 23
      2.3.5 Macro-economic modelling 25
      2.3.6 Applicability of the methods 27
   2.4 Outline of a new method 30
   2.5 Process energy analysis 31
      2.5.1 Selection of the energy service 31
      2.5.2 Theoretically lowest energy consumption 31
      2.5.3 Process exergy analysis 32
   2.6 Identification of technologies 34
   2.7 Characterization of the selected technologies 34
      2.7.1 Data gathering and handling 35
      2.7.2 Description of the technology and performance 35
      2.7.3 Energy efficiency improvement potential 35
      2.7.4 Costs and benefits 36
      2.7.5 Technical change 37
   2.8 Discussion and conclusion 41

3. SECTORAL POTENTIALS FOR ENERGY-EFFICIENCY IMPROVEMENTS IN THE NETHERLANDS 43
   3.1 Introduction 43
   3.2 Definitions 45

3.2.1 Potentials for energy-efficiency improvement 45
3.2.2 Cost-effectiveness of energy-efficiency improvement 46
3.2.3 Pay-back period of energy-efficient technologies 46
3.3 The database icarus 49
3.4 Energy-efficiency improvement potentials in all sectors 50
3.5 A taxonomy of energy-efficient technologies 54
3.6 Sectoral potentials for energy-efficiency improvement 56
3.7 The effect of combined financial instruments 58
3.8 Discussion 63
3.9 Conclusions 65

4. LONG-TERM ENERGY-EFFICIENCY IMPROVEMENTS IN THE PAPER AND BOARD INDUSTRY 67
4.1 Introduction 67
4.2 Energy analysis of a paper mill 67
4.2.1 Selection of the energy service 69
4.2.2 Process description of conventional paper making 69
4.2.3 Main production parameters in paper making 70
4.2.4 Enthalpy and exergy analysis 70
4.3 Identification of technologies 76
4.4 Characterization of technologies 76
4.4.1 Dry-sheet-forming 76
4.4.2 Innovative pressing and drying techniques 80
4.4.3 Latent heat recovery systems 84
4.4.4 Comparison of the technologies 86
4.5 Discussion 89
4.6 Conclusions and recommendations 90

5. FUTURE TECHNOLOGIES FOR ENERGY-EFFICIENT IRON AND STEEL MAKING 93
5.1 Introduction 93
5.2 Past technological development of iron and steel production 94
5.2.1 History of iron making 94
5.2.2 History of steel making 96
5.2.3 The current situation 100
5.3 Energy service and theoretical specific energy consumption 102
5.3.1 Description of the energy service 102
5.3.2 Calculation of the theoretically lowest energy demand 102
5.3.3 Heating and melting of iron 107
5.3.4 Iron ore reduction in the blast furnace 107
5.3.5 Comparison with practical processes 107
5.4 Exergy analysis of an integrated steel plant 109
5.4.1 The reference plant 109

5.4.2 Results of the exergy analysis . . . 114
5.4.3 Conclusions . . . 117
5.5 Identification and selection of long-term energy efficient techniques . . . 118
5.5.1 Gathering of information . . . 119
5.5.2 Selection of energy-efficient techniques . . . 119
5.6 Characterization of long-term energy-efficient techniques . . . 120
5.6.1 Smelting reduction processes . . . 120
5.6.2 Near-net-shape casting . . . 139
5.6.3 Scrap-based process . . . 145
5.6.4 Steel making at lower temperatures . . . 151
5.6.5 Waste heat recovery at high temperatures . . . 153
5.6.6 Conclusions on the long-term energy-efficiency improvement . . . 158
5.7 Discussion . . . 161
5.8 Conclusions and recommendations . . . 163

6. FIXING ATMOSPHERIC NITROGEN WITH LESS ENERGY . . . 167
6.1 Introduction . . . 167
6.2 The production of nitrogen fertilizers over the past 100 years . . . 168
6.3 State-of-the-art production processes . . . 171
6.3.1 Ammonia synthesis . . . 171
6.3.2 Urea production . . . 174
6.3.3 Nitric acid production . . . 174
6.3.4 Ammonium nitrate production . . . 176
6.4 Selection of the energy service . . . 177
6.5 Theoretical specific energy consumption . . . 179
6.5.1 Theoretically minimum SEC . . . 179
6.5.2 Theoretically minimum SEC for the main production processes of nitrogen fertilizers . . . 181
6.5.3 Actual exergy requirements of processes to produce nitrogen fertilizers . . . 183
6.5.4 Biological nitrogen fixation . . . 183
6.6 Exergy analysis . . . 184
6.6.1 Exergy loss in the production of urea and ammonium nitrate . . . 184
6.6.2 Exergy analysis of ammonia production processes . . . 186
6.6.3 Exergy analysis of a nitric acid plant . . . 192
6.7 Options to improve the energy efficiency of nitrogen fixation . . . 193
6.7.1 Selection of energy efficient options . . . 193
6.7.2 Collection of information . . . 199
6.8 Characterization of options to improve the energy efficiency . . . 199
6.8.1 Ammonia syngas production . . . 200
6.8.2 Ammonia synthesis . . . 207
6.8.3 Nitric acid production . . . 209
6.8.4 Alternative routes for atmospheric nitrogen fixation . . . 214

6.8.5 Comparison of the options to improve the energy-efficiency of nitrogen fertilisation 217
6.9 Discussion 220
6.10 Conclusions and recommendations 223

7. CONCLUSIONS 225

REFERENCES 231

PREFACE

This book does not give a prediction of what the efficiency will be of the energy use of industrial processes in the future. However, it does give an exploration of limits to the efficiency of current processes and an indication of what might be achieved if new technologies can be developed.
At the Department of Science, Technology and Society of Utrecht University research had been done to the opportunities for improvement of the energy efficiency in the short term since the 1980's. This had resulted in a comprehensive database on energy efficient measures. This database and a possible application are described in Chapter 3 of this book. The use of the database induced new research themes around efficiency improvement, e.g. concerning barriers for implementation of measures. It was around 1993 that I did a preliminary study to the potential for efficiency improvement in the long term. Historical analysis had shown us that the short term potential stayed constant over the years. It seemed to be replenished by the introduction of new technologies. This lead to the question whether there are limits to the efficiency, taking into account both thermodynamic considerations and ideas on the development and dissemination of new technologies.
By that time, the risk of climate change by an enhanced greenhouse effect was already an item on the political agenda. Since improvement of the energy efficiency is generally seen as the most cost-effective way to reduce the emission of the greenhouse gas carbon dioxide, the interest in this research was not only of scientific but also of societal importance.
The results of my research are published in this book. In another form most of the texts were part of my Ph.D. thesis, which I defended successfully in November 1998.

I would like to express my special gratitude to Kornelis Blok of Utrecht University. He was at the basis of the research to energy efficiency improvement in general and this research in particular. During many discussions on draft versions he showed his large knowledge of the topic and his ability to look at problems from a different angle, often resulting in surprising solutions. His input was of great importance for the final result.

Wim Turkenburg has read draft versions and provided valuable comments for which I am grateful. He co-authored Chapter 2 of this book. I would also like to thank Ernst Worrell - co-author of three chapters in this book - for the years we worked together and the useful comments he gave on earlier versions.

I would like to thank all my colleagues at the Department of Science, Technology and Society for creating a friendly and creative working climate. Special thanks go to Esther Luiten and Evert Nieuwlaar who were closely involved in this research. I would also like to thank the students who contributed in one way or the other to this research.

The Netherlands Organization for Scientific Research is acknowledged for its financial support. Finally, I would like to thank Sheila McNab for her linguistic assistance.

CHAPTER 1

INTRODUCTION

1.1 Impetus for improving the efficiency of energy use

All the energy we use is ultimately to satisfy human needs. We do not burn natural gas to heat the room we live in just for the sake of heating but we do it to feel comfortable in that room. We use energy to produce the materials that we need to make consumer goods, e.g. we produce steel to make cars, paper on which to print this book, and fertilizers to grow more food so that we do not have to feel hungry. Energy is not a need in itself. Energy is one of the goods we use to fulfil our needs. The human needs that can be realized by the use of energy are called energy services.

On the one hand, energy services contribute to human well-being. On the other hand, the use of energy for energy services can result in economic, social and environmental problems. These problems are very closely related to the pursuit of the sustainable development of society. Sustainable development is defined as "a development that ensures the needs of present generations without compromising the ability of future generations to meet their own needs". This definition is given in the report "Our Common Future", published by the World Commission on Environment and Development in 1987 [WCED, 1987]. To achieve sustainable development, the United Nations have formulated an action programme called Agenda 21 [UNCED, 1992]. Recently, the United Nations Development Programme (UNDP) published a report in which the relations between the production and use of energy and sustainable development are thoroughly examined [Reddy *et al.*, 1997]. We will keep to the UNDP classification in our discussion of these relations and will indicate the reasons why -within the context of sustainable development - we must strive to improve the efficiency of our energy use.

<u>Energy and economy</u>
Reduction in energy consumption is not a new item on the political agenda. In the 1970s, two oil crises confronted the industrialized world with its economic dependency on oil from the Middle East. This forced scientists and policy makers to think about a future that could draw on alternative energy sources and reduce its energy consumption. Although the industrialized world has reduced its dependency on Middle Eastern oil since the 1970s, about half of the oil exported in 1995 was still from the Middle East [BP, 1996]. This situation is unlikely to change drastically since more than half of the world's proven oil reserves are in this region [BP, 1996] and the production costs there are low [Reddy *et al.*, 1997]. Further improvement in the efficiency of our energy use is an important option for reducing the West's dependency on oil from the Middle East.

In the 1970s it was projected that on the basis of the economic growth at that time the proven reserves of many natural materials, including oil and gas, would be exhausted within several decades [Meadows *et al.*, 1972]. Nowadays, exhaustion of fossil fuel reserves is no longer seen as such an urgent problem [Nakicenovic and Grübler, 1993; Reddy *et al.*, 1997]. New detection and exploration techniques for fossil fuels have increased the proven reserves considerably [BP, 1996], and new categories of deposits hold the promise of much larger resources than the estimates made one or two decades ago [Nakicenovic and Grübler, 1993]. It can be concluded that at the moment the depletion of reserves is not one of the main reasons for striving for a reduction in the energy demand. However, to allow future generations to meet their needs as we meet ours today, it is still necessary to follow a strategy in which reserves are used as efficiently as possible.
In developing countries an economic problem associated with energy use is that high expenditure on energy imports hampers investments in other goods that could improve welfare. Furthermore, these countries may have difficulties in financing the infrastructure required for energy supply. A reduction in the energy demand, e.g. by improving energy efficiency, can improve the economic situation in developing countries.

Energy and security
The UN recognizes the link between energy and security as a separate issue that might give rise to problems that will obstruct sustainable development. A major problem is that world peace is threatened by nuclear weapon proliferation, which is a risk associated with the use of nuclear energy. Improvement in the efficiency of the energy use is one possible strategy for reducing this risk.

Energy and social issues
Social issues that are linked with the use of energy are, e.g. poverty and gender disparity. These issues play an important role particularly in developing countries, where a large proportion of household expenditure is on energy. In addition, in these countries much time is spent on obtaining biomass energy sources, like wood and dung. This task is performed predominantly by woman and children. The money and time spent on obtaining energy cannot be spent on issues that would improve the living standards, e.g. education. Improving the efficiency of energy use can help to solve this problem.

Energy and environment
The link between energy and environment involves issues like health, acidification, and climate change. Health problems may arise from poor indoor and outdoor air quality.
Acidification of soils and waters is caused by the deposition of acid compounds. These compounds are formed by chemical transformation of compounds like sulphur dioxide and nitrogen oxides, which are emitted from fossil fuel combustion processes. In many industrialized countries several abatement measures have been implemented, reducing the emission of both sulphur dioxides and nitrogen oxides [Nilsson and Johansson, 1994]. In a number of developing countries these emissions are increasing to serious levels, making acidic deposition a potentially serious problem in some regions [Reddy *et al.*, 1997].

Of all the environmental problems linked with energy use, the risk of a climate change has attracted the most attention in the developed world during the past decade. The combustion of fossil fuels leads to the production of carbon dioxide. Carbon dioxide accumulates in the atmosphere, causing enhanced radiative forcing. This in turn leads to a change in the earth's energy balance. There are still important uncertainties about the relationships between the accumulation of greenhouse gases, the energy balance of the earth and the climate on earth. Nevertheless, the Intergovernmental Panel on Climate Change (IPCC) concluded that "..the balance of evidence suggests that there is a discernible human influence on global climate " [Houghton *et al.*, 1996]. It also stated that due to human behaviour "climate is expected to continue to change in the future"[Houghton *et al.*, 1996]. Under the umbrella of the United Nations, the Framework Convention on Climate Change (FCCC) was established in 1992. The ultimate objective of the FCCC is "stabilization of greenhouse gas concentrations in the atmosphere at a level that would prevent dangerous anthropogenic interference with the climate system. Such a level should be achieved within a time-frame sufficient to allow ecosystems to adapt naturally to climate change, to ensure that food production is not threatened and to enable economic development to proceed in a sustainable manner" [UNFCCC, 1992]. The FCCC has been ratified by a majority of the UN-member states, including the Netherlands. In 1997 targets and timetables were set at the third Conference-of-the-Parties (that ratified the FCCC) in Kyoto [UNFCCC, 1997]. The member states of the European Union, for instance, have jointly committed themselves to a reduction of 8% in the emission of the six main greenhouse gases in the period 2008-2012 compared to the 1990-emissions [UNFCCC, 1997]. For the Netherlands this is translated into a reduction of 6%. Improvement in energy efficiency is considered to be one of the main opportunities for attaining this objective.

There are several methods for tackling the energy-related problems that impede the sustainable development of our society. One of these methods is to improve the energy efficiency by developing and disseminating efficient end-use technologies. Other methods are, for instance, making more use of clean, renewable energy sources, and applying end-of-pipe-techniques for low-emission fossil-fuel use. This book will concentrate on the potential for improving the efficiency of energy use in order to realize energy services.

1.2 Energy services and energy-efficiency improvement

Figure 1.1 shows a general representation of an energy system. An energy system comprises a sequence of conversion and transportation operations[1] that bring energy in the right form to the right place. An energy system is designed for realizing energy services. We distinguish two sub-systems: the energy supply system and the energy end-use system. In the energy supply system primary energy is converted to final energy. There may be an intermediary form of energy which we refer to as secondary energy. Final energy is

[1] An energy system can also include operations for extracting and mining of energy.

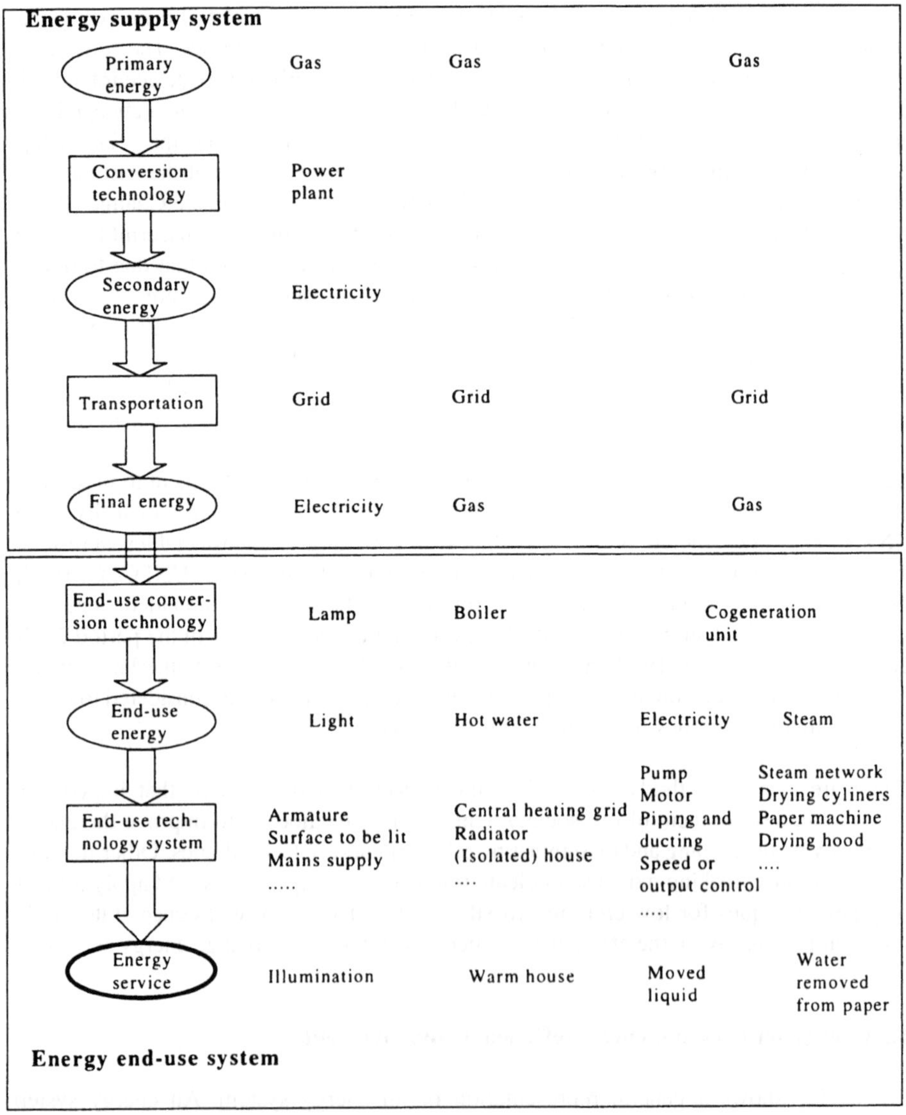

Figure 1.1: Representation of an energy system and some illustrative examples.

delivered to the end-user. In the energy end-use system final energy undergoes its final conversion into end-use energy. The end-use energy is delivered to an end-use technology system, which provides the energy service. Figure 1.1 gives some examples to illustrate the terms used.

Along the route from primary energy to end-use energy part of the initial energy is dissipated into the environment, mainly in the form of low-temperature heat. This energy can no longer be used to fulfil our needs and is therefore said to be lost. When the end-use energy has served its purpose, it is also said to be lost. However, according to the first law of thermodynamics, energy cannot be lost. The sum of all energy flows which we say that are lost equals the amount of initial energy. Nevertheless, the work that can be done with the lost energy is much smaller that the work that could be done with the initial energy. This notion is captured in the term 'quality' of energy carriers. The quality of energy carriers is a measure of the degree of ordering of energy. In almost all practical processes that convert one form of energy into another the degree of ordering of energy decreases[1]. The quality of the new form of energy is lower than that of the initial form, thus less work can be performed. At the end of the energy system the quality of the energy is so low that no more work can be performed. The concept of quality of energy is closely related to the concept of exergy. Exergy expresses the amount of work that can be obtained from an energy carrier. In Chapter 2 of this book we will give an exact definition of exergy and will pay more attention to the merits and drawbacks of using the concept of exergy. Throughout this book, exergy analysis will be used as a tool to evaluate the potential for energy-efficiency improvement.

This book deals with energy-efficiency improvement, i.e. reducing the consumption of energy per unit activity without affecting the level of these activities. The demand for energy can also be reduced by changing the human need underlying a given energy service. However, this option will not be considered in this book.

The energy efficiency of energy conversion processes is generally defined in terms of the ratio of useful energy output to energy input [Worrell, 1994]. This definition is applicable to all energy conversion processes. However, since human needs cannot be expressed in terms of energy, this definition is less useful for expressing the efficiency of energy use of an energy service as delivered.
In this book we will use the specific energy consumption (SEC) as indicator for the efficiency of the use of energy for an energy service. The SEC is the amount of energy required to realize the human activity associated with that energy service. The amount of energy can be expressed either in terms of primary energy carriers, or in terms of intermediary energy forms. Indicators for the human activity are, for instance, cubic metres of space heated, lumen per square metre, or tonnes of steel produced[2].

---

[1] In thermodynamics entropy is used as a measure of (the inverse of) the degree of ordering. The second law of thermodynamics states that the entropy of any natural process performed in an isolated system increases (degree of ordering decreases).

[2] It should be noted that one energy service can serve another energy service. For instance, steel is produced to make cars, which are turn required to fulfil our need for transportation. We will deal with this hierarchy of energy services and its consequences in chapter 2 of this book.

The lower the SEC of a plant the more efficiently that plant operates. Improvement in the energy efficiency can be translated into a reduction of the SEC. To determine the potential for energy-efficiency improvement the SEC of a contemporary plant should be compared to the SEC of a future plant. Both the contemporary plant and the future plant should be well-defined.
In this book, the currently most efficient plant is used as the contemporary plant in the assessment of the long-term potential for energy-efficiency improvement. The currently most efficient plant is defined as the plant with the lowest SEC that has been achieved in a complete plant that is currently in operation. For the future plant, the plant used is one in which a number of well-defined energy-efficient technologies have been implemented. he potential for energy-efficiency improvement can therefore be defined as:

$$\frac{\textit{SEC of the currently most efficient plant} - \textit{SEC of the future plant}}{\textit{SEC of the currently most efficient plant}} \times 100\%$$

Improvement in the energy efficiency can be attempted in all conversion and end-use processes in the energy system. Energy-efficiency improvement used to be predominantly aimed at the energy supply system, for instance, by making power plants more efficient. It has been recognized for many years that improvement in the energy efficiency of the end-use energy system is a very promising option for reducing the energy demand. Several sources state that increasing the efficiency of performing energy services is one of the greatest and most cost-effective opportunities for reducing the energy demand and for the pursuit of sustainable energy development, see e.g. ([Lovins and Lovins, 1991; Nakicenovic, 1995; Watson *et al.*, 1996; Reddy *et al.*, 1997; Worrell *et al.*, 1997a]). Nevertheless, a major part of the governmental energy R&D-budget of both Europe and USA is used to develop energy supply options [IEA/OECD, 1997; PCAST, 1997].

In this book we focus on options for improving the energy efficiency of the end-use energy system. Improvement of the efficiency of the energy supply system will be dealt with only when it is directly related to improving the efficiency of the end-use energy system. Moreover, in this book we will analyse promising research and development options for energy-efficiency improvement.

1.3 Energy for industry

Worldwide, the manufacturing industry is the largest energy-consuming economic sector. In 1990, the manufacturing industry accounted for 44% (136 EJ) of the global energy use [WEC, 1995]. This is shown in Figure 1.2. Although there are many sectors in the manufacturing industry, only five sectors are responsible for 45% of the industrial energy consumption: iron and steel making, chemical industries, petroleum refining, pulp and paper production, and cement production.

Between 1971 and 1995 the global industrial energy consumption grew at a rate of on average 1.9% per year, which is slightly less than the growth of the world energy consumption of 2.3% [Worrell *et al.*, 1997a]. The contribution by and the growth of the industrial energy use are not the same in all world regions. Figure 1.3 shows that the industrialized countries account for 45% of the 1995-industrial energy use, whereas developing countries and economies in transition used 32% and 23% respectively [Worrell *et al.*, 1997a]. The growth has been largest in developing countries, 5.5% per year between 1971 and 1992. China is responsible for half of the 1992 energy use in these countries. Whereas the contribution of heavy industry to the total energy consumption is growing in some rapidly-developing Asian countries with low labour costs, this contribution is declining in industrialized countries. This can be explained partly by a growing energy demand for transportation and a shift to less energy-intensive manufacturing industry in industrialized countries.

It is expected that in developing countries the energy requirement of the five largest energy-consuming sectors (see Figure 1.2) will continue to grow by about 3 to 5% per year up to the year 2020, if no measures are taken to encourage energy-efficiency improvement [WEC, 1995]. In industrialized countries and economies in transition the expected growth is much smaller. These facts stress the importance of aiming at improving the energy efficiency in developing countries.

As already indicated, the specific energy consumption for producing materials can be expressed as energy consumption per tonne of product made. It should always be mentioned which form of energy is used, for instance was it primary or secondary energy

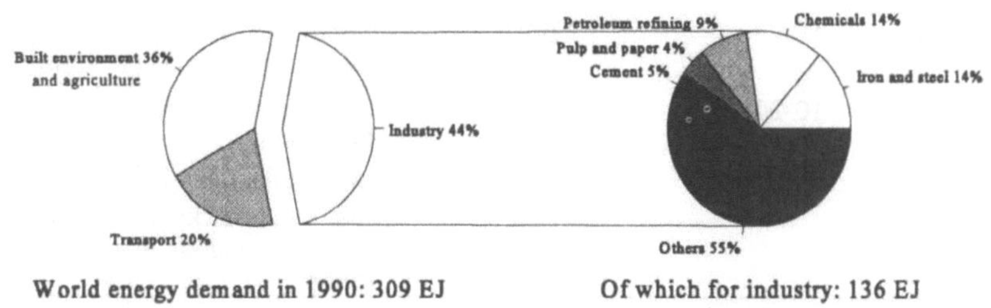

Figure 1.2: World energy demand and world industrial energy demand in 1990 [WEC, 1995].

carriers. In Table 1.1 the primary SEC for several products of the largest energy-consuming industrial sectors is given. The ranges are due to differences in raw materials, product range, processes applied, and energy efficiencies of the processes.

Table 1.1: World average primary SEC and 1990 production for several products from the five largest energy-consuming industrial sectors. Based on [WEC, 1995], unless stated otherwise.

	SEC GJ/tonne	1990 production (million tonnes)	Remarks
Steel	20-40[1]	770	Large differences in the efficiency of plants.
Petroleum products	3-6	1800	Based on energy consumption of a complex refinery configuration
Paper and board	22-30	45[2]	Pulp production accounts for about 10-15 GJ/tonne. Production of board requires about 10 GJ/tonne and of sanitary paper 17 GJ/tonne.
Cement	4-6	1160	The lower limit is for cement made from blast furnace slag, whereas the higher limit is for Portland cement.
Chemicals			Depends strongly on the product. All values exclude energy carriers used as feedstock.
Ammonia	6-20	95	Most efficient plant uses about 6 GJ/tonne and 22 GJ of natural gas per tonne as feedstock.
Ethylene	18-28	50	Feedstock energy use is about 43 GJ of crude oil per tonne.
Methanol	16-20	22	Feedstock energy use about 21 GJ/tonne.
Chlorine	41-45	27	Most efficient process is the membrane process, which uses about 10 GJ of electricity per tonne.

[1] Upper limit based on [Worrell *et al.*, 1997b];
[2] Data for 1992, based on [FAO, 1994]

This book focuses on possible ways for improving energy efficiency in the production of bulk materials in the manufacturing industry. We will analyse the production of three industrial products that are produced in sectors belonging to those with the highest energy demand, namely paper and board, iron and steel, and nitrogen fertilizers.

1.4 Scope and objective of this book

The potential for energy-efficiency improvement by implementing existing technologies has been studied extensively, see e.g. [ETSU, 1984; Farugui *et al.*, 1990; Ficket *et al.*, 1990; Melman *et al.*, 1990; Jackson, 1991; Koomey *et al.*, 1991; Mills *et al.*, 1991; Blok *et al.*, 1993; Beer *et al.*, 1994; ETSU, 1994a; Worrell, 1994; CADDET, 1997]. All these studies have a time projection of 10 to 25 years, and are focussed on improving the efficiency of the equipment already in place.

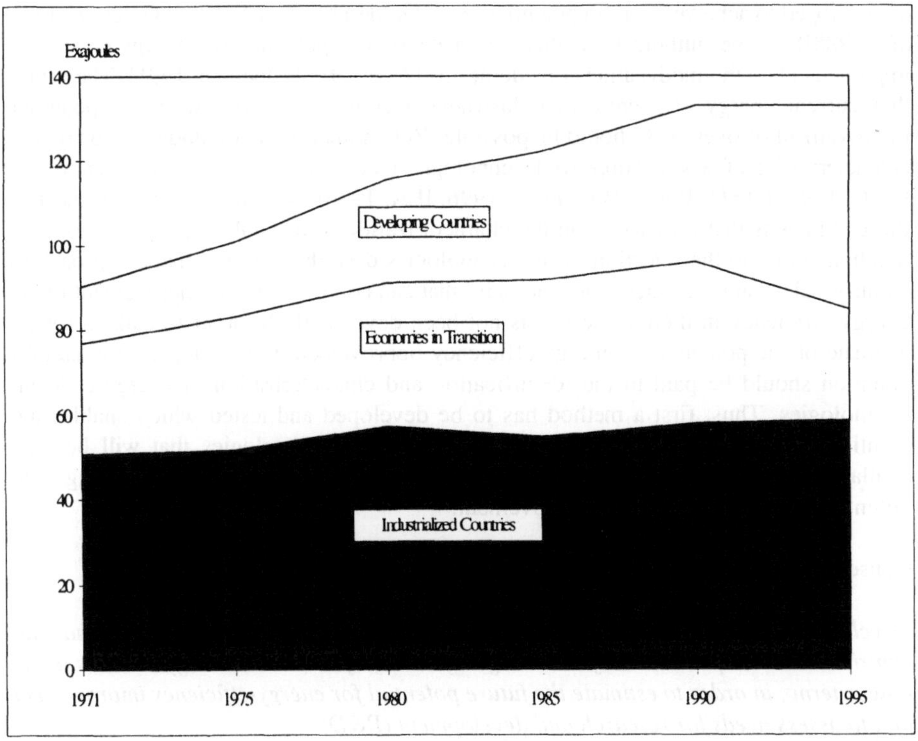

Figure 1.3: Global primary energy use for industry by region, 1971-1992 [Worrell, 1997b].

In a study by the International Energy Agency, which focussed on greenhouse gases, it is stated that the development and deployment of new and improved technologies will be essential if atmospheric concentrations of greenhouse gases are to be reduced significantly [IEA, 1994]. A study by the United Nations Development Program concludes that the opportunities for improving energy efficiency are far greater with new investments than with retro-fitting existing equipment [Reddy *et al.*, 1997]. Both sources indicate that the application of only existing technologies is not sufficient to attain the sustainable development of society. New technologies will also have to be developed and disseminated.

Currently few people have much insight into the potential for energy-efficiency improvement that can be attained in the longer term. Several studies have been performed to assess this potential. Most analyses assess this potential by estimating the development of the energy intensity, i.e. the ratio of the energy demand to economic activity, by assuming that it will be reduced by 0-2% a year. An overview of such studies is given by [Grubb *et al.*, 1993]. Only a few studies point out which technologies should be applied or developed to achieve a certain potential (e.g. [Goldemberg *et al.*, 1985; Goldemberg *et al.*, 1988]). Some authors base their estimate of the potential for energy-efficiency improvement on thermodynamic considerations [Ayres, 1988; Jochem, 1991]. They state that current energy efficiency in industrialized countries is so low that a potential improvement of over 80% should be possible. Yet another group of studies focus on the characterization of new and improved technologies that can improve energy efficiency (e.g. [CEC, 1992; ETSU, 1994b; Blok *et al.*, 1996; IEA, 1996]). A general feature of most of these studies is that they focus on the characterization of technologies, but do not give much attention to the selection of the technologies described. A systematic approach to identify, select and characterize technologies that can contribute to an improvement in the energy efficiency in the long term has not been developed[1]. In order to make a robust estimate of the potential for energy-efficiency improvement in the longer term, careful attention should be paid to the identification and characterization of energy-efficient technologies. Thus, first a method has to be developed and tested which enables the identification and characterization of energy-efficient technologies that will become available in the longer term. Only then will it be possible to estimate the long-term potential for energy-efficiency improvement.

Consequently, the central objective of this book can be formulated as follows:

Develop and test a method to identify and characterize promising technologies that can contribute to an improvement of the energy efficiency of industrial energy services in the longer term , in order to estimate the future potential for energy-efficiency improvement and to assess needs for research and development (R&D).

---

[1]See Chapter 2 of this book for an extensive overview of existing methods for assessing the long-term potential for energy-efficiency improvement.

This book does not give much attention to the integration of the energy flows between two or more industrial plants, or an industrial plant and the built environment. An example of such integration is the use of waste heat of one industrial plant to satisfy the heat demand of another plant. Integration between plants can also be considered at the level of material flows, e.g., using carbon dioxide from an ammonia plant as raw material in an urea plant. We realize that this option can result in additional energy saving. However, a thorough analysis that seeks optimal integration is very complex and beyond the scope of this book. Instead, we will indicate where energy or material flows that can be used in other processes become available. In most cases we will suggest some possible applications, and sometimes we will consider an option in more detail (e.g., electricity production by using as a fuel the process gas of smelting reduction processes for iron making). It should be noted, however, that our focus is primarily on the improvement of energy efficiency for realizing one specific energy service.

The question might be raised which level of energy-efficiency improvement should be strived at. What is enough to achieve sustainable development?
In 1994 a group of scientist and politicians, now know as the "Factor 10 Club" came with a declaration that stated, among others, that current resource productivity must be increased by an average of a factor of 10 during the next 30 to 50 years [Factor 10 Club, 1994]. This declaration was based on concern for the future of our economies as well as their ecological health. The Factor 10 Club challenges the notion that economic activity is tied tightly to consumptive use of materials. They believe that human welfare can be guaranteed by increasing the productivity of materials and energy use. However market imperfections, notably subsidies on activities that result in an environmental burden, currently hamper the exploitation of these opportunities. Furthermore, the time horizons that are currently used are too short.
An increase of material and energy productivity by a factor of 10 requires a long breath. Governments should provide the right incentives for change, e.g. by taxing resources instead of labour and cutting subsidies on environmentally harmful activities, according to the Factor 10 Club. It involves also a shift in thinking to services instead of products. What is really needed: the newspaper or the information.
In this book the goal of the Factor 10 Club will serve as guidance. We will evaluate to what degree energy-efficiency improvement can contribute to an increase of the energy productivity by a factor 10.

1.5 Outline of this book

In Chapter 2 we propose a method for achieving the central objective. We will define the demands we make on the method, for instance, with regard to the way information is gathered and technologies characterized. Then, we will review the existing methods for assessing future energy-efficiency improvement. For each of these methods, we will describe the approach, give some examples of studies that used that method, and discuss the applicability with regard to our requirements. On the basis of this evaluation, we develop a method that contains elements from existing methods and satisfies all our requirements. The method we propose consists of the following three parts:

- The first part is concerned with identifying the theoretical potential for improvement of the energy-efficiency compared to the currently prevailing process for realizing the energy service.
- The second part is dedicated to the identification of technologies that have the potential to reduce the amount of energy required to realize the energy service.
- The third part involves the characterization of the technologies so that the technologies can be compared with regard to their future impact on the energy service and so that R&D-needs can be assessed.

To be able to evaluate the potential for energy-efficiency improvement in the long term in terms of magnitude and attainability, it is useful to obtain insight into the potential that can be attained in the short term, and the policy instruments that can be applied to stimulate implementation of energy-efficient measures. Therefore, in Chapter 3, a study is presented of the potential for energy-efficiency improvement in the Netherlands for the period 1990-2000. This study uses one of the methods than will be discussed in Chapter 2, namely a technical-economic method using a bottom-up approach. This implies that data on saving potential and costs are collected at the level of energy services in processes. These data are cumulated to potentials on sectoral and national level. All data are collected in a database, which is called ICARUS. Potentials for energy-efficiency improvement in six sectors, namely heavy and light industries, households, services, transport and agriculture, are determined for the period 1990-2000. Finally, the possible effect of a combination of an energy tax and an investment grant is investigated.

Chapters 4 to 6 involve evaluations of future energy-efficient technologies used to make paper and board, to make iron and steel, and to fix atmospheric nitrogen. We will apply the method that is described in Chapter 2. The energy services are selected in such a way that the scope of the new energy-efficient techniques broadens from producing paper, via steel making to nitrogen fixation. The three industrial energy services are: the making of a flat sheet of paper, with specific properties, from an intermediate material (pulp or waste paper); making steel with certain well-described properties; and fixing atmospheric nitrogen as fertilizers. The analysis of paper production focuses on new processes for drying paper. The analysis of iron and steel production also takes recycling of steel scrap into account. Finally, the analysis of nitrogen fixation involves not only more energy-efficient chemical nitrogen fixation, but also biological nitrogen fixation. By using different scopes for the selection of the energy service we hope to be able to show the strengths and the weaknesses of the method.

The results are summarized at the end of this book. There, the method is also evaluated on its merits and shortcomings and a general conclusion is drawn regarding the potential for energy-efficiency improvement which can be achieved in the long term.

CHAPTER 2

TOWARDS A METHOD FOR ASSESSING LONG-TERM OPPORTUNITIES FOR ENERGY-EFFICIENCY IMPROVEMENT[1]

2.1 Introduction

There are many reasons for striving for an improvement of energy efficiency. Recently, the most important reason has been the possibility of reducing the emissions of the greenhouse gas carbon dioxide.
The potential for energy-efficiency improvement by retro-fitting existing equipment and implementing existing technologies has been studied extensively during the past decades. Numerous analyses have indicated that energy-efficiency improvements of about 30-50% are technically feasible over a period of one to two decades, see e.g. [ETSU, 1984; IEA, 1987; Ficket *et al.*, 1990; Jackson, 1991; Rosenfeld *et al.*, 1993; Beer *et al.*, 1994; Nilsson *et al.*, 1996; CADDET, 1997]. Currently few people have insight into the potential for energy-efficiency improvement that can be attained in the longer term by implementing new technologies. Nevertheless, several sources have indicated that the mere application of existing technologies is not sufficient to attain a sustainable development of society. In a study by the International Energy Agency, it is stated that the development and deployment of new and improved technologies will be essential if emissions of greenhouse gases are to be reduced significantly [IEA, 1994a]. A study by the UNDP concludes that the opportunities for improving energy efficiency are far greater with new investments than they are by simply retro-fitting existing equipment [Reddy *et al.*, 1997].

In this chapter we develop a method that can be used to identify and characterize new technologies that can contribute to improved energy efficiency, and that can investigate the potential for energy-efficiency improvement in the long term and to assess the associated needs for research and development (R&D). The method is developed so that it can be applied to a specific energy service. We define an energy service as a product of human activity obtained by the use of energy meant to satisfy a human need[2]. Improvement of the efficiency of energy consumption of an energy service means that the use of energy per unit product of human activity, expressed in physical terms, is decreased without

[1] Co-authors: W.C. Turkenburg and K. Blok, Utrecht University, Department of Science, Technology and Society.

[2] In economics, service is defined as *a product of human activity (e.g. transport, research) meant to satisfy a human need but not constituting an item of goods* [Webster, 1995]. In this book the term service is used in a broader sense, namely so that an energy service can also constitute an item of goods.

substantially affecting the level of this activity. The method is developed in the first place for application to industrial energy services.

We start by defining the requirements that have to be fulfilled by the method we intend to use in this study. Next, these requirements are used to evaluate various methods that have been used to investigate the potential for energy-efficiency improvement. In Section 3 the main existing methods are described. Also the applicability of these methods to our requirements is discussed. On the basis of our evaluation we propose a new method, an outline of which is presented in Section 4. Details on the method are described in the following sections. Section 5 deals with the analysis of the energy efficiency of energy services. Particular attention is paid to exergy analysis as applied in our method. In Section 6 another part of our method is described: the identification of potential energy-efficient technologies. Also we discuss how data on these technologies are gathered and handled. In Section 7 we discuss the parameters that characterize the technologies. Finally, in Section 8, we evaluate whether our method can satisfy our requirements.
The method is applied to three industrial sectors, described in three chapters [Beer and Blok, 1998; Beer *et al.*, 1998b; Beer *et al.*, 1998a], which are included in this book. An evaluation of the method, based on these applications, is presented in the Chapter 7 of this book.

2.2 Requirements of a method for assessing future opportunities for the improvement in energy efficiency

We are interested in the potential of technologies that can be applied in the longer term to improve the energy efficiency of industrial processes. The method to be applied to investigate this potential should fulfil the following requirements:

1. The method should start at the level of energy service. The development and application of energy-efficient technologies and energy carriers does not take place for its own merits but it has to satisfy specific human needs, thus to deliver a service (called energy service).
2. The method should allow an estimate of the potential for improving energy efficiency in the longer term. Consequently, the method should focus on technologies and options that are at present not commercially available.
3. The method should pay careful attention to the selection of technologies to realize an energy service in the best possible way. Not only should technologies be identified and evaluated which are being developed actively, new concepts should also be considered which have not yet been developed and which offer the opportunity to bring about a considerable improvement in the energy-efficiency. This provides us with an opportunity to select research areas in which investment in R&D could make a difference in the efficient use of energy in the longer term.
4. The characterization of the technologies should include a well-founded estimate of the potential efficiency of the energy consumption, diversified into the energy

carriers involved. It should also include an indication of the costs of applying these technologies when available on the market relative to the costs of the most efficient technology that is currently available. The characterization should also include a discussion of the performance of the technology with regard to the main process parameters, like quality of the product, production rate, and flexibility of the production. This is necessary to allow an assessment of possible barriers to the implementation of the technology and to identify research questions relating to R&D policy development.

5. The method should allow an assessment of the impact on the energy efficiency of applying two or more technologies that can perform the same energy service. It should be possible to identify whether this impact is smaller than, equal to or greater than the impact of the individual technologies. In the first case one can speak of synergy of the technologies, and in the latter of competition.
6. The method should make it possible to get an indication of the chance of successful development and the time required to allow market introduction of each technology. Information on these aspects is required to be able to estimate how long it will take before the potential improvement in energy efficiency can be attained, and to assess opportunities for actions to attain this potential sooner.
7. The basis for the estimate of the potential energy-efficiency improvement should be clear and the calculation of the improvement should be reproducible. Consequently, all sources and assumptions should be well-documented; it should allow verification by third parties. This requirement might seem trivial. However, not all publications on the subject of the potential for energy-efficiency improvement satisfy this requirement.

2.3 Existing methods for assessing future opportunities for improvement in energy efficiency

On the basis of the requirements formulated in Section 2.2, we will now review existing methods to assess the potential for energy-efficiency improvement. We categorized these methods into four groups:

1. In *thermodynamical methods* the ultimate potential for energy-efficiency improvement is estimated by comparing the current energy demand for delivering a specific energy service with the thermodynamical minimum demand.
2. The potential performance of new and existing technologies is assessed by *technology exploration*. In our case this results in an estimate of the potential for energy-efficiency improvement. Studies that use technology exploration can be grouped according to a set of data which is collected in order to identify the technologies and assess the potential improvement. We distinguish three groups of studies:
    a. Technical studies in which only the parameters that allow an assessment of the potential of energy-efficiency improvement are investigated.

b. Technical-economic studies in which not only technical but also economic parameters are investigated to allow economic evaluations of the potential for energy-efficiency improvement. Studies using this method are generally referred to as studies following a bottom-up approach [Krause, 1993].

c. Technology characterization studies in which data is collected on a set of the parameters that allow an assessment of the potential of the investigated technologies for improving energy efficiency. The parameter choice in these studies depends on the objective of the study. For instance, when the objective is to set priorities for R&D on energy-efficient technologies, data on e.g. the stage of development, R&D-needs, and R&D-environment are required.

3. *Trend extrapolation* is used to estimate the future potential for energy-efficiency improvement on the basis of an extrapolation of the historical trend in the development of the energy efficiency by which a specific service is delivered.

4. *Macro-economic modelling* is used to evaluate the potential future energy demand using parameters to describe the change in energy demand when economic activity changes. From these parameters information can sometimes be obtained about the (assumed) potential for energy-efficiency improvement. Studies using this method are generally referred to as studies following a top-down approach [Krause, 1993].

We will start the description of each method by elucidating the approach and the purpose it can serve. Then, we will give examples of studies in which a specific method has been used. We will focus on a few well-known studies. Finally, we discuss the merits and disadvantages of the method as far as our requirements are concerned.

2.3.1 THERMODYNAMICAL METHODS

The potential for energy-efficiency improvement can be defined with reference to the thermodynamical minimum energy demand for a given energy service. Several thermodynamical properties can be used to analyse and evaluate the amount of energy required to realize an energy service, e.g. *heating value, Gibbs free energy*, and *exergy*. The heating value of a compound is (in the case of a fuel) the difference between enthalpy between a compound (e.g. natural gas) and that of its combustion products under standard conditions. The heating value is frequently used in process energy analysis, because its value is easy to determine and because it indicates the value of a fuel for the application for which is mainly used, namely combustion. The use of the heating value to evaluate the full potential of an energy carrier is limited, since it does not take into account the quality of the energy carrier. Whereas the heating value includes only the change in enthalpy, the Gibbs free energy also includes the entropy. The Gibbs free energy of a compound is equivalent to the maximum amount of work that can be obtained when the compound is converted to its chemical elements. Exergy is comparable to the Gibbs free energy, the

difference being that exergy it is calculated with the most stable compounds in the environment as reference. We will deal in more detail with exergy analysis in Section 2.5. Thermodynamical methods can be used for process energy analysis. Overviews of methods are given by e.g. [Nieuwlaar, 1988; Kaiser, 1993]. *Exergy analysis* is used to locate operations and processes that use energy in an inefficient way and have large exergy losses. Two representative examples of publications in which the method of exergy analysis is developed and applied are [Kotas, 1985; Szargut *et al.*, 1988]. Some other examples of exergy analysis of processes are : a pulp and paper mill and a steel plant [Wall, 1988]; the chlorine-alkali industry[Morris, 1992]; production of synthesis gas from natural gas [Hinderink *et al.*, 1996]; ammonia production [Radgen, 1997]; and hydrogen production [Rosen, 1996]. Another thermodynamical method is *pinch analysis* that can be used to optimize heat exchanger networks and minimize heating and cooling demands. Since its introduction by Linhoff in the early 1980s [Linhoff, 1982], pinch analysis has been used to optimize the energy use of many processes. Some examples are: crude distillation units in a refinery [Sunden, 1988]; a chemical plant [Rossiter *et al.*, 1991]; and ammonia production [Radgen, 1996].
For some countries the thermodynamical potential for improving the energy efficiency has been determined at a national level. For instance, Ayres [Ayres, 1988] states that the overall efficiency of converting primary energy to useful energy in the US economy is only 2.5%. Wall performed an exergy analysis for the Swedish economy [Wall, 1987]. He found that of the 2539 PJ of exergy that went into the Swedish society in 1980, only 500 PJ was used by individuals for activities like space heating, and the consumption of food and materials.

Thermodynamical methods result in the maximum potential for energy-efficiency improvement of a specific energy service. In addition, they can give indications for those parts of the processes used to deliver a service which show the greatest potential for energy-efficiency improvement. Limitations of these methods are that they do not indicate which technology should be developed or used to achieve this potential and they do not indicate whether the potential can ever be realized. Pinch analysis is an exception, but its application is limited to heat exchangers and the optimal use of heating and cooling machines. It is not possible to assess specific R&D-needs and develop an effective policy to stimulate energy-efficiency improvement in the long term on the basis of the results of thermodynamical studies.

2.3.2 TECHNOLOGY EXPLORATION

In general, technology exploration refers to the investigation of the potential development of a specific technology, e.g. in terms of the efficiency of energy use. Often it means collecting and evaluating data on existing and new technologies, estimating the potential development of these technologies and assessing the potential impact of these developments on certain parameters, like energy efficiency. Evaluation of the effect of policy instruments on the development of the technologies can also be part of technology exploration methods.

Data on technologies are gathered using several approaches, e.g. by searching the professional and scientific literature, by consulting experts, or by using inventories of data on technologies. Searching the professional and scientific literature can nowadays be done using search engines for literature databases.

Experts can be consulted in various ways. For instance, experts can be asked to answer a list of questions. This may be take the form of a written questionnaire, but the investigator can also meet the experts. Another possibility is to ask for their views on future developments in a tight framework. Experts can also be interviewed personally, or asked to respond to views of other experts. The experts can be from all kinds of organizations, e.g. academia, industry, government, or NGOs.

Existing inventories on energy-efficient technologies often contain information on energy consumption, costs, and parameters like environmental emissions and the degree of implementation. Furthermore, they often give a description of the technology involved. These technology inventories are often based on the result of a governmental subsidy programme. Generally, the primary objective of these inventories is to disseminate information on the potential for energy-efficiency improvement to a larger audience. An example is the CADDET-database of the International Energy Agency [CADDET, 1997]; it includes worldwide data about demonstration projects on energy-efficiency improvement. Another example is the IDEE-database of the Dutch programme for industrial energy-efficiency improvement [Novem, 1998]. There are also technology inventories that contain other types of data that can be relevant for assessing the potential for energy-efficiency improvement. An example is Greentie, a database of the IEA that includes information on suppliers of technologies, services, research, data and literature, pertinent to the mitigation of greenhouse gases [Greentie, 1997]. Another example is MAESTRO, which is a database on information systems about environmentally sound technologies, developed by the United Nations Environment Programme [IETC, 1997].

As already indicated, we distinguish three groups of studies that use technology exploration as a method to investigate the potential development of technologies but each group gathers information via different types of studies: technical studies, technical-economic studies and technology characterization studies.

Technical studies

The approach used in technical studies is to collect data on energy-efficiency technologies that can be applied in a process and to cumulate the separate improvements to a total potential. The data collection is at the sectoral level, and is broken down into processes and operations, like space heating, pumping, or making steel. Since only the technical possibilities for improving the energy efficiency are considered, the results are referred to as the technical potential for energy-efficiency improvement.

Technical studies on the potential for energy-efficiency improvement have been carried out since the 1970s. The results have often been incorporated in scenario studies to assess potential future energy consumption. One of the first studies that made an inventory of technical possibilities to improve the energy efficiency and roused international interest was published by Lovins in 1977 [Lovins, 1977]. A few years before a similar study was performed in the Netherlands [Over and Sjoerdsma, 1974]. The study concluded that in the Netherlands the technical potential to improve the energy efficiency within the period 1972-1985 was 14%, or 1.2% a year. In 1979 the same approach was followed in a study by Leach *et al.* They developed a low energy strategy for the United Kingdom for the period 1976-2025 [Leach *et al.*, 1979]. Although in that study costs were not explicitly taken into account, the only energy-efficiency improvement measures which were adopted were those which representatives from industry considered to be economically and practically feasible. The result, which indicated an energy-efficiency improvement potential of 40-45% over the period 1976-2025 (about 1% per year), should therefore - strictly speaking - not be regarded as a technical potential. In 1980 Krause *et al.* published a comparable study for West Germany [Krause *et al.*, 1980]. The emphasis in this study was on renewable energy sources rather than on the identification of energy-efficient measures. At that time two studies were published in the Netherlands. Potma developed his so-called 'Forgotten Scenario' in 1977; it showed how the level of welfare could be maintained without production growth provided investments were made in energy conservation and renewable energy [Potma, 1977; Potma, 1979]. According to the 'Forgotten Scenario' the potential reduction in the energy demand in the Netherlands is 39% in 2000 and 60% in 2025 compared to 1975. The Foundation for Nature and Environment (SNM) made an inventory of energy-efficiency improvements and renewable energy options [Blok, 1984]. They concluded that a renewable energy system that includes most options can in the long term (2050) have an energy demand that is about 60% lower than the demand in 1980. The remaining demand can be supplied almost completely by renewable energy. Many studies followed, the studies by Goldemberg *et al.* receiving particular attention [Goldemberg *et al.*, 1985; Goldemberg *et al.*, 1988], because of the connection with the report "Our Common Future" published by the World Commission on Energy and Environment [WCED, 1987]. Another example is a study published by the International Energy Agency in 1987; it analyses past developments in energy intensity and future prospects for energy-efficiency gains [IEA, 1987]. Finally, a study commissioned by the Dutch Ministry of Economic Affairs and performed by the Netherlands Organization of Applied Scientific Research TNO in 1990 should be mentioned [Melman *et al.*, 1990]. This study concluded that the technical potential for energy-efficiency improvement in the Netherlands in the period 1986-2015 is 41%, or 1.7% per year.

Most technical studies focus on existing processes and technologies. As a result, it is usually only existing energy-efficient technologies which are considered. An advantage of the method is that the combined impact on energy consumption of applying two or more technologies can easily be assessed. A disadvantage of technical studies is that they give hardly any insight into the problem concerned with developing and implementing energy-efficient technologies, an important problem being the associated costs. In addition, information on the stage-of-development of the energy-efficient technologies and the R&D needed to develop these technologies is usually not included.

Technical-economic studies
Technical-economic assessment studies are an extension of the technical studies. The most important difference is that not only data on the technical potential for energy-efficiency improvement are collected, but also data on the present or potential costs of these technologies. In the literature these studies are referred to as studies that use a bottom-up approach. In this context bottom-up means that, starting at the level of unit operations in all sectors of society, a picture is obtained of the potential for energy-efficiency improvement and the associated costs of exploiting this potential for society as a whole. Since bottom-up constructed databases contain data on costs and the energy-efficiency improvement potential of a number of options, it is possible to construct marginal cost curves for all options to improve the energy efficiency at system level. The system could be, e.g. an apparatus, a factory, an industrial sector, or society as a whole. An example of such a cost curve, also called a supply curve for energy-efficiency improvement technologies, is shown in Figure 2.1. The vertical axis gives the marginal costs of the identified technology, expressed as costs per unit of energy saved. The horizontal axis gives the cumulative reduction in energy consumption. This can be expressed either in absolute terms or relative to the energy demand of the initial situation. In this example, the costs per unit of energy saved have been calculated by analysing the net present value of the potential investments in energy-efficient technologies. Consequently, in Figure 2.1 part of the curve lies below the line of zero specific costs. This part of the curve represents investments with positive net present values. These investments pay themselves back over their lifetime because of saved energy purchase costs. The total energy-efficiency improvement that can be achieved at negative or zero specific costs is called the economic potential for energy-efficiency improvement. Supply curves can be used to gain a preliminary estimate of the potential economic costs for society if a certain potential for energy-efficiency improvement were to be realized. The curves can also be used to obtain insight into the effect of financial policy instruments which can be used to simulate energy-efficiency improvement.

Since the end of the 1980s, a whole series of technical-economic studies have been published with supply curves of energy-efficiency improvement options as one of the outcomes. For instance, in the USA several studies have been published on the potential for improving the efficiency of electricity use in buildings (see e.g. [Farugui *et al.*, 1990; Ficket *et al.*, 1990; Koomey *et al.*, 1991]). This has led to studies that compare the outcomes and discuss the differences (see e.g. [Lovins and Lovins, 1991; Rosenfeld *et al.*, 1993]). The results show that sometimes large differences in economic potentials are found, mainly because of the differences in technologies and in the data that are taken into account.
In 1991 Jackson proposed a method for constructing supply curves for CO_2-abatement options [Jackson, 1991]. He also included abatement of the greenhouse gas methane in his

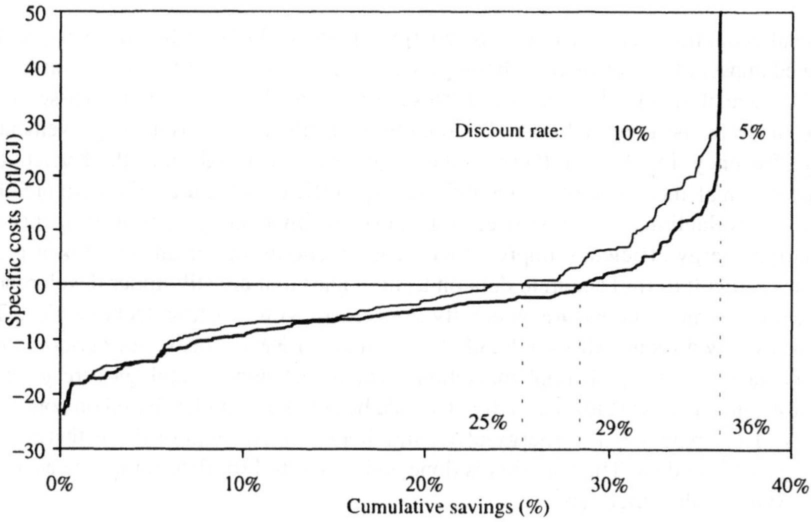

Figure 2.1: Supply curve of about 450 measures to improve the energy-efficiency of the economy of the Netherlands between the years 1990 and 2000. The curve is constructed from the database ICARUS-3 by calculating the net present value as well as the energy savings of each measure [De Beer *et al.*, 1994]. Details are presented in chapter 3 of this book.

analysis. Rubin *et al.* extended the method by including other greenhouse gases as well [Rubin *et al.*, 1992]. To allow comparison between abatement options he uses the concept of the Global Warming Potential of each greenhouse gas. At that time other studies were published with supply curves for greenhouse gas emission reduction, especially as a result of energy-efficiency improvement (see e.g. [Mills *et al.*, 1991; Blok *et al.*, 1993]).

The collected data vary per database. In general all databases contain data on the costs, the energy-efficiency improvement and the lifetime of the technology. Sometimes they also give additional information, for instance, on the penetration in the reference year and the expected implementation path. The costs of energy-efficiency improvement measures can be determined at various levels of detail. The simplest approach is to make a calculation based on a constant figure of the investment costs as well as the annual costs of the measure. In more advanced approaches, other types of costs are included as well, like losses due to the decrease in productivity during installation and test runs of the technology, training costs of employees, replacement costs of equipment before complete depreciation, and costs associated with the decision-making process. Investment costs can also be incorporated as a function of the level of application of the technology. These functions are based on the notion that production costs of a technology reduce with increasing experience and scale of production.

Technical-economic studies can use the energy service as the basis for the analysis. The combined impact of two or more technologies on the same energy service can be assessed. The advantage of technical-economic studies over technical studies is that an assessment can be made of the potential costs for society for achieving a certain improvement in energy-efficiency. In addition, these studies give enhanced insight into the barriers that may hamper implementation of the identified energy-efficient measures. Furthermore, they permit an evaluation of the cost-effectiveness of financial policy instruments for stimulating energy-efficiency improvement, e.g. an energy tax or an investment grant. Since it is difficult to find accurate data on technologies that are still under development, technical-economic studies are generally concerned with existing technologies. The outcome usually does not allow a detailed assessment of R&D-needs. Furthermore, non-economic factors that impede implementation of energy-efficient technologies are generally not investigated in these studies. Finally, it should be noted that studies based on bottom-up analysis of the potential for energy-efficiency improvement require the gathering and handling of many data. The way this is done and accounted for determines the accuracy and the value of the outcomes[1].

Technology characterization

The last group of technology exploration studies we discuss is referred to in the literature as technology characterization studies. The purpose of these studies is to characterize and evaluate specific technologies with regard to their future performance, e.g. potential for energy-efficiency improvement. Besides information on the potential for energy-efficiency improvement potential and the associated costs, fairly detailed information is given on other aspects, e.g. non-energy benefits, emissions, stage of development, developers, and required adaptation of the process. The gathered information can be used in decision making on R&D priority setting and on policy instruments to stimulate energy efficiency. There have been many technology characterization studies that focus on energy-efficient technologies. We give some examples. In the USA the Department of Energy commissioned a review of the status of energy technologies that aimed at the identification of R&D opportunities [Fulkerson *et al.*, 1989]. The California Energy Commission regularly publishes reports that review the status of new and existing energy technologies (see e.g. [CEC, 1992]). These reports lead to recommendations for the energy policy of the state of California. In the UK ETSU publishes on a regular basis a comprehensive review of the technical, economic and environmental status of existing and new energy technologies in order to formulate recommendations for the UK Research, Development,

[1]In chapter 3 of this book a study of the potential for energy-efficiency improvement using a technical-economic method is presented. In this chapter more details about the development and use of supply curves are presented and discussed.

Demonstration and Dissemination-policy, see for instance [ETSU, 1984; ETSU, 1994]. In 1994 IEA/OECD published a study that evaluated energy and environmental technologies to respond to climate change [IEA, 1994a]. In 1996 a study of the potential contribution of new technologies to sustainable development of the energy supply for the Netherlands was published [Olthof and Muradin, 1996]. In this study, called Syrene, specific attention is given to technologies that can improve the energy efficiency. The European Commission commissioned a detailed characterization of selected innovative technologies for both energy supply and energy demand, known as the ATLAS-study [ETSU, 1997]. In 1997 a study was performed to advise the US Department of Energy on the spending of R&D-budgets for energy technologies; this study reviewed the R&D-activity and needs of many energy-efficient technologies [PCAST, 1997].

Technology characterization makes it possible to formulate a policy for R&D priority setting. It is possible to determine how much a particular technology will contribute to energy-efficiency improvement, but it is difficult to estimate the total effect for a specific energy service. It is not possible to obtain good insight into the combined impact of different technologies that can perform the same energy service. In addition, technology characterization does not permit the evaluation of the effect of changing parameters, for instance the energy price. Another disadvantage is that it is often not clear on what basis the technologies that are included in the assessment are selected, or whether certain technologies have been omitted from the assessment.

2.3.3 TREND EXTRAPOLATION

The potential for energy-efficiency improvement can also be estimated by extrapolating the historical trend in the development of the energy efficiency of an energy service. For example, Marchetti uses historical trends in the efficiency of energy use in the supply of specific services – like illumination, iron making, electricity production and ammonia production – to estimate future development [Marchetti, 1979]. On the basis of his analyses, Marchetti concludes that: (1) energy efficiency does increase in time and (2) the increase in energy efficiency is extremely regular. Consequently, Marchetti also concludes that "this evolution (of the energy efficiency) has very little to do with the price of energy" [Marchetti, 1979]. Historical trends in energy efficiency are sometimes used for long-term predictions. According to Marchetti, these predictions can be considered to be dependable [Marchetti, 1979].

In Figure 2.2 we present three examples of historical trends, based on data we have collected. The figure shows the development of the specific energy consumption for the production of pig iron and aluminium and of industrial nitrogen fixation. The data for pig iron production go back to the 18th century. Since 1760 the average decrease in specific energy consumption has been 1.4% a year. The industrial production of aluminium and industrial nitrogen fixation started around 1900. Since then the specific energy consumption of aluminium production has decreased on average by 1.0% a year and of nitrogen fixation on average by 2.7% a year.

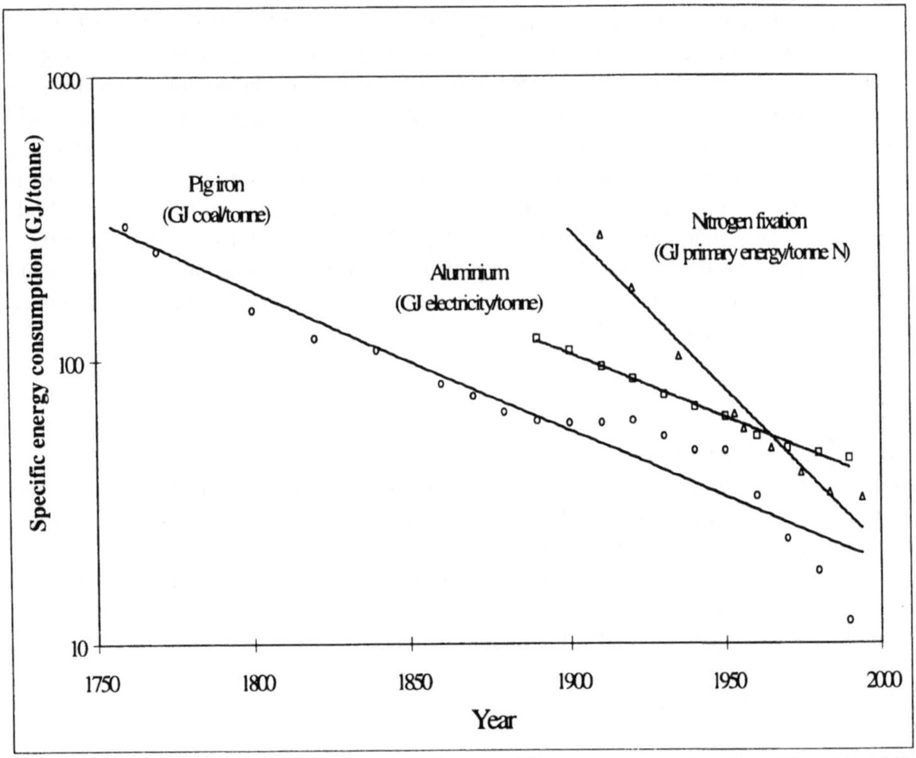

Figure 2.2: Historical development of the specific energy consumption of some bulk materials. Data for pig iron are based on [Hammersley, 1973] and [IISI, 1996]; data for aluminium are based on [Eichhammer *et al.*, 1995]; data for nitrogen fixation on [Honti, 1976; Tamaru, 1991; Radgen, 1996]. Note that the vertical axis uses a log-scale.

Historical analyses have also been performed at a more aggregated level. One example is the comparison of historical trends of the energy efficiency (see e.g. [IEA, 1987; Nakicenovic, 1990]) and of the energy intensities in different countries[1] (see e.g. ([Howarth *et al.*, 1991; Nilsson, 1993]). Other examples are detailed studies of the development of the energy efficiency and energy intensity in specific countries. Within this category, Farla *et al.* studied the development of the energy efficiency in the Netherlands [Farla *et al.*, 1998], Schipper *et al.* did the same for a number of other countries, e.g. Norway [Schipper *et al.*, 1992] and USA [Schipper *et al.*, 1990]. Finally, detailed studies

[1]Energy intensity is defined as the amount of energy needed to execute a certain economic activity expressed in monetary terms, e.g. value added or gross domestic product [Worrell *et al.*, 1997]. Energy-efficiency improvement is one of the factors that affect energy intensity, other factors being a shift in the mix of economic activities and a change in the level of activities. Some studies decompose energy intensity into these factors, e.g. [Farla *et al.*, 1998].

have been made of the development of the energy efficiency in specific industrial sectors, see e.g. [Farla *et al.*, 1997; Worrell *et al.*, 1997].
Historical analysis is a suitable method for obtaining a picture of the development of the efficiency of the way in which an energy service is performed, independent of the technology used. Sometimes these analyses can be used to evaluate the effect of introducing a new technology. However, they cannot be used to assess the chance of future technological breakthrough. Therefore, they cannot play a role in formulating an effective R&D-policy. In fact, Marchetti's modelling of historical trends implies that there is no direct policy influence on the rate of energy-efficiency improvement. However, this outcome is not as solid as it may seem, because until the 1970s there were never any long periods during which policies were in force to stimulate R&D to develop energy-efficiency technologies.

2.3.5 Macro-economic modelling

Macro-economic modelling has been widely used to evaluate the potential demand for energy in a future year. In the past decade macro-economic modelling has been used to assess the costs of limiting CO_2-emissions. Forecasting energy demand is an important element of these assessments. Macro-economic models often make it possible to evaluate the effects that certain price-related policy instruments, e.g. an energy tax, will have on the economy and on energy demand. Many assumptions and parameters underlie these models. It is beyond the scope of this study to discuss them all. We will concentrate on the way in which the parameter we are interested in is modelled: the potential for energy-efficiency improvement.
In most macro-economic models the change in energy efficiency over time is not incorporated as a separate parameter. In general, macro-economic models incorporate parameters for the change in energy intensity, that is the ratio between energy demand and economic activity of a specific industry, an economic sector or a country as a whole. The change in energy intensity is determined by several factors. One of these factors is improvement of the energy efficiency is. Other factors are structural change of the economic system (like an economic shift from more energy-intensive to less energy-intensive activities), a change in the level of human activities (influencing the implementation rate of new technologies), and changes in the value added per unit of physical production.
The change in energy intensity is usually modelled as a combination of a non-price factor, referred to as the autonomous end-use energy-intensity improvement[1], and a price-induced energy-intensity improvement, see e.g. [Grubb *et al.*, 1993]. The autonomous end-use energy-intensity improvement indicates the rate at which the ratio of economic activity, e.g. gross domestic product, and energy demand, would change in the absence of price changes

[1] In some studies this parameter is called autonomous energy-efficiency improvement, see e.g. [Manne and Richels, 1992]. However, in these studies too it is not the improvement of the energy efficiency that is indicated but the reduction in energy intensity.

[Grubb *et al.*, 1993]. The range of values of this parameter used in the different studies varies considerably, from 0% to 1.5% a year [Grubb *et al.*, 1993]. The price-induced energy-intensity improvement represents an additional improvement in the energy-intensity that is obtained when energy prices increase. This can be modelled using a figure for the price elasticity of the demand for energy.
Parameters for the energy-intensity improvement can in principle be determined by analysis based on historical data series of the ratio between economic activity and energy demand or between energy price and energy demand. There are many national and international studies that use this approach to estimate the potential energy demand in a future year. Examples of macro-economic models and studies that project the global energy demand are Global 2100 [Manne and Richels, 1992] and the model of Edmonds and Reilly [Edmonds and Reilly, 1983]. Studies using macro-economic modelling at other geographical levels have also been performed. For instance, in 1979 a study commissioned by the European Commission evaluated a low-energy growth scenario of the nine member states of the Union up to the year 2030 [Colombo and Bernardini, 1979]. Later, the European Commission published a number of other studies. One example is the 1996 scenario study that projected the 2020 energy demand of the member countries [European Commission, 1996]. An example of a macro-economic model at national level is CENECA, developed by CPB (Netherlands Bureau of Economic Policy Analysis) [Stoffers *et al.*, 1984].

An interesting way of determining parameters to evaluate the potential energy demand is used in macro-economic models that extract data on energy intensity and energy efficiency from technical-economic studies. Some models incorporate explicitly energy-efficiency improvement by penetration of (cost-effective) energy-efficient technologies, based on a pattern of penetration over time, see e.g. SAVE [Dril *et al.*, 1995] and the Global Energy Perspectives of WEC and IIASA [Nakicenovic, 1995]. Data on these technologies are taken from technology databases that are the result of technical-economic studies. SAVE uses data from the database ICARUS [Beer *et al.*, 1994]. ICARUS contains cost data per technology which do not change over time. The database used for the Global Energy Perspectives incorporates the effect that production costs of a technology usually decrease with ongoing technological development. This study also uses the energy intensity as a function of economic development, based on historical analyses. Technology databases derived from technical-economic studies can also be used to obtain data on sectoral energy-intensity trends and price elasticities of energy demand. This approach is used in NEMO, an adaptation of the Dutch model CENECA, also developed by CPB (Netherlands Bureau for Economic Policy Analysis) [Koopmans and Velde, 1996]. NEMO incorporates information extracted from the database ICARUS. Although trends in energy intensity are used in NEMO, the difference between the reduction in energy intensity and the improvement in energy-efficiency is small, because parameters are defined at a detailed sectoral level. The shift from more energy-intensive to less energy-intensive sectors does not affect the energy intensity at this level of analysis. However, changes in the product mix of a sector can influence both the energy use and the economic output, thus the energy intensity. This does not result in a change in energy efficiency.

Macro-economic models have some advantages over other types of methods. First, macro-economic models are very suitable for evaluating the possible development of the demand for energy if no policy measures are taken to stimulate energy-efficiency improvement. Second, with most models the potential effect on the energy demand of changing the energy price can also be evaluated. Finally, macro-economic effects of investing in energy-efficiency improvement can be assessed in principle, for instance the effect on employment.
The potential for energy-efficiency improvement can be assessed on the basis of macro-economic models that explicitly incorporate parameters for modelling the development in energy efficiency. However, most macro-economic models focus on energy intensity as a parameter. These models make it difficult to assess the potential for energy-efficiency improvement. Furthermore, these studies and models do not include specifically the development of technology that has an impact on the energy efficiency. Therefore, with these models it is not possible to foresee changes in energy-efficiency improvement due to specific technological breakthroughs. Generally speaking, it is not feasible to formulate an effective R&D-policy on energy-efficient technologies on the basis of these studies.

2.3.6 APPLICABILITY OF THE METHODS

In this section we evaluate the methods on their ability to satisfy the requirements we formulated in section 2.2. Then, we assess which elements of the existing methods can be used in the development of a new method.

Table 2.1 gives an indication about whether a method can be used to meet a specific requirement. The indication gives the potential use of the methods. Each study that use a specific method does not necessarily satisfy these requirements.

With regard to each requirement we now discuss why a specific method can or cannot be used:

1. Analysis at the level of energy services is in principle possible with all methods. Thermodynamical methods can be used to assess the minimum energy demand to realize an energy service. Technology exploration can yield the potential for reduction of energy use for an energy service, which is obtainable by specific technologies. Trend extrapolation yields also a potential but does not specify technologies. Macro-economic modelling, finally, can be used to, for instance, assess the effect of changing energy prices on the energy use of the energy service. However, information about the technical potential for energy-efficiency improvement should be taken from other sources, e.g. technology exploration studies.

Table 2.1: Comparison of existing methods for estimating the future potential for energy-efficiency improvement. When the demands we make on a new method (see section 2) can be met potentially by the method this is indicated by ●. If this is not the case this is indicated by O.

Requirements:	Thermo-dynamical methods	Technology exploration: Technical studies	Technology exploration: Technical-economic	Technology exploration: Technology characterization	Trend extrapolation	Macro-economic modelling
1. Energy service analysis	●	●	●	●	●	●
2. Potential for energy-efficiency improvement in the longer term	●	O	O	●	●	●
3. Careful selection of technologies	O	O	O	O	O	O
4. Detailed characterization of:						
energy consumption	O	●	●	●	O	O
costs	O	O	●	●	O	O
performance	O	O	O	●	O	O
5. Combined impact analysis	O	●	●	O	O	O
6. Assessment of R&D-parameters	O	O	O	●	O	O
7. Clear and reproducible	●	●	●	●	●	●

2. An estimate of the potential for energy-efficiency improvement in the long term cannot be obtained by technical studies and technical-economic studies, since these methods focus on existing technologies. Thermodynamical methods yield the ultimate potential, whereas trend extrapolation yields a potential in which barriers to the implementation of a technology are implicitly taken into account. Both potentials can be seen as long-term potentials. Some studies using macro-economic modelling can estimate the long-term potential for energy-efficiency improvement. However, most of these studies evaluate only the development of the energy intensity. Because technology

characterization can consider new and improved technologies, this method is suitable for estimating the long-term potential for energy-efficiency improvement.
3. The requirement that the method should pay careful attention to the selection of the technologies is met by none of the methods. The only method that considers technologies is technology exploration. Studies using these method collect information on existing and new technologies but do consider new concepts that have not yet been developed and are likely to lead to a considerable improvement in energy efficiency.
4. The requirement for detailed characterization can best be met by technology characterization. Technical studies take only the energy consumption into account, whereas technical-economic studies also consider costs. Other methods do not consider technologies at all.
5. Assessment of the combined impact on the energy efficiency of two or more technologies can only be achieved by technical and technical-economic studies. These studies generally use a calculation model to determine the total potential. Calculation rules to assess the combined impact are important to ensure that the potential is not over- or under-estimated. Such a model is not used in technology characterization.
6. Technology characterization is well-suited to assess R&D-parameters and to obtain an indication of the chance of successful development and the time required for the market introduction of each technology. Other methods cannot be used to meet this requirement, although macro-economic modelling can be used to evaluate the effect of a changing energy price over time on the cost-effectiveness of energy-efficiency improvement. Unless specific information is included, e.g. the costs development with increasing experience, this does not give an indication of the chance of successful development and of the time required to the market introduction of each technology.
7. The basis for the estimate of the potential for energy-efficiency improvement can be clear and the potential can be calculated with all methods, provided that they are performed properly.

It can be concluded that none of the methods can satisfy all requirements. Therefore, we will develop a new method, based on elements from existing methods. New elements will be added to the method, to satisfy requirements that cannot be met by existing methods,

From the above discussion, it can be concluded that technology characterization can satisfy 5 of the 7 requirements. Technology characterization is therefore a useful first step in the development of a new technology. However, two requirements are not satisfied by technology characterization. The first is that it pays too little attention to the selection of technologies. The second is that the combined impact of two or more technologies is difficult to analyse. There are two further requirements that can be satisfied in theory but are not satisfied in practice by most technology characterization studies. The first of these requirements is that the study should be performed in a clear and reproducible way. This requirement is seldom satisfied because many technology characterization studies rely on expert consultancy for the collection of data which give only limited insight into uncertainties in the data and the considerations that led to the prioritizing of technologies.

The second requirement, seldom satisfied in practice, is that the method should be applied starting from the concept of energy services.
The requirement regarding the combined impact analysis can in principle be met by using elements from technical and technical-economic studies. All technologies should be evaluated separately on the basis of their competitiveness with other technologies. This evaluation should include the expected year of commercialization.
The requirement that the technologies should be selected carefully cannot be satisfied by existing methods. However, exergy analysis, a thermodynamical method, can be used to form the basis for the selection of the technologies.
Finally, to ensure that the analysis is performed in a clear and reproducible way, an account should be kept of the way in which data were gathered and handled.

2.4 Outline of a new method

We will now outline a method that takes advantage of the merits of existing methods and fills some gaps left by these methods. The method is divided into three parts.
The first part of the method is concerned with selecting the energy service and identifying the theoretical potential for improvement of the energy-efficiency compared to the process that is currently providing the energy service. It starts with the selection of the energy service. Next, the theoretically lowest amount of energy needed to perform this energy service is determined. Then, the state-of-the-art process used to perform the energy service is described briefly. The energy flows in the process are analysed and a process exergy analysis is performed to investigate in which part of the process the most important opportunities for energy-efficiency improvement are to be found. Then, it is evaluated whether exergy losses can be avoided or reduced either by adapting the existing process or by applying another process to achieve this.

The second part of the method is aimed at making an inventory of technologies that have the potential to reduce the amount of energy required to realize the energy service. Possible areas for energy-efficiency improvement are identified on the basis of the results of the first step. An inventory of technologies under development for each area is drawn up on the basis of literature searches, and by contacting developers of technologies and processes in the field of the sector under investigation. For areas for which no technologies are under development or for which technologies cover only part of the potential, an analysis is performed to find out whether technologies for improving the energy efficiency are conceivable but are not yet under development. This analysis is made on the basis of our own expertise, developments in other sectors, and consultation with experts.

The third part of the method involves the characterization of the technologies. The aim of this part is to add information about the identified technologies so that the technologies can be compared with regard to their potential impact on energy consumption and to assess R&D-needs. Technologies are characterized on the potential performance regarding energy efficiency, costs and relevant operational parameters (e.g. productivity). Also an assessment is made of the time it could take to develop the technology into a commercial product. Furthermore, the chance of successful development is assessed. All information on the technologies is brought together and presented in a comprehensive way. Finally, conclusions are drawn concerning the potential for energy-efficiency improvement, taking into account the combined impact of two or more technologies, and the needs for RD&D.

In the following sections we will elaborate on the three parts of the method.

2.5 Process energy analysis

The first part of the method is concerned with identifying the theoretical potential for improving the energy-efficiency compared to the currently prevailing process used to realize the energy service. There are three stages in the process energy analysis: (1) selecting the energy service; (2) determining the theoretically lowest specific energy consumption to perform the energy service; and (3) performing an exergy analysis of the currently prevailing process that performs the energy service.

2.5.1 Selection of the energy service

The first stage is to select the energy service. In the introduction to this chapter we defined an energy service as the product of a human activity obtained by the use of energy meant to satisfy a human need. Energy services can be defined at various hierarchical levels [Blok *et al.*, 1994]: one energy service can satisfy a human need that has to be fulfilled to realize another energy service. For instance, energy is required to satisfy the need for transportation. However, it also requires a transportation medium, like cars, the manufacture of which does require energy. To satisfy the need for cars, steel is consumed, the production of which again requires energy. In this example the levels of delivering energy services are steel production, car manufacturing and transportation.
Since we are interested in the potential for energy-efficiency improvement in the industrial sectors, we will focus on energy services in terms of the products made by the industrial sector under investigation. It should be realized that the selection of the energy service that will be the basis for the analysis determines the scope of the technologies for improving the energy efficiency. As a consequence, the assessment of the potential for energy-efficiency improvement depends on the selection of the (hierarchical level of) energy service.

2.5.2 Theoretically lowest energy consumption

The next stage is to determine the theoretically lowest energy consumption needed to perform the energy service on the basis of thermodynamical principles, regardless the process or technology. The efficiency of energy use to perform an energy service can be determined by the specific energy consumption (SEC). The SEC is the amount of energy required to perform an energy service and is expressed as the amount of energy (final or primary energy carriers) per unit product of human activity measured in physical terms, e.g. tonnes of product, persons transported per kilometre, or cubic metres of heated space [Worrell, 1994].

2.5.3 Process exergy analysis

The last stage of the first part of the method to perform an exergy analysis of the current process used to realize the energy service. First, the process for detailed analysis is selected. The process should be based on the current state-of-the-art method used to perform the energy service. The current method is generally determined on the basis of global production capacities. Within this context the most energy-efficient process is selected as the starting point of our analysis.

Our objective is to find where in a process the main energy losses occur. Also we want to determine the maximum potential for improving the energy efficiency of this process. For these purposes, the best method is to make an exergy analysis, see e.g. [Gool, 1995], [Kaiser, 1993], and [Morris, 1992]. An exergy analysis is preferred to an energy analysis (on the basis of heating values of energy carriers only), because it takes into account the quality of energy flows. Furthermore, exergy analysis gives an indication of the potential for energy-efficiency improvement by giving insight into the degree to which energy losses can be avoided, at least in theory.
In section 3 we mentioned some studies in which exergy analysis was used. Here we will briefly discuss the principles of exergy analysis. For a more detailed discussion on exergy analysis we refer to, for instance, [Kotas, 1985], [Szargut and Morris, 1987], [Wall, 1988] and [Gool, 1995].
Exergy is the shaft work or electrical energy necessary to produce a material in its specified state from materials common in the environment, in a reversible way, heat being exchanged only with the environment at the temperature of the environment [Riekert, 1974]. It can be determined by calculating the Gibbs free energy with enthalpy and entropy defined with respect to an environmental reference system (ERS) (see Table 2.2). Expressed as a formula, the exergy B of a compound at pressure p and temperature T is

$$B\,(p,T) = H_{base}\,(p,T) - T_0 \times S_{base}\,(p,T) \qquad (1)$$

where:

$H_{base}\,(p,T)$	=	the enthalpy of that compound at pressure p and temperature T with respect to the ERS:
$S_{base}\,(p,T)$	=	the entropy of the flow at pressure p and temperature T with respect to the ERS;
T_0	=	environmental reference temperature;

In this book the exergy analysis is performed with the software package ENERPACK [Nieuwlaar, 1996]. ENERPACK uses the environmental temperature (in our analysis: 298.15 K) and pressure (101.325 kPa) as a reference for hot and cold flows and flows with an elevated or reduced pressure. It also uses the most stable compound that occurs in the natural environment as a reference for chemical elements. The exergy of an element can be determined by considering the process in which these reference compounds under environmental reference conditions are converted to the elements under thermodynamical reference conditions. The chemical elements nitrogen, oxygen, carbon and the noble gases have their reference compounds in the atmosphere. Nitrogen, oxygen and the noble gases are themselves reference compounds. To determine the exergy the difference between the environmental and the thermodynamical reference temperatures and the difference between the partial pressure in the atmosphere and the environmental reference pressure should be accounted for. Carbon dioxide is the reference compound for the element carbon. For the remaining elements the reference compounds are taken from the lithosphere, the hydrosphere or a combination of both. Liquid water is taken as the reference compound for the element hydrogen. Once the exergy of the elements is determined, the exergy of all other compounds can be calculated. The ratio of exergy and enthalpy is called the quality, or exergy factor. The quality of electricity is by definition unity. Chemical fuels have a quality that is also approximately unity. Heat, e.g. hot air or water, has a quality that is less or far less than unity, depending on the temperature of the heat. Table 2.2 gives an overview of the ERS-values of some chemical elements and compounds. For some compounds the quality is given as well.

To start an exergy analysis one selects parts of the process, the so-called unit operations that are considered in the analysis. For each unit operation the material balance and the enthalpy balance are made complete. Then, the exergy of all input and output flows is calculated. The difference between exergy input and useful exergy output is called exergy loss. Exergy losses can be divided into external and internal losses. External loss is the exergy of all flows that cross the boundaries of the process and have no commercial value. Examples are the purging of exhaust gas to the atmosphere and waste water flows at elevated temperatures. Internal loss is the difference between the exergy of the input flows and the exergy of all output flows. Internal loss may have various causes, all related to gradients of temperature, pressure or chemical potential. An example is the exergy loss that occurs in burning a fuel due to the irreversible nature of the combustion process.
Once all exergy losses are registered, one assesses whether or not these losses can be avoided or reduced. Some external losses can be avoided or reduced by recycling or heat handling. An example is the recovery of heat from exhaust gases in order to preheat fuel or combustion air. Some losses are inherent in the process and cannot be avoided without changing the method in which the energy service is performed. An example is the internal exergy loss due to combustion of a fuel for the purpose of electricity generation. One way of reducing this loss is to convert the chemical exergy of the fuel directly into electricity using a fuel cell.

Table 2.2: ERS-values of selected chemical elements at T_0=298.15 K and p=101.325 kPa [Nieuwlaar, 1996].

Element/ compound	Reference compound	H_{base} (kJ/mole)	S_{base} (J/K/mole)	B (kJ/mole)	Lower caloric value (KJ/mole)	Quality B/LHV
H_2	H_2O	285.69	169.34	235.20	241.82	0.99
C	CO_2	393.52	-56.82	410.46	393.52	1.00
N_2	N_2	0.00	-2.31	0.69	-	-
O_2	O_2	0.00	-13.26	3.96	-	-
Fe	Fe_2O_3	412.10	147.31	368.18	-	-
CH_4	-	890.05	201.24	830.05	802.31	1.03
CH_3OH	-	764.23	145.52	720.84	676.49	1.06
NH_3	-	382.64	153.91	336.75	316.83	1.06

2.6 Identification of technologies

The goal of the second part of the method is to identify as many future technologies as possible which might reduce the exergy losses as identified by the process exergy analysis. These can be technologies that are improvements on the current process or technologies that lead to alternative processes that perform the same energy service. So that this overview will be as complete as possible a list has been compiled of technologies or opportunities for improving the energy efficiency. This list contains two types of technologies: (1) technologies that are already under development; and (2) technologies that are conceivable but not (yet) under development. Both types of technologies are identified by scanning the scientific and professional literature, by consulting experts and producers, and by screening technologies that are applied in other sectors on the grounds of their ability to improve the energy-efficiency of the process analysed.
At this stage, the data on all the technologies that have the potential to realize the energy service with a lower energy consumption than the current process are not yet sufficient for us to characterize those technologies that could satisfy our requirements.

2.7 Characterization of the selected technologies

The aim of the third part of the method is to add information to the technologies identified and to compare the technologies with respect to their future performance.

2.7.1 DATA GATHERING AND HANDLING

In the first place, data are gathered on the energy consumption and the costs and the benefits of each technology. A second category of data that are collected concerns the chance of successful innovation and the time period to commercialisation.
Some of these data are obtained during the identification of the technologies. The set of data is supplemented by contacting experts and producers and by searching the literature. This process is executed in an iterative way. When no data are available an estimate is made based on the authors' own expertise, e.g., derived from evidence from other sectors. The figures on energy-efficiency improvement are verified by checking the balance of energy and material flows to and from the process. The reliability of the data is verified by comparing data for comparable technologies. Finally, in a review process experts are asked to check the information on all technologies with regard to completeness and accuracy and to check whether it is up-to-date.

2.7.2 DESCRIPTION OF THE TECHNOLOGY AND PERFORMANCE

Each characterization of a technology starts with a description of the technology. A history of the development is given as well as an overview of the parties involved in the development. Insight into the actors involved in the development is required in order to make recommendations concerning further R&D. The principle behind the technology is explained. Parameters that are important for assessing the performance of the process, e.g. production rate, and parameters that express the quality of the product, are identified and quantified. If no quantitative information is available, the effect of the new technology on the performance is assessed in a qualitative way [1].

2.7.3 ENERGY EFFICIENCY IMPROVEMENT POTENTIAL

The potential specific energy consumption (SEC) of a process using a new technology is determined on the basis of the collected data on energy use. The potential for energy-efficiency improvements is presented by comparing the SEC of the selected state-of-the art process with the SEC of a process that uses new technologies. Similarly, if appropriate, the specific heat consumption (SEC_h) and the specific electricity consumption (SEC_e) of the current and new technology are compared. When secondary energy carriers, like electricity and steam, have to be expressed in units of primary energy, the method and assumptions are specified.

[1] It should be emphasized that the impetus for developing new technologies is often better performance and not an improvement of the energy efficiency. Important driving forces for the development are improvement of the reliability and the productivity of the process and an improved product quality.

2.7.4 Costs and benefits

Several types of costs of energy efficient technologies can be distinguished. First, there are the costs associated with the research, development and demonstration. It is assumed that, in the case of a successful innovation, *research and development (R&D) costs* are small compared to the ultimate investment costs. Therefore, this type of cost is not considered It should be noted, however, that high R&D costs may be a bottleneck for further development.

Second, there are the costs related to the purchase and installation of the technology. Besides the direct *investment costs,* there may be costs associated with collecting information, decision-making, gaining expertise and monitoring, referred to as transaction costs. The transactions costs for implementing new technologies to improve the energy efficiency in industry in the Netherlands have been estimated at 3-8% of the total investment costs [Hein and Blok, 1995]. Finally, there may be costs due to productivity loss caused by the temporary shut-down of the process or a change in investment cost for other parts of the production process. These indirect costs may vary strongly from plant to plant.
In this book only the direct investment costs are considered. However, it should be noted that the absolute investment costs of long-term technologies are often hard to estimate. If insufficient information is available, an indication is given as to whether a new technology is more expensive, just as expensive or less expensive than the current technology.

Third, there are the *operation and maintenance (O&M)-cost.* O&M-costs include variable costs, labour and fixed costs [Chauvel and Lefebvre, 1989]. Variable costs are costs associated with raw materials, other materials (e.g. chemicals, solvents, and filters), and utilities (e.g. steam, electricity, process water). Labour is usually expressed in number of operators per shift. Fixed costs include maintenance, taxes, insurances and overheads [Chauvel and Lefebvre, 1989]. Well-established figures for the O&M-costs of possible ways of improving the energy efficiency in the long terms are not always available. In such cases a qualitative estimate is given.

Two types of benefits of energy-efficiency technologies are distinguished. First, the *saved energy purchase costs.* The benefits of energy efficiency improvement are often expressed as a reduction in the costs of energy consumption. It should be noted that these benefits depend on future energy prices. Since future energy prices depend on several unpredictable factors, e.g. policies concerning energy taxes, the saved energy purchase costs are not mentioned separately. A relative assessment of these benefits is obtained by comparing the potential future SECs of the technologies.
New technologies often have *other benefits*, manifested in an improvement in performance. These improvements can lead to lower O&M-costs (e.g. less maintenance or interruptions) or a higher value added (product quality). For instance, the working conditions in a workshop can be improved by recovering heat that otherwise would have been vented into

the room. The benefits cannot always be expressed in monetary units. Therefore, other benefits are described in a qualitative way

2.7.5 TECHNICAL CHANGE

In our method we estimate both the time period up to commercialization of the technology and the chance that the technology will be commercialized. To this end more insight is required into the processes that underlie the development of a technology and the replacement of an existing technology by a new technology. Many parameters influence technical development and technical change, e.g., the expected market, the RD&D-environment in the firm, sector and country, the specific technical problems, competitiveness and governmental regulations. Ever since the industrial revolution economists have attempted to find a relationship between economic growth and technical change (see e.g. [Schumpeter, 1928] and [Schmookler, 1962]). In 1977 Nelson and Winter [Nelson and Winter, 1977] initiated the development of a framework for a new theory for explaining the timing and occurrence of technical change. Since then this theory has been adapted and improved several times (see e.g. [Dosi, 1982] and [Sahal, 1985]). At the heart of these theories is the understanding that technologies are developed along certain trajectories [Nelson and Winter, 1977]. The concept of a 'technological trajectory' is nowadays used to describe the continuous process of change along a typical path (see e.g. [Dosi, 1982] and [Kemp *et al.*, 1994]). The new theories and concepts help us to understand why a given technology is developed at a certain point in time, and is preferred to another possible technology. Unfortunately, the theories on technical change are very qualitative and, so far, have been used only to explain technical change in retrospect. However, the underlying ideas of these theories can be used to identify to what extent each technology requires a deviation from the current trajectory of technical development, i.e. the degree of technical change. In the method the degree of technical change is assessed. Furthermore, the stage of development of the new technology is identified.

Degrees of technical change
Technologies can be categorized according to the degree to which they imply a change in the current technology. There is a big difference between technical changes that adapt or improve a technology based on an existing technical principle and a change that introduces an entirely new technology to an industry [Kemp *et al.*, 1994]. In between these two types of innovations more categories can be distinguished.
In our method the changes are divided into three categories. However, it should be noted that these three types of change represent points in a continuum.

1. Evolutionary change can be characterized by:
   - a logical continuation of current trends to perform the energy service;
   - a small change in performance (expressed in e.g. production rate, or energy demand), meaning a continuation of the trend;
   - no effects on the quality or the nature of the products;

- no important effects on the main process parameters, e.g. temperature, pressure;
- there are no effects on the supplying or the purchasing industries;
- small adaptations to the process;
- small organisational needs and small changes associated with the implementation of the technology.

An example is the implementation of blast furnace gas recovery in the iron and steel industry instead of the gas purging.

2. Major change can be characterized by:
   - the application of a new principle in the performance of at least part of the energy service;
   - a leap forward in performance compared to the trend;
   - significant changes in the quality or nature of the product;
   - significant effects on the main process parameters;
   - small or negligible the effects on the supplying or the purchasing industries are;
   - considerable adaptations of the production process. Main components of the process have to be replaced, but (in the case of a retrofit application) it is not necessary to build a complete new plant;
   - the organisational needs and changes associated with the implementation of the technology are considerable.

An example, again from the iron and steel industry, is the basic oxygen furnace that has replaced the open hearth process for steel making.

3. Radical change can be characterized by:
   - the application of a new principle in the performance of the energy service, or the introduction of a new energy service;
   - a leap forward in performance compared to the trend;
   - considerable changes in the quality or nature of the product, or the introduction of a new product;
   - large effects on or changes in the main process parameters;
   - significant effects on the production processes of the supplying and purchasing industries, and even on industries further up or down in production chain;
   - the need for a complete new production plant.
   - large adaptation of the organisation of the firm that implements the technology and in that in the supplying and purchasing industries.

Changing from primary steel production from iron ore in an integrated steel plant to secondary steel production from scrap in an electric arc furnace is an example of a radical change.

Since specific technologies can contain elements of more than one category, it is not always possible to make an unambiguous categorisation. In that case the technology is placed between two categories.

Stage of development

An indication as to the stage of development is the type of ongoing research, development and demonstration (RD&D) activity. The Frascati Manual, the proposed standard practice for surveys of RD&D used by OECD-countries, distinguishes three types of research activity: basic research, applied research and experimental development [OECD, 1994]. Basic research is undertaken without any view to a particular application or use. For our purpose this type of research activity is not of direct importance since we have an application-oriented approach.

In our method three stages in the development of a technology are distinguished. The first two categories are identical to the last two categories given in the Frascati Manual. A third category is demonstration. The Frascati Manual distinguishes only governmental demonstration programmes, and not demonstration at the level of firms. We use a definition of the IEA for demonstration [IEA, 1994b].

1. Applied research includes original investigation aimed at new knowledge directed towards a practical objective [OECD, 1994]. Experiments at laboratory level are part of applied research.
2. Experimental development is systematic work directed to the production or improvement of materials, products or devices [OECD, 1994]. Knowledge and expertise from research and practical experience are used. Part of this stage can be the construction and building of a prototype or a pilot plant. A prototype is an original model that includes all technical characteristics of the new products [OECD, 1994]. A pilot plant is a small plant constructed for the purpose of obtaining experience and compiling engineering and other data [OECD, 1994].
3. Demonstration is the realization of projects which take place on a large scale but which are not expected to operate on a commercial basis [IEA, 1994b]. Both pilot plants and demonstration units can switch to operating as a normal production unit, albeit on a small scale [OECD, 1994].

The degree of technical change and the stage of development both affect the time period to commercialization and the chance of market introduction. We will discuss these relationships on the basis of Figure 2.3. The degree of technical change is shown on the horizontal axis and the stage of development on the vertical axis. The time period to commercialization generally decreases from the bottom left corner to the top right corner. Radical change often requires the development of new knowledge and experience, whereas evolutionary change can build on existing knowledge and experience. Radical change also requires changes in the supplying and purchasing industries. Normally, new technologies have also to be developed in these industries. For these reasons, it generally takes longer to commercialize a radical change than an evolutionary change. The time period is also affected by the stage of development. The time period from applied research to demonstration will be shorter. However, the period may differ considerably from technology to technology and from sector to sector.

The chance of commercialization is affected by both the degree of technical change and the stage of development. Radical change has less chance of success than evolutionary

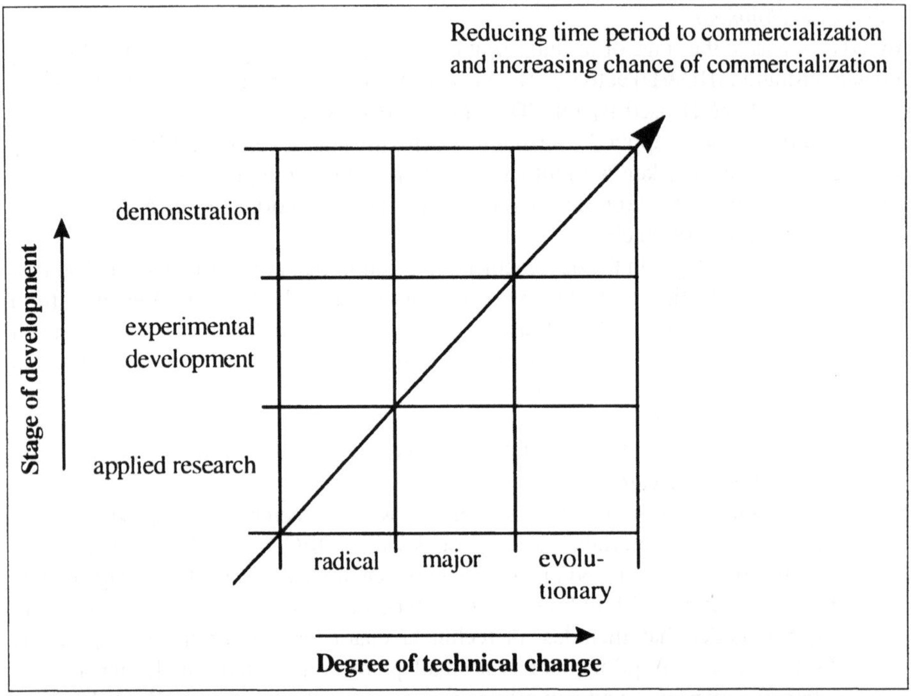

Figure 2.3: Diagram to illustrate the effects of the degree of technical change and stage of development on the time period to and the chance of commercialization.

change. The latter type of change can often be developed along a standard path. The uncertainties are smaller than for a radical change for which a new path of development has to be found. Finally, since many innovations are needed to obtain only a few commercial successes, the chance of commercialization is smaller in the early stage of development.

The technologies identified are compared by positioning them in the figure. It should be emphasized that the parameters selected are only the first indicators of technical development. The chance of commercialization, for instance, is also influenced by the other indicators we have defined like the costs and the benefits of the technology, and by external influences, like governmental policy. However, the combination of the two parameters quickly gives an idea about the likelihood that new technologies will become available and within what time frame, or - by contrast - gives an indication of the governmental effort required to stimulate technologies.

2.8 Discussion and conclusion

There are several methods for assessing the potential for energy-efficiency improvement, but no one method can be used to achieve the objective of this book: to identify and characterize technologies that can contribute to an improvement in the energy efficiency for realizing energy services, in order to estimate the future energy-efficiency potential and to assess RD&D needs. We propose a method that contains elements of a number of the existing methods. The method consists of three steps:

1. The first step is concerned with identifying the theoretical potential for improvement compared to a current state-of-the-art process that is used to perform a specific energy service and to investigate which parts of the process contain the main opportunities for energy-efficiency improvement.
2. The second step involves identifying alternative and mutually competitive technologies that have the potential to improve the energy efficiency along the road to the theoretical maximum value.
3. The third step is aimed at a characterization of the technologies. This consists of determining the expected performance concerning energy consumption, costs and relevant operational parameters, and an assessment of the time it may take to develop the technology to a commercial product and the chance that this development will be successful.

Does this method meet the requirements we set out in section 2.2? We will answer this question in relation to each requirement. The first step in the analysis is to select the energy service. With the help of an exergy analysis, all areas are evaluated that can contribute to an improvement in the energy efficiency of the current process used to realize the energy service. This allows an estimate to be made of the potential for energy-efficiency improvement that can be attained in the longer term. The selection of the technologies is based on this evaluation and on information on the status of technological development obtained from producers and experts in the field of the sector under investigation. This approach ensures that both technologies currently under development and new opportunities to improve the energy efficiency of the way the energy service is realized are taken into account. The characterization allows the assessment of the impact of each technology on future energy consumption and costs. In addition, parameters are identified for assessing impediments to successful development and recommendations are made for R&D-priority setting. Since all technologies apply to the same energy service, and sufficient information is available, the combined impact of two or more technologies on that energy service can be evaluated. Finally, if carefully executed, the method results in a well-documented and reproducible set of data on future technologies that should be able to perform a given energy service with less energy.

We can conclude that this method holds the promise to meet all the requirements we make on a method to determine the future potential for energy-efficiency improvement.

In this book the method will be applied to three industrial processes, namely paper and board production, iron and steel production and nitrogen fertilizer production. To test the feasibility of the method under different conditions we will vary the selection of the energy service in such a way that the range of technologies to identify and characterize will differ. In the study concerning the paper and board industry we will focus on one unit operation: drying of paper. The study to the iron and steel industry will also include energy-efficiency improvement by recycling of steel scrap. Finally, the scope of the study to nitrogen fertilization will be that biological fixation of atmospheric nitrogen is also included. In the last chapter of this book the method will be evaluated on the basis of the results of the case studies and recommendations will be given for adaptations and further research.

CHAPTER 3

SECTORAL POTENTIALS FOR ENERGY-EFFICIENCY IMPROVEMENTS IN THE NETHERLANDS[1]

3.1 Introduction

For about ten years now the assessment of the possible risks of climate change due to an enhanced greenhouse effect has been an important topic on the global political agenda. Scenarios that depict the possible response of the global and regional climate have been constructed, e.g. by the Intergovernmental Panel on Climate Change [Houghton *et al.*, 1990]. A large number of options for reducing the emission of carbon dioxide, the greenhouse gas that has received most attention, have been proposed and are included in these global response strategies. Improvement of the efficiency in energy use is widely regarded as the main technical option for reducing the emission of carbon dioxide.
In the Netherlands the governmental goal is a 3% reduction in CO_2 emissions by the year 2000, compared to the 1989/1990 levels. Energy-efficiency improvements will contribute most to this reduction. The Ministry of Economic Affairs states that an annual improvement of 1.7% in energy efficiency in the period 1990-2000 is sufficient to achieve the goal. This goal is set in the Sequel to the Memorandum on Energy Conservation [Ministry of Economic Affairs, 1993], published by the Dutch government in 1993. In this Memorandum the efficiency improvement is broken down into six separate sectors on the basis of technical and economic feasibility. Table 3.1 sets out the governmental goal for efficiency improvement in each sector. To achieve its objectives the government will rely on a mixture of instruments. For instance, information campaigns to increase public awareness are being organised, subsidies to promote renewable energy and energy efficient technologies are being introduced, and the government is working out covenants with particular industries. The energy distribution companies have also proposed a programme to reduce pollution and increase energy efficiency. For effective and efficient energy-efficiency policy-making it is essential to have insight into the sectoral distribution of the potentials for energy-efficiency improvements.

The objective of this chapter is to provide insight into the structure of energy-efficiency improvements in the Netherlands. To this end we determine per sector the energy-efficiency improvement potential for the period 1990-2000. We distinguish technical, economic and profitable potential. In section 2 we deal with the definitions of these potentials. All calculations are done using a database containing potentials and economic

[1] This chapter is an adapted version of: de Beer, J.G., Worrell, E. and Blok, K. (1996) 'Sectoral Potentials for Energy Efficiency Improvement in the Netherlands", *Int. J. of Global Energy Issues,* Vol. 8, Nos 5/6, pp. 476-491.

data of more than 400 energy conservation measures. A description of this database, called ICARUS, is given in section 3. In section 4 the technical potential for energy-efficiency improvements is considered. To obtain more insight into the technologies underlying this potential we categorize the energy improvement technologies in ICARUS according to the kind of energy loss that is reduced. The technical potential for energy-efficiency improvement in each category is given. The categories are described in section 5.

Table 3.1: Efficiency improvement goals per sector for the period 1990-2000.

Sector	Primary energy consumption in 1990 [a]		Governmental energy-efficiency improvement goal [b]
	(PJ)	(%)	(%)
Heavy Industry [c]	1054	37	12 [d]
Light Industry [c]	88	7	19
Agriculture	174	13	26
Services	274	10	23
Households	459	16	23
Transport	375	10	10
Power plants	317	11	26
Total	2840	100	17

a The sectoral consumption is derived from [CBS, 1991] and based on primary energy demand. Electricity is converted to primary energy using the average national electricity generation efficiency (40.4%).

b Source: [Ministry of Economic Affairs, 1993]. The goal reflects the aimed energy-efficiency improvement in the year 2000 relative to 1989.

c In the sequel to the memorandum on energy conservation [Ministry of Economic Affairs, 1993] the industry is treated as one sector. The industry can be subdivided into heavy and light industry. In this study we define heavy industry as: the basic metal industries; paper and board industries; petroleum industries; petrochemical and inorganic chemical industries; refineries; coke production; and building materials industries. Light industries are: the food and beverages industries; the textile industries; the metal manufacturing industries; other chemical industries (e.g. pharmaceutical industries); and other small industries.

d The goal for industry is 19% conservation excluding feedstocks. Almost all feedstock energy is used in the heavy industries. In the petrochemical industries approximately 60% of the energy input is used as feedstock in the form of crude oil. In the fertilizer industry the percentage is about 55% (natural gas). In total the feedstock input is 46% of the total energy input. Consequently, the efficiency improvement goal, including the use of energy carriers as feedstock, is calculated to be 19% in light industry and 12% in heavy industry. The goal excluding feedstock remains 19% for both sectors.

In section 6 the sectoral potentials are presented. Implementation of energy improvement technologies may be stimulated by introducing financial measures, for instance an energy tax or an investment grant. In the seventh section of this chapter the possible effect of a combination of these two financial instruments on the profitable potential for energy-efficiency improvement is assessed and discussed. We conclude with a discussion of the methodology and the results.

3.2 Definitions

In this section we start by giving definitions of the potentials for energy-efficiency improvement. Subsequently, we describe two methods for evaluating investments in energy-efficient technologies. The first method considers the investments from a national point-of-view. The second method uses the point-of-view of firms.

3.2.1 POTENTIALS FOR ENERGY-EFFICIENCY IMPROVEMENT

In this book we refer to energy-efficiency improvement when the energy used to perform a certain energy service is reduced without any effect on the level or nature of the activity. An example could be more efficient engines in cars. When the level or nature of activity does change, e.g. people take the train instead of the car, we talk of energy conservation. One of the starting points of our analysis is that the level and the nature of all activities remain unchanged. Consequently, in this book we limit our discussion to energy-efficiency improvements and energy-efficient technologies.

We distinguish several types of potentials for energy-efficiency improvement:

- The *technical potential* is the difference between the projected energy demand for a given year in a baseline development and the energy demand in that year assuming that all technically achievable energy-efficiency improvements have been implemented by then. It should be made clear whether the technical potential is limited to end-use measures or whether efficient conversion technologies are also taken into account. In our analysis we take both efficient end-use and conversion technologies into consideration.
- The *economic potential* is that part of the technical potential which has a net positive economic effect, i.e. the benefits of the measure are higher than the costs (including interest, depreciation, and operation and maintenance costs) [Blok and Jager, 1994]. A national perspective is used, which implies that the benefits of the investment are calculated over the entire lifetime of the measure and that a low discount rate (i.e. 5 or 10%) is used.
- The *profitable* potential is that part of the technical potential which satisfies an investment decision criterion as used by firms (e.g. characterized by pay-back period, internal rate of return).

The *market* potential can be distinguished too. This is that part of the technical potential which one expects to be realized in practice [Blok and Jager, 1994]. This potential takes into account possible barriers to the implementation of energy-efficient technologies, due to both market and non-market failures (see e.g. [Jaffe and Staffe, 1994]). Since this is beyond the scope of this book, we will not deal with the market potential.

3.2.2 COST-EFFECTIVENESS OF ENERGY-EFFICIENCY IMPROVEMENT

To determine the economic potential we have to calculate the specific costs per measure. In our analysis we define the specific costs (C_{spec}) as the annual costs per unit of energy saved:

$$C_{spec} = \frac{\alpha \times I + OM - SEPC}{PES} \tag{3.1}$$

where:

α = an annuity factor depending on the interest rate and the technical lifetime (y^{-1});
I = investment costs (Dfl)
OM = annual operation and maintenance costs (Dfl/y)
SEPC = annual saved energy purchase costs (Dfl/y)
PES = annual amount of primary energy saved, expressed as the amount of primary energy saved per unit of final energy consumption saved (GJ)

The real energy prices of 1990 are used to determine SEPC (1 Dfl is approximately US$ 0.55 (1990)). One can assume that the measures will be implemented throughout the period from 1990 to 2000. So average prices will give the best reflection of reality. However, this means one is dependent on a price scenario. History teaches us that the development of energy prices is hard to predict. For this reason known 1990-prices are used.
Saving on electricity is converted into primary energy saving using the expected average efficiency of the power plants in 2000 (42.5%) and is corrected for distribution and transmission losses.
As can be seen from formula 3.1 C_{spec} is influenced by the discount rate. It is not obvious what discount rate should be used [Burg, 1992]. A minimum is the real interest rate (generally in the range of 3-6%), but for certain sectors this might be higher. In our analyses C_{spec} is calculated using a discount rate of both 5 and 10%.

3.2.3 PAY-BACK PERIOD OF ENERGY-EFFICIENT TECHNOLOGIES

A number of economic instruments are available to assess the probability of profit from an investment, e.g. the return on investment, the pay-back period, and the net present value.

In this study we will use the simple pay-back period (PBP) as a measure to judge the attractiveness of an investment. The PBP is often used in industrial practice to make preliminary evaluations, because it is simple to use and easy to calculate. It is calculated according to:

$$BP = \frac{Total\ investments}{Annual\ revenues} = \frac{I}{SEPC - OM} \tag{3.2}$$

where the symbols have the same meaning as in formula 3.1.

All measures with a pay-back period that satisfies a pre-set criterion (e.g. 5 years) are assumed to be financially attractive. However, the pay-back period criterion can vary largely, depending on e.g. the nature of the firm or the size of the investment. Thus, it is not clear beforehand what criterion one should use.

A survey carried out in the Netherlands in 1983/84 among a restricted number (16) of industrial companies in order to assess the effectiveness of stimulating measures showed that the simple PBP is used in almost all companies to make a preliminary evaluation of investments [Koot *et al.*, 1984]. Although that survey was carried out for another purpose and the number of companies is limited, we can use the results to obtain insight into the pay-back period used in industrial practice. The survey showed that the pay-back criterion used varies widely and does not depend on the size of the firm. Only in some cases a sharply defined criterion was used. Table 3.2 shows the results of this study. The weighted average is 3.8 years.

Table 3.2: Fraction of companies applying pay-back period criteria as indicated according to Koot *et al.* [1984] and Gruber and Brand [1991]. The figures used in this study are also given.

Pay-back period (years)	Percentage of firms that use the indicated pay-back period as criterion to assess investments		Percentage of firms willing to invest when a pay-back period criterion is met
	Koot *et al.*	Gruber and Brand	Figures used in this study
< 2	19%	17%	100%
< 3	31%	27%	89%
< 4	19%	13%	56%
< 5	19%	27%	39%
> 5	12%	15%	15%

This distribution was more or less confirmed by a German study performed in 1989 among 500 industrial firms [Gruber and Brand, 1991], see Table 3.2. The average pay-back period criterion is 4.1 years. This distribution holds for general investments. For energy-saving investments 69% of the firms require the same pay-back period as for other investments,

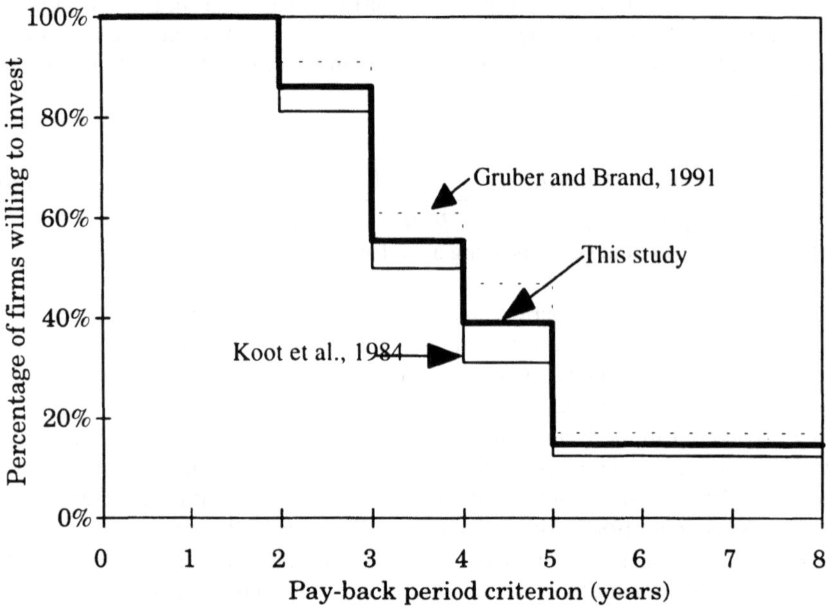

Figure 3.1: Percentage of firms willing to make an investment with a given maximum pay-back period. On the horizontal axis the pay-back period criterion is given.

8% requiring a shorter pay-back period and 10% a longer one.

Figure 3.1 shows both distributions in a graphical form. We can use this to weight the energy-efficiency improvements with a factor that reflects the probability that the firm will decide to make the investment on the basis of economic rational grounds. The thick line is the distribution we will use in this study. We assume that a pay-back period of more than five years is accepted by only 15% of the firms. When a pay-back period criterion is used of less than five years, the acceptance by firms grows to 39%. The acceptance continues to increase as the criterion is further relaxed. We assume that a pay-back period of less than two years is accepted by all firms. We will call this potential the *weighted profitable potential.*

In summary, we will present the profitable potentials according to a 3- and a 6-year pay-back period criterion and weighted using the distribution shown in Figure 3.1.

3.3 The database ICARUS

ICARUS is a database containing information on the potential and details of the costs of a large number of both demand-side and supply-side technologies for improving efficiency in all sectors of the Dutch economy. ICARUS is an acronym for: "Information system on Conservation and Application of Resources Using a Sector approach". The calculations presented in this chapter are made using ICARUS version 3 [Beer *et al.*, 1994], which is an update of earlier versions. In earlier publications [Blok *et al.*, 1990; Worrell *et al.*, 1992; Blok *et al.*, 1993] the structure of the database and the methodology followed to acquire data have been described extensively. Therefore, we will restrict ourselves to the main features of the database.

Data for ICARUS are collected using a bottom-up approach. This means that the analysis starts from the unit operations that the consumer ultimately desires, such as space heating, lighting, or blast oxygen furnaces.

ICARUS has three components:

1. A database on energy-efficient technologies;
2. The 1990 energy balance of the Netherlands and sectoral growth figures;
3. A database containing additional data, like carbon dioxide emission factors and energy price figures.

Table 3.3 shows the basic data per measure included in the database on energy-efficient techniques. With these input data and data from the auxiliary databases, it is possible to calculate, for instance, the associated carbon dioxide emission and economic characteristics, like the specific costs of energy conservation and the pay-back period.

The energy balance for 1990 is based on figures published by the Netherlands Central Bureau of Statistics [CBS, 1991], and have been corrected for the influence of ambient temperature on the energy consumption. This energy balance is given in Table 3.4. As can be seen from Table 3.4 the total climate-corrected domestic primary energy consumption in 1990 was approximately 2800 PJ.

The sectoral growth figures are based on the European Renaissance scenario, constructed by the Netherlands Central Planning Bureau (CPB) [Albers, 1993]. However, the original set of growth figures which comes with this scenario is based on the growth in value added. The use of value added as indicator for the growth of energy demand has some drawbacks. Changes in value added of a sector can occur without there being any change in the sectoral energy demand. The development of energy demand is more strongly related to the growth of physical production. Changes in this relation indicate an improvement in the efficiency. In ICARUS we use a set of growth figures based on growth of the production in physical terms. Furthermore, we assume that in the (hypothetical) reference scenario the efficiencies of the unit operations remain unchanged. Using these growth figures the primary energy consumption in the year 2000 is calculated to be 3510 PJ, if no efficiency improvement occurs (efficiency frozen at the 1990-level). The sectoral growth factors are also given in Table 3.4.

Table 3.3: Input data for each of the energy efficiency improvement measures included in ICARUS

- the conservation potential (on fuel or electricity);
- the part of the fuel or the electricity consumption of the sector to which the conservation technique applies;
- the degree to which the conservation technique leads to an increase in the other type (i.e. electricity or fuel) of energy carrier;
- the sector to which it can be applied;
- the capital investment;
- the operation and maintenance costs;
- the average lifetime of the equipment.

3.4 Energy-efficiency improvement potentials in all sectors

The range of cost-effectiveness figures valid for energy-efficient technologies can be expressed in the form of a supply curve. A supply curve of the techniques for the period 1990-2000 is depicted in Figure 3.2. On the vertical axis the C_{spec} of each measure is given, expressed in Dutch guilders (Dfl) per GJ primary energy that can be saved annually by the measure. On the horizontal axis the cumulative saving on primary energy demand in the year 2000 is given. The technical potential for energy conservation in all sectors is 36% of the projected primary energy demand in the year 2000. When a 5% discount rate is used the economic improvement potential amounts to 29% of the projected energy demand without savings. If a 10% discount rate is used, the economic potential reduces to 25% of the projected energy demand in the year 2000.

Table 3.4: Energy balance for the Netherlands for the year 1990. The energy figures are based on data provided by the Netherlands Central Bureau of Statistics [CBS, 1991], corrected for the influence of ambient temperature on the energy consumption. The energy consumption is given in PJ, and includes use of energy carriers as feedstock. Growth factors express expected growth of physical output [RIVM, 1991; Albers, 1993].

	Energy consumption (PJ)						Growth factor	
Sector	Gas	Oil	Coal	Elec-tricity	Other	Total	Fuel	Elec-tricity
End-use sectors								
Greenhouse horticulture	141.6	3.0	0.0	2.0	0.0	146.6	1.28	1.55
Other agriculture	4.9	18.0	0.0	4.7	0.0	27.6	1.05	1.05
Food & drugs industry	40.9	2.6	3.1	16.7	18.6	81.9	1.15	1.15
Textile industry	4.2	0.1	0.1	1.6	0.6	6.6	1.24	1.24
Paper & board industry	7.6	0.1	0.0	8.5	13.4	29.6	1.18	1.21
Fertilizer industry	111.2	0.2	1.4	4.0	6.1	122.9	1.06	1.06
Inorganic chemicals	17.6	6.5	11.8	12.3	13.7	61.9	1.16	1.16
Petrochemicals	52.5	270.2	7.1	20.0	50.1	399.9	1.27	1.27
Other chemicals	18.2	2.0	0.0	2.7	0.0	22.9	1.66	1.66
Building materials	25.2	6.2	3.3	5.4	0.8	40.9	1.03	1.12
Basic metals (ferrous)	11.9	0.0	69.4	11.3	0.0	92.6	1.21	1.21
Basic metals (non-ferrous)	3.2	10.2	0.0	15.9	4.3	33.6	1.00	1.05
Other metals industry	20.3	3.4	0.3	12.9	0.7	37.6	1.45	1.45
Other industries	11.8	1.7	0.1	6.9	0.1	20.6	1.17	1.17
Building & construction	4.6	27.6	0.0	1.6	0.0	33.8	1.12	1.12
Commercial services	45.3	7.2	0.0	22.4	0.0	74.9	1.38	1.44
Catering	15.7	2.2	0.0	5.0	0.0	22.9	1.37	1.45
Health care services	16.3	1.3	0.0	4.1	1.5	23.2	1.27	1.32
Non-commercial services	60.8	0.1	0.0	21.6	7.0	89.5	1.03	1.03
Households	383.7	10.2	0.6	59.4	4.7	458.6	1.13	1.45
Passenger transport	0.0	245.0	0.0	4.0	0.0	249.0	1.23	1.23
Freight transport	0.0	125.6	0.0	0.6	0.0	126.2	1.33	1.23
Other	-12.0	0.6	1.3	10.6	0.6	1.1	1.00	1.00
Total final consumption	985.5	744.0	98.5	254.2	122.2	2204.4		
Energy supply sectors								
Cogeneration CHP	128.2	30.3	6.5	-32.1	-98.2	34.6		
Waste incineration	0.0	0.0	0.0	-2.6	15.4	12.8		
Refineries	8.9	123.3	0.3	6.3	17.2	156.0	1.16	1.16
Coke production	0.0	0.1	12.8	0.4	0.2	13.6	1.21	1.21
Electricity supply	246.2	29.7	250.0	-213.9	-3.0	309.0		
Other types of energy conversion	25.5	-1.3	0.0	12.1	2.4	38.7		
Total energy consumption	1394.3	926.1	368.1	24.4	56.2	2769.1		

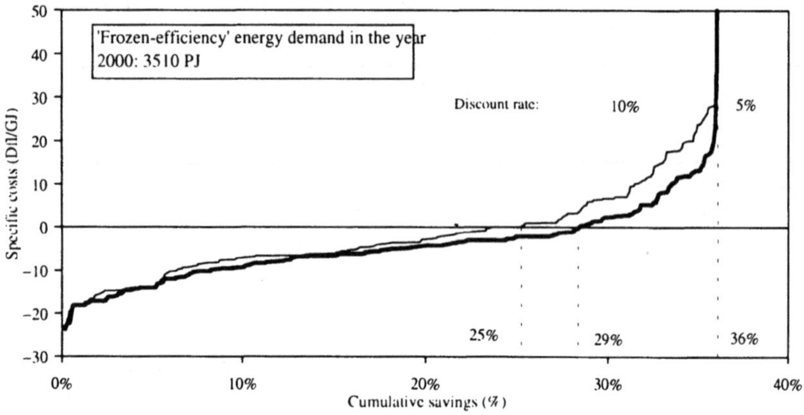

Figure 3.2: Supply curve of energy conservation measures for the Netherlands for the period 1990-2000. On the horizontal axis the cumulative energy efficiency improvement in the year 2000 is given. Vertically the specific costs of the last measure is depicted. These costs are calculated with real energy prices of 1990 and for a discount rate of both 5% and 10%.

These conservation potentials are also given in Figure 3.3 in the form of bar graphs. In figure 3 the profitable potentials according to pay-back period criteria of three and six years are also shown. When only the measures with a pay-back period of six years at maximum are implemented, the energy conservation potential is 25%. If the pay-back period criterion is a maximum of three years, the potential reduces to 14%. The weighted profitable potential is 19%.

Figure 3.3 also shows the potential for efficiency improvement that can be achieved with measures that do not involve investment costs (e.g. good housekeeping). An example is not heating a room that is unoccupied. Note that these measures might entail additional operation costs. This potential is 4% for the whole Dutch economy.

Table 3.5: A taxonomy of energy efficiency improvements

Application	Category	Examples	Share[a]
End-use	Good housekeeping	Reduction of uncontrolled ventilation in buildings Improve behaviour of building users Prevent equipment running idle	8%
	Process control and management	Building management systems Daylight- or occupancy-dependent lighting control Industrial energy management systems	5%
	Reduction of heat losses through surfaces	Insulation of solid building surfaces Double glazing Insulation of kilns and process-equipment Better insulation of refrigerators	24%
	Heat recovery	Heat recovery in building ventilation systems Application of heat exchangers in general	3%
	Process integration	Optimization of heat exchanger networks, using pinch-technology or exergy analysis	5%
	Energy recovery other than heat	Power recovery by expansion turbines for natural gas Gas recovery at BOF-furnace	2%
	Improved lighting systems	Compact fluorescent lamps Better lamp fittings High frequency lighting	4%
	Reduction of friction losses during movement	Reduction of aerodynamic drags of cars Apply wider piping in ventilation systems Better piping design	4%
	More efficient conversion of power to movement	Adjustable speed drives for pumps, fans and compressors More efficient pumps and fans Reduce over-sizing of electric motors	10%
	New process technologies	New separation technologies, e.g. membranes Apply new catalysts Membrane process in chlorine production Converted blast furnace LCD screens	9%
Conversion	More efficient boilers and furnaces	Condensing boilers Weather dependent boiler control	7%
	Heat upgrading	Heat pumps Heat transformers Mechanical vapour recompression	2%
	Combined generation of heat and power (CHP)	Gas turbine/waste heat boiler Gas engine total energy systems Combined cycle district heating systems	18%
	Efficiency improvement of fuel to power conversion	More efficient gas turbines Improve car engines Fuel cells	1%

[a] The relative contribution of each category to the total technical potential for energy-efficiency improvement of 36% compared to a frozen-efficiency level.

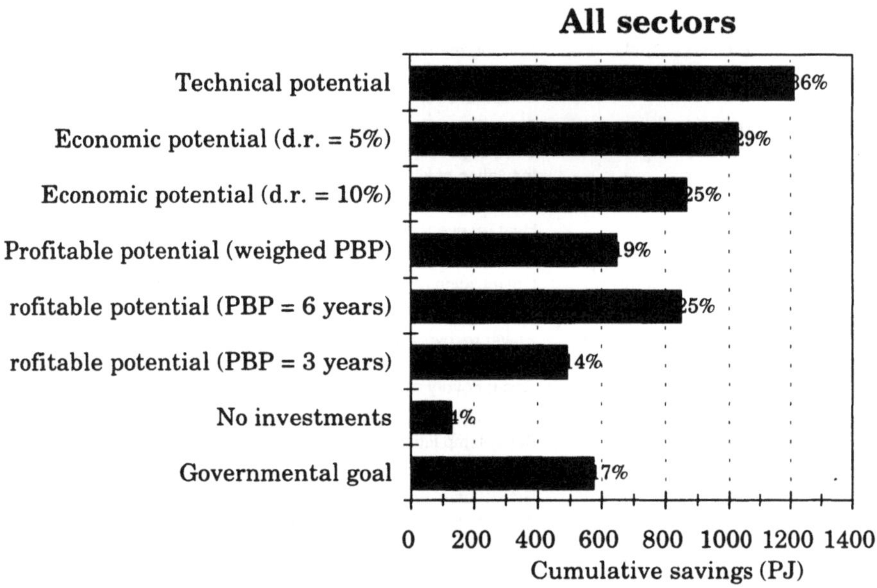

Figure 3.3: Energy efficiency improvement potentials for the whole Dutch economy. Three different potentials are shown: the technical potential, the economic potential, for discount rates of 5 and 10%, and the profitable potential, according to two pay-back period criteria (three and six years). Also the potential of measures without investment costs is given. Finally, for comparison, the Dutch governmental goal is shown. On the horizontal axis the cumulative saving on primary energy demand in the year 2000 is shown. To the right of each bar this potential is expressed as a percentage of the unabated 'frozen-efficiency' energy consumption in the year 2000.

3.5 A taxonomy of energy-efficient technologies

To obtain insight into the division of the potential for improvement over several types of improvement measures we made a categorization of energy-efficient technologies. The basis was the kind of energy loss reduced by the measure. The aim of energy-efficient technologies is to reduce the losses associated with the performance of a given energy service. This resulted in categories like 'reduction of friction losses during movement' and 'heat recovery'. This list is supplemented with categories consisting of measures relating to the kind of behaviour (e.g. good housekeeping) or technology (e.g. process integration).

The resulting taxonomy of energy-efficiency improvements is shown in Table 3.5. Each category is illustrated by some examples. Using ICARUS, we determined the relative contribution made by each category to the total technical potential. We will briefly describe the categories with the largest potential savings.

The fact that about one quarter of the improvement potential can be realized by the *reduction of heat loss through surfaces* emphasizes the importance of insulation. More than half of this potential can be realized in households. Insulation is also an important measure in the services sector and in light industry. Light industry includes many offices and workshops, where there is a large potential for insulation. In heavy industry the insulation of kilns or furnaces also leads to a reduction in heat loss through surfaces.

A second important category is *combined generation of heat and power (CHP)*. CHP is an important measure in the Dutch policy to achieve the CO_2-emission reduction goals. In our calculations we assume that CHP will only be implemented after all end-use measures have been taken. In reality CHP will be installed in parallel with end-use measures, or will even be given priority. This implies that the relative contribution of CHP will become even larger. CHP can be applied in all sectors, to provide either space heating or process heating. The sectors with the largest potential for CHP are the heavy industries (mainly chemical industries), agriculture (glasshouse industry), and households (mainly district heating).

A third important category is *more efficient conversion of power to movement*. Large energy losses are associated with this conversion. The efficiency of, for instance, a liquid pumping system, consisting of a motor, coupling, pump, valves and piping, can be as low as 42% [Larson and Nilsson, 1991]. These losses can be reduced by better sizing of the system, adjustable speed drives, enlarging the diameter of the piping etc. The largest potential for this category lies in the industrial sectors.

The last category we deal with is *good housekeeping*. Measures in this category generally do not involve investment costs. Operation and maintenance costs, however, may increase. The services sector and light industry show the largest potential for energy savings by good housekeeping. Good housekeeping relies on the behaviour of users of equipment or buildings. They can turn off the light when not in a room, or prevent equipment from running idle.

3.6 Sectoral potentials for energy-efficiency improvement

In this section we determine the sectoral energy-efficiency improvement potentials. We distinguish the six end-use sectors mentioned in Table 3.1. Figure 3.4 presents the technical improvement potentials per sector. It can be seen that the absolute potentials in households and heavy industry are both in the range of 300 and 325 PJ primary energy. The four other sectors all have a potential ranging from 100 to 200 PJ. Note that these potentials include the application of CHP. In all sectors, except for agriculture, the produced electricity can be used within the sector itself.

Now we will go into more detail about the sectoral potentials. All results are also presented in the form of bar graphs, see Figure 3.5.

Heavy industry has a technical potential of 25%. When CHP is excluded the potential reduces to 20%[1]. The economic potential is approximately 25%, both for a discount rate (d.r.) of 5% and of 10%. The profitable potential using a weighted pay-back period criterion (see section 2) is 16%.

The technical potential in *light industry* is larger, namely 40%. Without CHP this potential reduces to 36%. With a six-year criterion the potential is still 23%, but with a three-year criterion the potential is only 12%. If we consider light and heavy industry together the technical potential is 29%. The total industrial economic potential (d.r = 5%) equals 22%. If a pay-back criterion of three years is used the total industrial potential is only 13%. The weighted profitable potential is 16%.

The technical potential of *agriculture* is 73% including CHP. The technical potential without central CHP plants is 47%. Note that these plants supply heat to the horticultural sector and most of thier electricity to the grid. The electricity production by central CHP exceeds the electricity demand of this sector by a factor 6.

Households have a technical potential of 51% including CHP and 42% excluding CHP. However, in this sector a large number of expensive measures are identified, which explains the reduction in the economic potential: 43% using a discount rate of 5% and only 30% using a discount rate of 10% (including CHP). It must be concluded that unlike the commercial sectors, households rarely use sharp investment criteria to judge the profitability of their investments. Only a few studies have investigated the investment behaviour of households. Train [Train, 1985] gives a review of the literature concerning implied discount rates up to 1985. A comparison showed that the implied discount rates were highest for refrigerators (39-100%) and second highest for other appliances (18-67%).

Estimated average discount rate for other actions (thermal integrity, space heat, air

---

[1] The technical potential includes about 22 PJ savings on feedstock. If we do not account for this savings the technical potential is 24% (incl. CHP) of the energy demand.

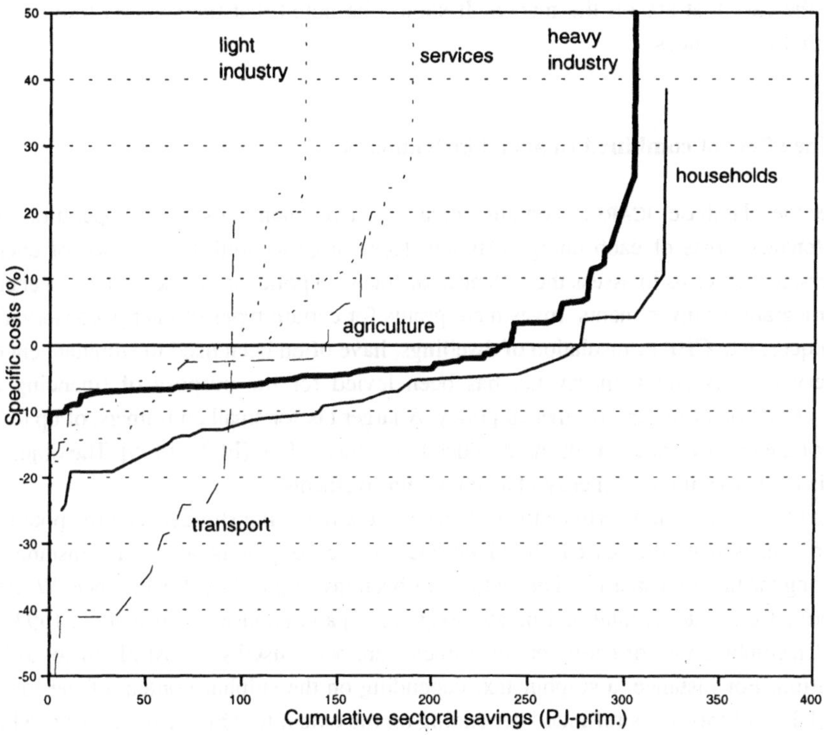

Figure 3.4: Supply curves of energy-efficiency improvement technologies per sector.

conditioning) are in the same general range (5-30%). One possible explanation for this difference is that the energy usage of different refrigerators, water heaters, stoves etc. was less well known to consumers than the relative operating costs of space heating and insulation. Anyway, it means that the simple pay-back period criterion does not have the same meaning for households as for firms.

The *services* sector has a technical potential of 42% and a much smaller economic potential (28%). This large difference between the technical and the economic potential is due to the large number of expensive insulation measures. CHP accounts for only 3% of these savings.

Finally, *transport* shows a small difference between the technical potential and the economic potential, both of which are about 17%. On the basis of these results it might be concluded that financial instruments have little effect on the saving potential. But the party that has to make the investment, the car manufacturer, is not the party that pays the energy bill. Even when e.g. an improved Otto-engine is cost effective, the user can only choose from the cars that are on the market. It should be noted that the same is true for e.g. household appliances.

3.7 The effect of combined financial instruments

Since the database ICARUS contains data on investment costs and operation and maintenance costs of each energy-efficient technology as well as the price of energy carriers, it enables us to assess the influence of financial policy instruments, like an energy tax or a grant for investments. Investment grants for certain types of energy conservation techniques, like CHP or insulation of dwellings, have often been used to stimulate energy conservation. A small energy tax has been levied for the purpose of financing the implementation of the environmental policy. A larger tax for small consumers, of up to 20-25% of the energy price, has been introduced in January 1996 [ECN, 1995]. The objective of this tax is to stimulate energy-efficiency improvements.
Using ICARUS one can determine the influence of tax recycling strategies on the potential. The revenues from the tax can be given back to the tax-paying sector, for instance by lowering the taxes on salaries. This option has been investigated by, for instance, Wolfson [Steering Committe Regulation Energy Taxes, 1992] and Krause [Krause *et al.*, 1993]. In Sweden combinations of economic instruments are being used successfully to control air pollution. For instance, a sulphur tax, depending on the sulphur content of the fuel, is levied. Part of this tax is refunded depending on the extent to which emissions have been reduced by flue gas desulphurization. This has led to a decrease of the sulphur content of the fuel (especially coal) used as well as to an increase in the efficiency of sulphur removal [Lövgren, 1993].
In this analysis we will assume that the revenues from an energy tax will be used to finance subsidies for investments in energy improvement measures. In this way the government can subsidize investments in energy-efficiency improvement technologies without extra costs or at only small cost. Moreover, the individual company can recover part of the costs of the energy taxes by investing in (cost-effective) energy conservation techniques.

We will use the following method. First, the investment grant is subtracted from the capital investment and the energy tax is added to the energy costs. Next, the pay-back period of each measure is calculated. Then, the saving is multiplied by the fraction of the investments that is expected to be made with this pay-back period (see Table 3.2) to obtain the profitable potential under the grant/tax regime. Next, the total costs of an investment grant for all profitable measures are calculated. We have a view period of ten years (1990-2000). Therefore, the annual costs of the grant are calculated by dividing the total costs by ten.

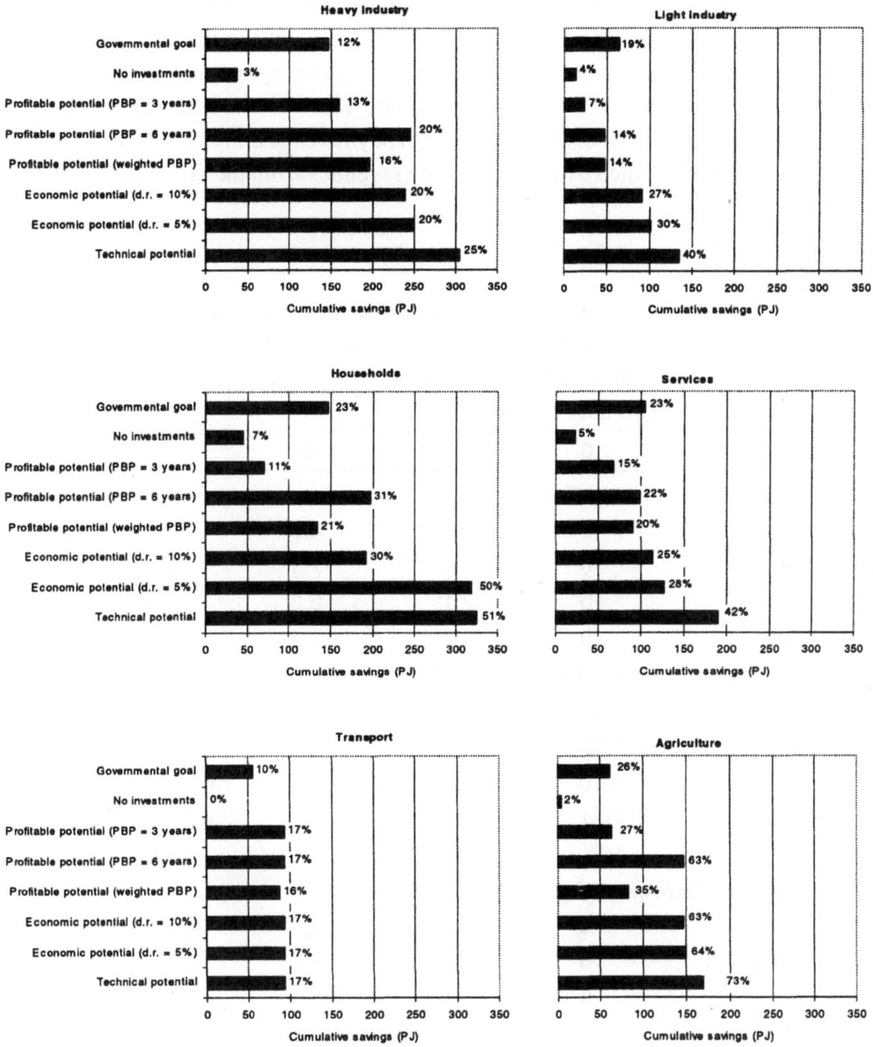

Figure 3.5: Sectoral potential for energy-efficiency improvement. See caption of Figure 3.3 for an explanation.

Then, the annual revenues from the tax are determined. This is done by first determining what the revenues will be if no efficiency improvements occur in the period 1990-2000 (frozen efficiency scenario), then by determining the lost income due to reduced energy consumption as a result of energy conservation. The revenues are determined by subtracting the lost income from the frozen-efficiency revenues. Finally, the net costs for the government are calculated by subtracting the revenues from the tax from the costs of the grant.

Table 3.6: Tax required to break even at the sectoral level with different investment grants, and the effect of this grant/tax combination on the profitable potential. The amount of money associated with this combination is also shown. At the break-even point the revenues from the tax equal the costs of the grant.

grant (%)	0	20	40	60	80
Tax required to break even (%)					
heavy industry	0	1.9	4.5	7.4	11.7
light industry	0	2.6	7.1	14.1	23.0
services	0	2.1	5.3	11.8	28.2
households	0	4.0	11.2	23.0	50.2
Primary energy savings at the break-even point (PJ/%)					
heavy industry	190.1/16	221.1/18	241.8/20	254.9/21	273.9/23
light industry	62.4/19	71.5/21	89.0/27	105.8/32	116.1/35
services	89.6/20	98.2/22	111.0/25	131.9/29	163.9/36
households	132.2/21	178.0/28	219.4/35	257.7/41	309.5/49
average	18%	22%	25%	29%	33%
Revenues from tax/costs of grant (million Dfl)					
heavy industry	0	113.3	266.3	435.3	689.3
light industry	0	58.7	155.0	295.5	487.5
services	0	82.4	203.1	426.4	992.8
households	0	220.3	575.6	1131.1	2456.5

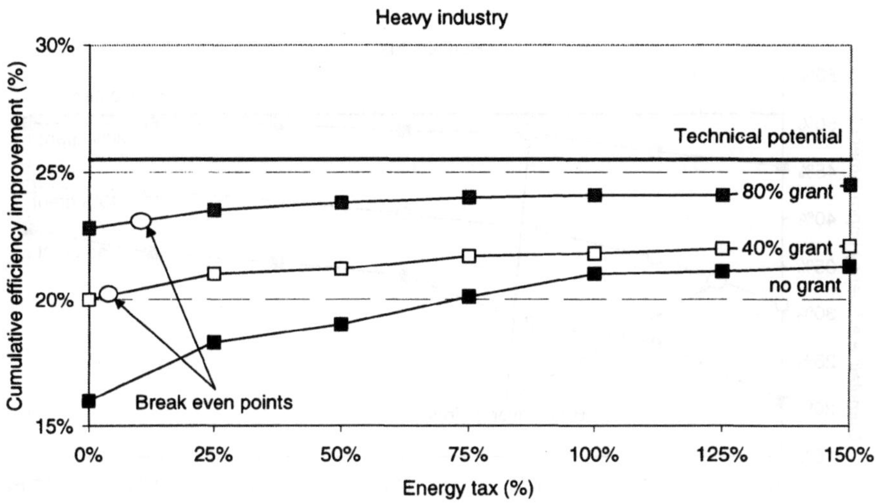

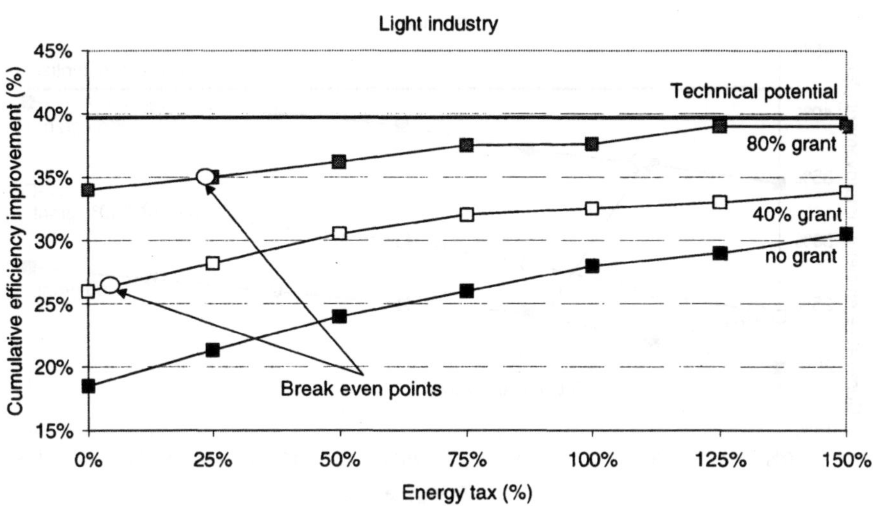

Figure 3.6: The possible influence of a combination of an investment grant and an energy tax on the weighted profitable potential of four sectors: heavy industry, light industry, households and services. (caption continued overleaf)

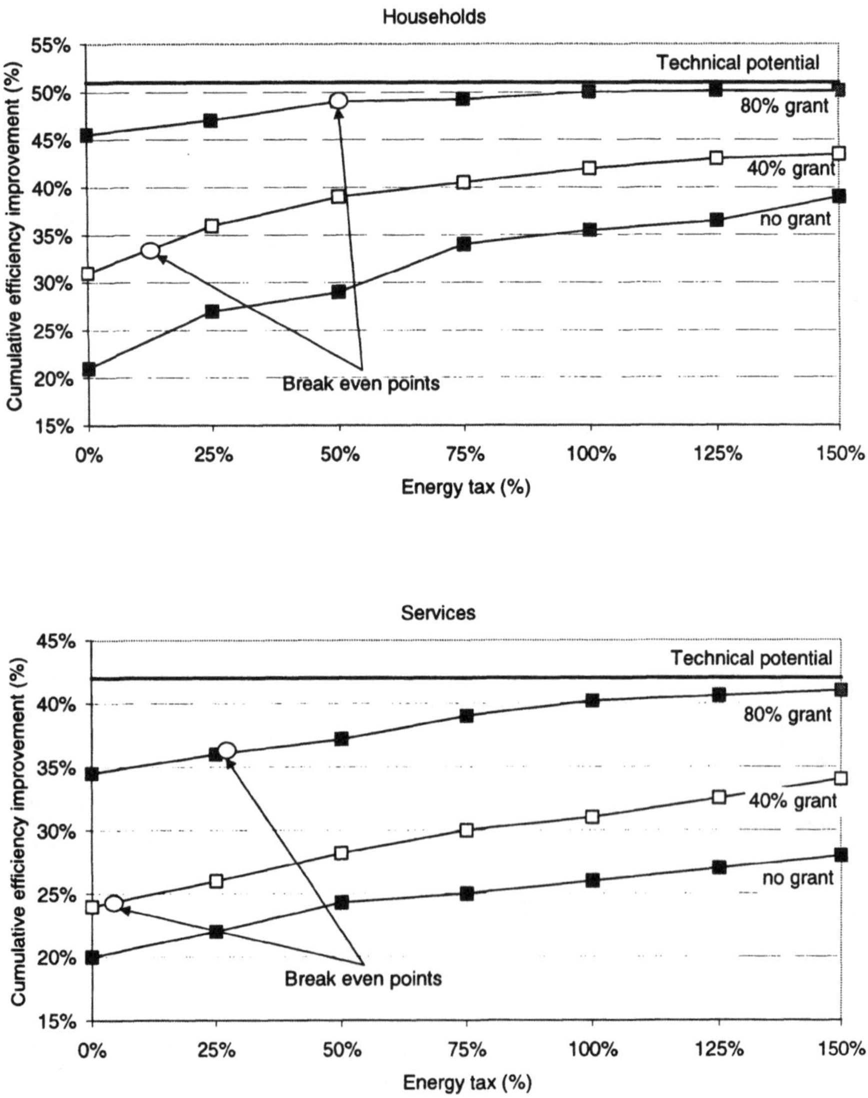

Figure 3.6 (continued): The effect of a tax only can be read from the vertical axis. Also the break-even points are indicated. At these points the revenues of the tax equal the costs of the grant.

The results for four sectors (heavy industry, light industry, households and services) are shown in Figure 3.6. On the horizontal axes the energy tax is presented. The vertical axes show the potential for energy-efficiency improvement. Two different levels of investment grants are assumed: 40% and 80%. A grant of 40% is the maximum that has been used in the Netherlands in the past. Also the curves for the potential without investment grant are shown. The bottom curves give the influence of a tax only. The influence of a grant only can be read from the vertical axes. The break-even points are the combinations where the costs of the grant equal the revenues from the tax. Table 3.6 presents more detailed information about these break-even points. It can be seen that in heavy industry a tax of only 2% (also on feedstock) would be sufficient to finance a 40% grant. In this case the profitable potential increases from 16 to 20%. The flow of money associated with this combination is about 115 million Dfl (85 million US$ (1990)). In the light industries the tax required to finance a 40% subsidy is 7% and in the services sector 5% is required. These combinations of tax and grant result in an increase in the weighted profitable potential from 20 to 25% in the services sector and from 19 to 27% in the light industries. In households an 11% tax is required to finance a 40% grant.

The profitable potential increases from 21 to 35%. In households the effect of a tax only is larger than in the commercial sectors. The large difference in the effect of a combined tax/grant in the commercial sectors and in households supports the choice of a sectoral approach.

3.8 Discussion

We start this section by discussing the validity of the potentials determined in this study. Next, we comment on the method of calculating the possible influence of financial instruments.

The potentials determined in this study are dependent on the completeness of the database. Therefore, the technical potentials probably represent a lower limit. Additional measures that we have not yet identified or which have not been explored at all are conceivable. However, after two updates of the database for the period up to the year 2000 we believe that this additional potential is very small.
A second comment relates to the costs of energy-efficiency improvements. The investment costs included in the database account only for direct investments. Additional costs, usually referred to as transaction costs, may occur. These costs are associated with acquiring information, selecting a supplier, dealing with the contract etc. It is difficult to estimate the amount of these additional costs. They will not be equal for all measures and they may also

depend on the way in which implementation is enforced. For large industries, with relatively low investments, the transaction costs are estimated to range between 2 and 8% [Hein and Blok, 1995]
A third point is that not all implementation barriers are taken into account. A barrier that is taken into account in our calculation of the profitable potential is that firms use a lower pay-back period criterion than required. The implementation of energy-efficient measures is also influenced by non-economic factors, for instance, lack of knowledge concerning energy-efficient technologies, or a company being not familiar with the technology required by the measure. Moreover, other economic factors that are not taken into account in the profitable potential may influence the implementation. Examples are: the energy bill is too low relative to the total production costs; energy-intensive equipment has not yet been marked down by depreciation; and companies use an upper limit for investments [Berkhout *et al.*, 1991]. Gillissen [Gillissen and Opschoor, 1994] en Velthuijsen [Velthuisen, 1995] performed a survey in 300 Dutch firms to analyse the investment behaviour. Results indicate the existence of some barriers (e.g. uncertainty due to fluctuations of energy prices), denying the existence of others (e.g. distance to core activities).
For a further discussion on the possibilities and limits of ICARUS the reader is referred to [Blok *et al.*, 1993].

Now we will turn to the method of calculating the possible effect of financial instruments. In this article we have studied the possible effects of levying an energy tax and subsidizing investments, without going into the possible side-effects of such instruments. Although this represents a study complete in itself, we will make some comments on this topic. Levying an energy tax might not leave the macro-economic structure of the society unchanged. For instance, it is possible that large companies will close their Dutch plants and continue their production abroad. It is important to assess these possible effects carefully before implementing a tax. However, this assessment cannot be made using ICARUS.
A second complicating factor is the so-called 'free-rider' effect. The objective of an investment grant is to make investments in energy efficiency more attractive by financing part of it. However, it has been observed that investors who had already decided to invest in a technology, whether or not they received a grant, simply raked in the money [Gruber and Brand, 1991]. Although there are costs for the government, there is no additional effect of the grant. In our analysis this effect is implicitly taken into account.
A third aspect that should be considered are the additional costs. The combination of a tax and a grant brings with it the 'pumping around' of large amounts of money. An analysis of the 'energy bonus', a large subsidy scheme that existed between the 1980 and 1985 for stimulating investment in items such as energy-efficiency improvements in the Netherlands showed that the administrative costs are of the order of 1-2% of the direct costs of the

bonus [Farla and Blok, 1995]. This does not mean that the costs of pumping around money are negligible. However, they are small compared to the total money flow.

3.9 Conclusions

The objective of this study was to obtain insight into the characteristics of energy-efficiency improvements in the Netherlands. A taxonomy of energy-efficiency improvements has been presented. It has been shown that the reduction of heat loss through surfaces is the most important measure. Large saving can be achieved by insulation, specially in households.
Technical, economic and profitable potentials of six economic sectors have been calculated. The technical potentials vary considerable from sector to sector, ranging from 17% in transport to 51% in households. Comparison with the governmental goals shows that these potentials are technically feasible in all sectors. The governmental goal cannot be reached with profitable measures only: in the services sector and the households the profitable potential is lower than the governmental goal.
Financial instruments can be proposed which will increase the profitable potential. It is shown that a increase of 4.5 to 7% in the energy price is sufficient to finance a 40% subsidy on investments in the industry and services sector. The average profitable potential increases from 17 to 22%. In households a tax of 11% is required to finance a 40% subsidy. However, in that case the profitable potential increases from 21 to 35%.
The differences in potentials and effects of financial instruments underline the necessity of a sectoral approach in energy-efficiency improvement policy.

CHAPTER 4

LONG-TERM ENERGY-EFFICIENCY IMPROVEMENTS IN THE PAPER AND BOARD INDUSTRY[1]

4.1 Introduction

Since the first oil crisis in the seventies, potentials for energy-efficiency improvements in the short and medium term have been studied extensively. Numerous analyses have indicated that energy-efficiency improvements of about 30-50% are technically feasible in industrialized countries over a period of one to two decades [ETSU, 1984; Maier, 1986; Lovins and Lovins, 1991; Beer *et al.*, 1994].

There is less consensus about the efficiency-improvement potential for the longer term. Studies relating to a longer time frame are usually less detailed than studies with a short time frame. Most analyses assess this potential by estimating the development of the energy intensity, i.e. the ratio of the energy demand to economic activity, by assuming that it will be reduced by 0-2% a year. An overview of such studies is given by [Grubb *et al.*, 1993]. Only a few studies point out which technologies should be applied or developed to achieve a certain potential (e.g. [Goldemberg *et al.*, 1988; Grübler *et al.*, 1995]). Some authors base their estimate of the potential for energy improvement on thermodynamic considerations [Ayres, 1988; Jochem, 1991]. They state that current energy efficiency in industrialized countries is so low that a potential improvement of over 80% should be possible. However, they do not indicate by what technologies or system optimization this potential can be achieved. Yet another group of studies focus on the characterization of new and improved technologies that can improve energy efficiency (e.g. [CEC, 1992; IEA, 1994; Blok *et al.*, 1995; ETSU, 1994; IEA, 1996]. The main impetus for these studies was the need to be able to prioritize limited RD&D resources and develop new RD&D-programs. A general feature of most of these studies is that they focus on the characterization of technologies, without paying much attention to which ones should be selected.

This chapter aims to test the method (described in Chapter 2 of this book) to assess the potential for energy-efficiency improvement in the long term and will focus on the selection and characterization of technologies.

4.2 Energy analysis of a paper mill

---

[1] This chapter is an adapted version of: Beer, J.G. de, E. Worrell and K. Blok (1998), Long-Term Energy-Efficiency Improvements in the Paper and Board Industry, Energy - The International Journal, Vol. 23, No. 1, pp. 21-42.

Paper can be manufactured in an integrated pulp and paper mill or in a mill where paper is produced from imported pulp. Integrated mills are located close to their main raw material input, namely wood. In the US, integrated mills are concentrated in a few wooded states, such as Wisconsin, Main, Mississippi, and Georgia [Boyd, 1996]. Non-integrated mills may be located closer to the market. In The Netherlands, with only a small forest area, only one of the 32 paper mills is an integrated mill [VNP, 1995].
The annual primary energy consumption of the pulp and paper industry worldwide is estimated at 8 EJ (1990), of which 2.3 EJ are non-conventional energy, i.e. from wood waste and pulping chemicals [WEC, 1995; Farla *et al.*, 1997]. The pulp and paper industry is the fourth largest consumer of primary conventional energy in the industrial sector worldwide; its share of industrial energy consumption is about 4% [WEC, 1995]. On the basis of the average specific energy consumption (SEC) per paper grade and a distribution of world paper production over paper grades, the global primary energy consumption of paper making excluding pulp making is estimated to be about 3.1 EJ (Table 4.1). The SEC of the paper-making process is hardly affected by whether the paper mill is integrated or not. The SEC is affected by the type of energy carrier (e.g. wood chips or natural gas) and the method of energy generation (e.g. black liquor recovery boiler or cogeneration plant). The specific heat consumption (SEC_h) and specific electricity consumption (SEC_e) differ considerably per type of paper, and even for one type of paper the range can be large. A review of ranges based on values reported in the literature is presented in Table 4.1 for some types of paper. The differences are caused by paper-specific operations, like coating and glueing, by the efficiency of operations, e.g. heat transfer in drying cylinders, and by the operational practice.

The efficiency of pulp and paper making has been improved in the past few decades. The annual decrease of the SEC in the pulp and paper industry in the OECD countries between 1973 and 1991 was on average 1.1% [Farla *et al.*, 1997]. In the short term the SEC can be further improved. Savings on the SEC_h of paper making can be achieved by e.g. the long nip press, energy management, avoiding steam use in stock preparation, and more efficient steam distribution. Savings on the SEC_e can be realized by energy management, direct drives and more efficient appliances. It is estimated that savings of about 30% on both heat and electricity are technically possible in the period 1990-2000 in The Netherlands [Beer *et al.*, 1994]. It should be emphasized that this is a technical potential, i.e. it takes no account of economic constraints and other barriers to implementation.

Table 4.1: Ranges in SEC_h and SEC_e and average values for SECs based on primary energy carriers (in GJ per mt of paper) [Reese, 1989; Melman, 1990; Komppa, 1993a; Mulder and Sinon, 1994; Nilsson, 1996], global production of paper in 1992 [FAO, 1994], and the estimated global primary energy consumption.

Type of paper	SEC_h (GJ/mt of paper)	SEC_e (GJ/mt of paper)	Average SEC (GJ primary per mt of paper)[a]	Global production in 1992 (million mt)	Estimated global primary energy consumption (EJ)
Newsprint	2.3-8.6	1.3-2.9	8.7	32	0.3
Printing/writing	2.9-8.6	1.9-3.2	15.4	71.4	1.1
Sanitary	2.6-7.0	2.4-3.6	16.9	14.5	0.2
Packaging	2.3-7.7	1.3-2.9	12.1	106.3	1.3
Other paper	5.0-7.0	1.3-1.8	9.5	21.3	0.2
TOTAL				245.4	3.1

[a] Assuming all heat is generated in boilers with an efficiency of 90%, and all electricity is generated in power plants with an efficiency of 33% [WEC, 1995]. The average SEC is based on the authors own judgement and includes production of heat and electricity from non-conventional fuels, like wood and chemicals.

4.2.1 SELECTION OF THE ENERGY SERVICE

The energy service of paper making that we will use in this analysis is making a flat sheet of paper, with certain desired properties, out of intermediate material (pulp or waste paper). Thus, pulp making is not considered here. We will limit our analysis to non-integrated paper mills, and focus on the paper-making process. Theoretically, only a small amount of energy is required to perform this energy service. The only energy needed is for aligning the fibers and for obtaining fiber-to-fiber bonding.

4.2.2 PROCESS DESCRIPTION OF CONVENTIONAL PAPER MAKING

In conventional paper making, the energy service is performed by mixing the fibers with about 100 times their weight of water and subsequently removing the water. The processes in a conventional non-integrated paper mill can be categorized into three sequential clusters: stock preparation, paper machine and finishing operations. In the stock preparation, pulp and waste paper are screened and de-inked and then mixed with water. Fibers are refined and additives are brought in. The mixture is called stock. The percentage of dry solids (ds), also called the consistency of the sheet, is about 1%. The operations in the paper machine are conducted in three steps: forming, pressing and drying. Forming means dispersing the stock over a wire screen to form a sheet, and subsequently removing most of the water by gravity and suction. The consistency of the sheet at the end of the forming section is about 20% ds. Next, the consistency is increased to 40-45% ds; water is removed by passing the sheet, supported by a felt, through three or four pairs of press cylinders. Finally, when no more water can be removed mechanically, the sheet is passed over 40-50 steam-heated cylinders (drying section), the final consistency being about 90-95% ds. Finishing operations are calendaring to smooth the paper surface, winding on wheels, cutting etc.

4.2.3 MAIN PRODUCTION PARAMETERS IN PAPER MAKING

For each new technology the impact on the production rate, the runnability, and the paper quality should be considered. The production rate of modern paper machines paper ranges from lower than 1000 m/min for board to 1600 m/min for newspaper [Tissari, 1994]. In 1980, the production rate for newspaper was still only about 1100 m/min [Tolonen, 1994], whereas the current tendency is towards rates above 2000 m/min [Kerttula, 1994]. The production rate is determined largely by the drying rate in the drying section, expressed in kg water removed per m^2 contact surface between paper and heated surface per unit of time. Increasing the drying rate makes it possible to increase the production rate without having to extend the drying section. The drying rate in a traditional drying section with 55 drying cylinders operating at a speed of 1500 m/min is about 15-30 kg water/m^2h. The term runnability is used to indicate how the paper goes through the paper machine. A technology with poor runnability decreases the reliability of the process. Important aspects are: does the paper stick to the supporting felt, how easily is the transfer from one cylinder to the next, does much paper get damaged, and is the paper dried equally over the whole surface? The main parameters used to determine the paper quality are the basic weight (in g/m^2), the thickness, the optical properties (brightness, color, opacity, transparency and gloss), stiffness, moisture content, and strength (tensile, bursting and tearing). These parameters are determined largely by the types of fiber, the machine speed, the pressing load, and the drying temperature [Baum *et al.*, 1981].

4.2.4 ENTHALPY AND EXERGY ANALYSIS

In the paper-making process heat is required mainly in the form of low-pressure steam. When heat and electricity are produced in a cogeneration plant, it is common to generate high pressure steam first, and let this expand to low-pressure steam in a back-pressure turbine to generate electricity. Paper drying consumes about 90% of the steam demand. Some steam or hot water is used in the stock preparation. In some paper mills a small amount of steam is used by steam showers in the pressing section. The electricity demand is more evenly distributed over the various unit operations than the steam demand. Electricity is used to drive the pumps that handle the huge flows of water and the fans for removing the damp air from the paper machine, and to drive the paper machine. Furthermore, refining the fibers also requires electricity. During the numerous wet processes the temperature of the process water increases by 10° to 20°C.

In order to make an enthalpy and exergy analysis of a paper mill, we constructed a standardized paper mill. The assumptions we made are shown in Table 4.2, while Figure 4.1 is a schematic of the unit operations and the material and energy flows. The material, energy and exergy balances of the flows in the standardized paper mill are shown in Table 4.3. The enthalpy values, calculated with respect to an environmental reference system [Nieuwlaar, 1988], are not necessarily valid for other paper mills. The general approach and analysis, however, do hold for other mills.

Table 4.2: Assumptions about the standardized paper mill.

- Three unit operations are distinguished: the stock preparation, the paper machine and a Combined Heat and Power (CHP)-unit.
- The steam consumption is 5.0 MJ per kg of paper and the electricity consumption is 1.8 MJ_e per ton of paper. These values were reported for the production of wrapping paper [Melman *et al.*, 1990]].
- The raw materials for paper production are fibers in the form of wood pulp and/or waste paper, water and additives, like clay. In our analysis additives are left out of consideration.
- For determining the exergy of paper, we assume it consists completely of cellulose. The exergy of cellulose is calculated to be 23.6 MJ/kg.
- 1.03 kg of dry pulp is required to produce 1.00 kg of paper with 7% water [Brown *et al.*, 1985]].
- 10% of the input fibers is lost during the process; 5% is lost to the environment, 5% is recycled [Brown *et al.*, 1985].
- Assuming a stock is made with a 1% consistency, 102.3 kg of water per kg fibers is utilized. After application the bulk of the water is cleaned and re-used. About 1% is lost in the drying section.[Own estimates]
- The temperature of the process water flow is 40°C [Schareman, 1995].
- 6.7 kg of air is required to remove 1 kg of water vapor in the drying section [Stubbing, 1990].
- All energy flows used in the paper machine stem from the CHP-unit, which consist of a gas turbine and a waste heat boiler. The CHP-unit is fired with NG using an excess air ratio of 3.5 [Huigens, 1984]. The capacity of the CHP-unit is such that the steam requirement can be met.
- Saturated steam of 7 bar is generated with an efficiency of 50.7%; Electrical efficiency is 36.7 % [Huigens, 1984]; condensate is returned to the waste heat boiler.
- We assume all electricity required in the process is generated in the CHP-unit; no electricity is used from the electricity grid. Excess electricity is delivered to the electricity grid.

Table 4.3: Material, enthalpy and exergy flows in a standardized paper mill. All values are normalized to an output of 1 kg of paper.

INPUT (MJ)				OUTPUT (MJ)			
material	mass (kg)	enthalpy (MJ)	exergy (MJ)	material	mass (kg)	enthalpy (MJ)	exergy (MJ)
CHP-UNIT							
natural gas	0.27	10.80	10.20	electricity[a]		3.60	3.60
air	7.11	0.10	0.00	steam (7 bar, 140°C)[a]	1.80	4.80	1.50
water/condensate	1.80	0.00	0.00	exhaust (105 °C)	7.38	1.50	0.20
				heat loss (500 °C)		0.20	0.00
				energy balance correction[b]		0.90	0.00
STOCK PREPARATION							
steam	0.19	0.50	0.20	wet pulp[a]	103.33	19.80	24.40
pulp/waste paper	1.03	19.80	24.40	condensate (100°C)	0.19	0.10	0.00
water	102.30	0.00	0.00	energy balance correction[b]		1.30	0.00
electricity		0.90	0.90				
PAPER MACHINE							
wet pulp	103.33	19.80	24.40	paper[a]	1.00	18.10	22.00
steam	1.61	4.30	1.30	broke[a]	0.05	1.00	1.20
electricity		0.90	0.90	lost	0.05	1.00	1.20
atmospheric air	6.82	0.10	0.00	water	101.21	0.00	0.00
				damp air (55 °C)	7.84	3.20	0.20
				condensate	1.61	0.50	0.10
				energy balance correction[b]		1.30	0.00

[a] useful flow

[b] The enthalpy balance is corrected with a term H_{lost} to deal with unrecorded losses. The entropy balance must be corrected with a term $-H_{lost}/T_o$ which makes the exergy due to this lost term zero.

The exergy input for this mill is 34.6 MJ/kg of paper, of which natural gas (NG) contributes 10.2 MJ/kg, and pulp and waste paper 24.4 MJ/kg. The useful exergy output is 25.0 MJ/kg of paper: 1.8 MJ/kg of electricity and 23.2 MJ/kg of paper and broken paper. The total exergy loss of 9.6 MJ/kg is shown in Table 4.4, divided into internal and external losses. The largest internal exergy loss occurs in the CHP-unit and is caused by the conversion of a high-quality fuel (NG) to low-quality steam. The steam used in this mill has an exergy factor of only 0.3, whereas the exergy factor of NG is about 1. This loss would have been even larger if only steam had been generated (in a boiler) and no electricity had been produced. The exergy loss in the paper machine is about 35% of the total exergy loss. About one third of this loss is caused by lost fibers. The exergy of the steam and electricity input is largely lost internally in the process; a small amount is found in the damp air and the condensate. Reduction in the steam requirement not only reduces this loss, but it also reduces the losses that occur in the CHP-unit. Therefore, it is worthwhile considering the processes in the paper machine in more detail.

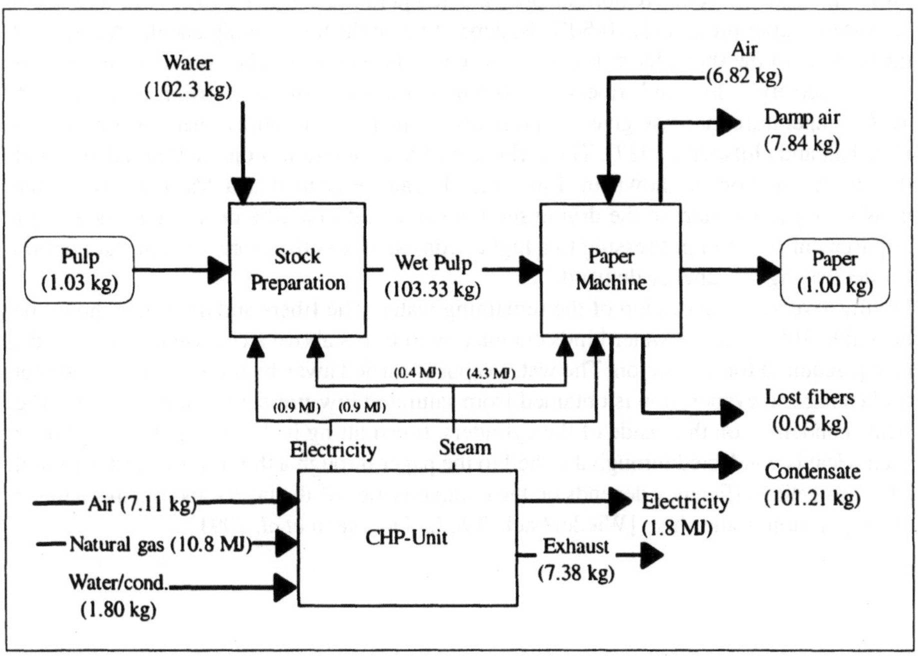

Figure 4.1: Schematic representation of the unit operations taken into account in our analysis. The material and energy flows are also given.

Table 4.4: Absolute exergy losses (GJ/mt of paper).

Unit operation	Exergy loss		
	Interna	External	Total
CHP-unit	4.9	0.2	5.1
Stock preparation	1.1	0.0	1.1
Paper machine	1.9	1.5	3.4
Total	7.9	1.7	9.6

Figure 4.2 shows the relationship between energy use and the amount of water removed in the three processes occurring in the paper machine. The bulk of the water is removed in the forming section, using the smallest amount of energy. Drying uses by far the most energy per mt of water removed.

Pressing involves squeezing water out of the voids and the cell walls. This water drains away through the felt. At normal pressing temperatures, i.e. 40-50°C, the maximum consistency after pressing is 40-50% ds, depending on the type of pulp and the density and the porosity of the sheet. Increased temperatures aid water removal by pressing because the water viscosity is lowered, fibers are softened and water surface tension is reduced. A 10°C temperature increase gives a minimum of one percent improvement in consistency [Cutshall and Hutspeth, 1987]. The influence of the consistency on the steam demand of the drying section is shown in Table 4.5. It can be seen that a 5% increase of the consistency at the start of the drying section can result in a 20% decrease in the energy requirement for drying. Pressing to a higher consistency is therefore an important option for decreasing this energy demand.

Drying involves evaporation of the remaining water. The fibers and the water should be heated to 100°C. Since water binds chemically to fibers above a consistency of 70% ds, heat is required for desorption. The water vapor is carried away by pre-heated air. The heat for heating and evaporation is obtained from saturated low-pressure steam (3-8 bar). The steam condenses on the inside of the cylinders, transmitting its latent heat to the cylinder shell. Heat is conducted through the shell to the paper through a thin layer of dirt, rust and air. The heating efficiency depends on the conductivities of the layers and the mechanism of evaporation in the sheet [Wiedenback, 1987; Nederveen *et al,* 1991].

Table 4.5: Minimum energy requirement for water evaporation from paper, expressed in GJ/mt of water evaporated and GJ/mt of paper [Gavelin, 1982]. It is assumed that the final consistency of the paper is 93% ds. In the case of the energy requirement per mt of water evaporated, the ingoing consistency is 45% ds.

	(GJ/mt of paper)		(GJ/mt of water evaporated)
	Ingoing consistency		
Minimum energy requirement for:	40%	45%	
Heating and evaporation water from 50°C to 100°C	3.25	2.63	2.46
Heating of fibers from 50°C to 100°C	0.07	0.07	0.07
Desorption heat	0.02	0.02	0.02
Total	3.34	2.72	2.55

The minimum energy requirement for water evaporation from paper is 2.55 GJ/mt of water [Gavelin, 1982]. In practical situations at least 0.15 GJ additional heat per mt of water evaporated is required to preheat the air and to compensate for condenser and radiative losses [Gavelin, 1982]. Actual values for the steam consumption of newsprint dryers in Canada range from 3.5-6.7 GJ/mt of paper, with the average value being 4.5 GJ/mt of

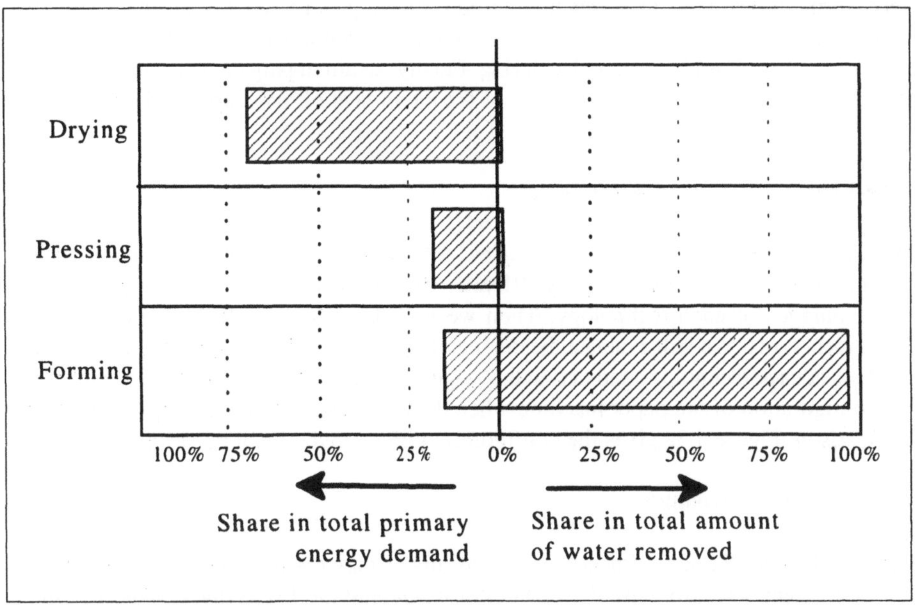

Figure 4.2: Final energy requirement vs amount of water removed for the three operations forming, pressing and drying in a paper mill.

paper [Nilsson *et al.*, 1996]. In Sweden, the steam consumption ranges between 2.4 and 5.5 GJ/mt of paper, with an average value of 3.4 GJ/mt [Nilsson *et al.*, 1996]. The consistency after the press ranged from 39% to 47% ds. Values for the Netherlands range from 1.7 to 8.0 GJ/mt of dried paper, the average being about 5 GJ/mt [Mulder and Sinon, 1990].

It can be concluded that improvement of the energy efficiency of paper making in the long term should be directed primarily at reducing the energy requirement for drying.

4.3 Identification of technologies

In order to identify technologies that have the potential to reduce the energy requirement for drying, we scanned recent volumes of *Pulp and Paper International, Paper Technology* and the *Technological Association of the Pulp and Paper Industry (TAPPI) Journal*. We extended the list of technologies by checking the references of the relevant papers. In addition, we used the proceedings of TAPPI seminars and conferences. Finally, information was obtained from experts in the field. The results of our search were also reviewed by these field experts.

The list of technologies can be divided into three groups according to the approach that is followed to improve the energy efficiency. The most far-reaching step is to avoid the use of water. In theory, this can be accomplished by processes generally referred to as dry-sheet-forming. The amount of steam required for drying can also be reduced by a number of innovative pressing and drying technologies: air impingement drying, press drying, condensing belt drying, and impulse drying. Finally, steam impingement drying and airless drying are two processes that make use of the latent heat of the evaporated moisture.

4.4 Characterization of technologies

We will characterize seven technologies. We will give a brief description of the technology, describe the ongoing RD&D, and indicate the main advantages and disadvantages of each technology. Then we will determine the SEC and estimate the investment costs. We will evaluate the degree of technical change required to develop and implement each technology and define the stage of development. Figures 4.3b-g illustrate the configuration of the paper machine when the new technologies are implemented. For comparison, Figure 4.3a gives the configuration of a conventional paper machine.

4.4.1 DRY-SHEET-FORMING

The idea behind dry-sheet-forming is the making of paper without the addition of water. The first patent dates from the 1930's, but it was not until the 1970's that dry-sheet-forming became a commercial technology [Attwood, 1996]. Dispersing the dry fibers to a flat sheet can be achieved by applying either a carding technique or an air-

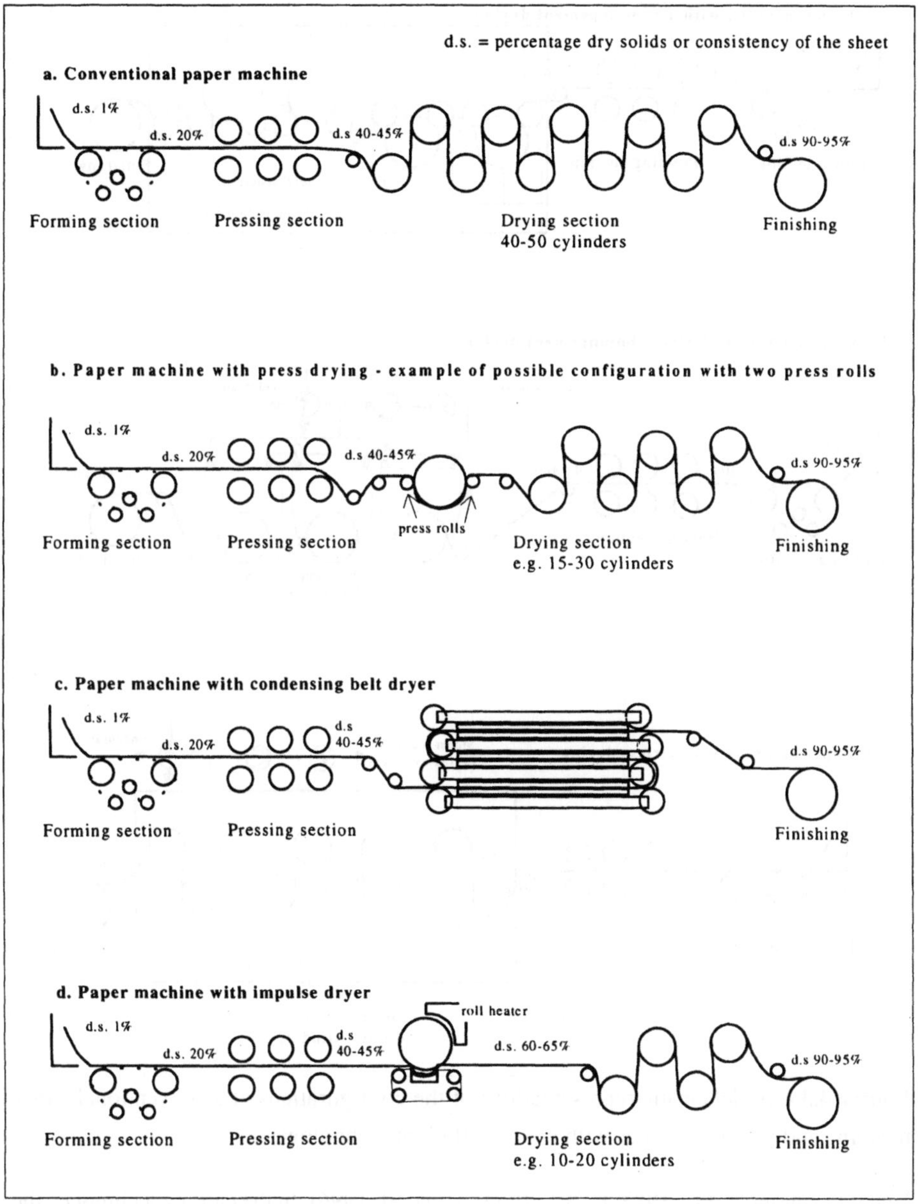

Figure 4.3 a-d: Captions overleaf.

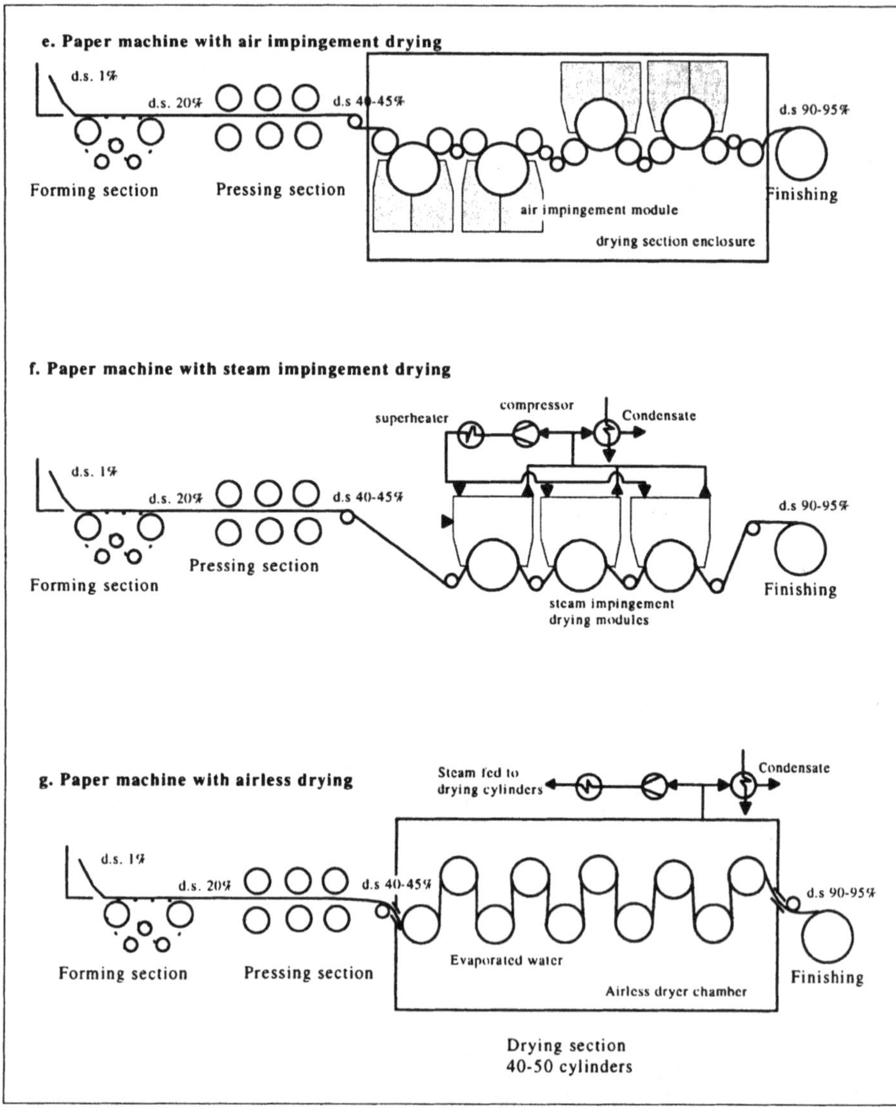

Figure 4.3 e-g: Schematic representations of the configurations of a conventional paper machine and paper machines with energy efficient technologies.

laying technique [CEC, 1983; ETSU, 1984; Attwood, 1996]. In the carding techniques the fibers are dispersed by mechanical means. In the air-laying technique the fibers are suspended in air and the paper is formed from this suspension. Fiber-to-fiber bonding is obtained by adding resins to the fibers or spraying a polymer latex on the web formed. In the first case hot pressing or air-heating is applied to polymerize the resins. In the second case, the web is dried in an air-heated drying chamber. The air-laying techniques permits a higher production rate. Moreover, by control of the air stream the characteristics (fiber direction, strength) of the paper can be better adjusted [ETSU, 1984]. Worldwide about 15 dry-sheet-forming process are in operation, with a global capacity of less than 100,000 mt/year in 1990 [CEC, 1983]. The majority use the air-laying technique and latex bonding [CEC, 1983]. Typical capacity is about 7,000 mt a year.

Commercial dry-sheet-forming plants use typically about 7.5 GJ/mt product for drying the paper that is wetted with the latex solution [Attwood, 1996]. The SEC_e for maintaining the air stream and driving equipment is 5.4 GJ_e/mt [Attwood, 1996]. Consequently, these plants are less energy efficient than conventional paper mills. The thermal-bonding process eliminates the need for drying. On the basis of a comparison with the latex bonded process, we assume that when the thermal bonding is performed by air-heating, a saving of 50% of the SEC_h is possible. When the thermal bonding is performed by passing the paper through an electrically-heated hot press, the SEC_h can even be reduced by 100%. We estimate the increase in SEC_e to be 0.6-1.0 GJ_e per mt of product. This estimate is based on the electricity requirement for heating the roll used for impulse drying (see later). This is additional to the electricity requirement for maintaining the air stream and driving the equipment, which is comparable to the latex bonded process. The SEC_e of current dry-sheet-forming plants might be reduced by conventional energy-efficient techniques, like adjustable speed drives.

The direct investment costs are one third to one half of the investment costs for a conventional non-integrated paper mill [Attwood, 1996]. Operation and maintenance costs are also lower [Attwood, 1996]. Contrary to all other selected technologies, dry-sheet-forming is already commercial, although it is used only for the production of specialties. The products can be grouped into soft products (e.g. napkins, sanitary towels, and diapers) and hard products (e.g. insulation board, industrial filter, roofing type material). It has been demonstrated that the dry-sheet-forming of corrugated medium and molding board is possible. Attempts to produce folding board were not successful [Attwood, 1996].

There are four major producers of dry forming systems: M. and J. (Denmark), Dan Web (Denmark), United Paper Mills (Finland), and Honshu (Japan). All have pilot plant facilities for developing new, special, products [Attwood, 1996]. At the moment there is no ongoing RD&D to adapt the process for paper types that are produced in large quantities [Attwood, 1996; Komppa, 1993]. If this could be achieved, this would mean making paper according to a new principle, but the new process would require the construction of a complete new production plant. Whether or not this will affect the paper quality is at present a hypothetical question, because this technology has not yet been used for the production of other paper types. Therefore, dry-sheet-forming will be either a major change or a radical change.

4.4.2 Innovative pressing and drying techniques

Research has been conducted into new pressing techniques since the early seventies. The research is based on the application of elevated temperatures (above 100°C) and higher pressures [Back, 1991]. The initial motive for this research was to improve the paper quality and to achieve higher production rates by replacing part of the conventional drying section by components with higher drying rates. Pressing at higher temperatures is not new. Since the 1950's steam showers have been used to increase the temperature of the paper in the pressing section [Cutshall and Hutspeth, 1987]. The new techniques we will describe are press drying, condensing belt drying, impulse drying and air impingement drying.

In press drying (Figure 4.3b), the sheet is pressed between two hot surfaces (100-250°C), e.g. two heated pressing cylinders. Many research institutes have conducted research into press drying; this has resulted in many configurations of the pressing and drying cylinders [Back, 1991; Gunderson, 1991]. In most designs the press dryer is located between the conventional pressing and drying section; the number of cylinders in the conventional drying section can be reduced significantly. There are probably no restrictions regarding the types of paper that can be produced and it is possible to improve the paper quality. Several concepts have been tested at pilot plant scale [Gunderson, 1991]. Although press drying seems to be over the top of its research phase, it has not developed into a commercial process [Back, 1991; Gunderson, 1991]. Drying rates in the press-drying section can be 2 to 10 times faster than the rates in conventional drying cylinders [Polat and Mujumdar, 1995].
Data on the SEC of press drying are scarce. One source reports the SEC for press drying to be 2.5-3.5 GJ/mt water removed [Polat and Mujumdar, 1995]. Assuming that paper is being dried from 45 to 93% ds, we can estimate the reduction of the SEC_h to be in the range of 5-30%. Less electricity is needed for driving the ventilation air fans and the paper machine. Consequently, we estimate that the SEC_e can be reduced by about 5%.
Press drying means that the paper machine can be smaller. The direct investment costs may therefore be smaller than for a conventional dryer. The improved paper quality results in higher revenues. The smaller number of cylinders reduces the demand for maintenance. This effect will probably be balanced by additional O&M-costs for the new equipment.
Press drying can be seen as a continuation of pressing at higher temperatures using steam showers, although the technology differs considerably. The new technology affects the nature and the quality of the paper. On the other hand, the production process requires only slight adaptations. We consider press drying as an evolutionary technical change.

In condensing belt drying (or condebelt drying, Figure 4.3c), the paper is dried in a drying chamber by contact with a continuous hot steel band, heated by either steam or hot gas [Lehtinen, 1993; Unkila and Lehtinen, 1991]. On the other side of the sheet there are three layers: a fine wire gauze, a coarse wire gauze, and an externally cooled steel band. Water is evaporated by the heat from the hot steel band. The evaporated water passes through the wire gauzes and condenses on the cold steel band. The condensate is removed by pressure and suction. The pressure on the sheet can range from atmospheric to 10 bar and the temperature of the hot steel band to a maximum of 180°C. The method is applied in the absence of air between the steel bands and requires [Polat and Mujumdar, 1995] a long pressing time (0.25-10 sec). The length of the drying zone is more than 20 m. Three different configurations are being developed so that the process will be applicable to almost all grades of paper. In fact, condensing belt drying can be seen as a specific configuration of press drying. However, unlike press drying, condensing belt drying can completely replace the drying section of a conventional paper machine.

With a condensing belt dryer newsprint paper can be dried to a consistency of 94-95% ds, starting from 44-45% ds [Lehtinen, 1993]. The drying rate is 5-15 times higher than in conventional steam drying [Lehtinen, 1995b]. The higher drying rate is achieved by a lower thermal resistance via the steel band than via the conventional cylinder shell, better thermal contact between the hot surface and the paper, and lower thermal and diffusional resistance in the paper because of the absence of air [Lehtinen, 1995a].

As far as energy consumption is concerned, we have to rely on data obtained from pilot plant tests performed by Valmet-Tampella [Lehtinen, 1993]. It appears that the largest steam losses occur at the seals of the drying chamber. For narrow machines (2.5 m), where the seals are relatively large compared to the drying surface, the steam consumption is more or less equal to conventional machines of the same size. For larger machines a saving on steam requirement of approximately 10-20% is possible compared to a conventional dryer [Lehtinen, 1993; Lehtinen, 1995b]. The SEC_e is expected to be more or less equal to the SEC_e for conventional paper making [Lehtinen, 1995b]. Several schemes with heat pumps have been proposed to recover the heat of the cooling water and the hot condensate. Calculations have shown that by the application of heat pumps it is possible to produce all the steam that is needed from the discharge heat. However, 0.75-1.0 GJ of electricity is required to drive the compressor of the heat pump [Lehtinen, 1993].

Because of the higher drying rate, a condebelt dryer is considerably smaller than a conventional dryer. Presumably, the direct investment costs are lower. A benefit of condensing belt drying is improved paper quality, due to the softening of lignin as a result of the long drying time. O&M-costs are assumed to be comparable to conventional paper making.

Condensing belt drying is being developed by the paper machine manufacturer Valmet-Tampella (FIN). RD&D concerning the condensing belt dryer has been conducted since 1975. Since 1992 pilot plant scale tests have been performed [Cox, 1992]. The first production machine went into operation in Finland in the spring of 1996. The unit has a maximum production speed of 230 m/min (later 450 m/min), the width of the sheet is 2.3 m, and the length of the drying zone is 22 m [Lehtinen, 1996]. As the production speed and the width of the sheet are a factor 2-5 and 2-3.5 smaller respectively than these factors for current paper machines, we consider this unit as a demonstration plant. The total investment costs were 12.5 million US$ (1996) [Lehtinen, 1996].

Condensing belt drying has characteristics of evolutionary change and major change. It implies drying paper according to the principle of pressing at elevated temperatures, and is therefore a continuation of the current trend. It also involves drying with steam, as in conventional drying. However, the technology differs considerably. It involves a complete replacement of the conventional drying section. The performance of the paper machine will change in that a higher production rate can be achieved. There may be a positive effect on the paper quality. We consider condensing belt drying as a change that is in between an evolutionary change and a major change.

Impulse drying (Figure 4.3d) involves pressing the paper between one very hot rotating roll (150-500°C) and a static concave-shaped press (also called the nip) with a very short contact time (15-100 ms). The pressure is about 10 times higher than that in press and condensing belt drying. The theory is that the steam layer formed along the hot roll can displace water from the sheet without evaporating it. However, this theory is still not proven, and is even questioned by some authors [Back, 1991; Wahlstrom, 1991].
Impulse drying dries paper 50-500 times faster than conventional drying [Polat and Mujumdar, 1995]. Furthermore, paper with different and improved properties can be produced [Wahlstrom, 1991]. Two important drawbacks are that the sheet can be delaminated especially if it is of a heavy grade, and that it can stick to the roll.
The consistency of the sheet after the impulse dryer can be increased drastically to about 55% ds for board and to 78% ds for lightweight paper, if the consistency of the ingoing sheet is 40% ds [Wahlstrom, 1991]. This means that from paper with these consistencies 35% and 85% less water has to be removed in the drying section. Comparable savings with regard to the steam requirement should be possible. These figures have been confirmed by other experimental studies. For instance, the Institute of Paper Science and Technology (IPST) in the US reports that the consistency of newsprint paper can be increased from 46-52% to 60-65% [Smith, 1993]. Vincent [1993] states that a 50% saving is possible. When we compare values ranging from 0.6 to 1.4 MJ/kg evaporated water reported by Nilsson *et al.* [1996] to the 2.7-5.1 MJ/kg water evaporated required for conventional drying, we can conclude that the saving of steam consumption lies in the range of 50-75%. Since conventional drying is still required to dry the sheet further, the saving on SEC_h is smaller. Assuming an initial consistency of 45% and a final consistency of 95%, the reduction of SEC_h is 35% when the consistency after the impulse dryer is 60%. When the consistency after the impulse dryer can be increased to 70%, the reduction on SEC_h is about 50%[1]. Model calculations by IPST showed that the additional electricity requirement for heating the roll is of the order of 10-15% of the saving of steam consumption [Smith, 1993]. Since the electricity requirement for driving fans and the paper machine is reduced, the resulting SEC_e will be 5-10% higher than the SEC_e in conventional paper making. In terms of primary energy consumption, further savings may be feasible by heating the roll with steam, but we could not find any reports of research on this subject.
The development of impulse drying started in 1980 at IPST in cooperation with the paper machinery manufacturer Beloit [Smith, 1993]. The objective was to increase the drying

[1] In the original publication (Energy, vol. 23, (1), 1998, pp. 21-42) the potential reduction of SEC_h by impulse drying is said to be 50-75%. However, this is the saving on steam consumption per kg of water removed. As explained in the text, the reduction of the SEC_h of the total drying section is lower: 35-50%.

rate, and so to decrease the size and costs of the paper machine. At IPST, laboratory scale experiments have been performed to try to reduce sheet delamination [Smith, 1993]. These experiments showed that ceramic coated rolls can be operated at higher temperatures and pressures ranges without inducing sheet delamination. Other institutes are also working on the development. For instance, PAPRICAN (CAN) is doing research on impulse drying. A pilot plant has been built at the estimated costs of CAD 17.5 million installed (approximately US$ 14.2 million). These costs are offset by savings on equipment (shorter drying section) and increased production [Anonymous, 1994]. The reduced drying section will result in lower investment costs [Wahlstrom, 1991]. The additional O&M-costs for the impulse dryer are assumed to be balanced by the reduced O&M-costs for the shorter paper machine.

Impulse drying is at the pilot plant stage. Development of this technique is in the first place directed at newsprint and liner board. However, the development might be expanded to other grades of paper.

Impulse drying may have an impact on the quality of the paper. Considerable adaptations have to be made to the paper-making process: one or more impulse dryers have to be installed and the drying section can be shortened. Contrary to press drying and condensing belt drying, impulse drying involves drying according to a new principle, although the debate on the formulation of this principle is still ongoing. Therefore, we consider impulse drying as a major change.

Air impingement drying (Figure 4.3e) involves blowing hot air, heated to 300°C in gas burners, at high velocity against the wet paper sheet. Combination with existing technologies is possible. For instance, Valmet-Tampella proposed an air impingement module that consists of a large diameter roll covered by the air impingement hood, followed by a steam-heated cylinder, a small roll and another steam-heated cylinder [Sundqvist, 1994]. Both the large and the small roll are vacuum-assisted. About 70% of the drying is carried out by the impingement hood. The remainder of the water is removed by the steam-heated cylinders and suction at the small roll. Since the water evaporated on the steam-heated cylinders has to be removed as well, the complete drying section is covered. Humidity, temperature, and impingement speed of the drying air can be optimized to achieve a low SEC and high drying rate. Some of the heat in the exhaust air can be recovered. Drying rates can be beyond 100 kg/m^{2}h, depending on the temperature and velocity of the air stream [Sundqvist, 1994].

Model calculations show that the SEC_h ranges from 1.9 to 4.3 MJ/kg evaporated water, with the average being about 3 MJ/kg water evaporated [Sundqvist, 1994]. The minimum value is achieved at high exhaust air humidity and low drying air temperature and velocity, and a low drying rate [Sundqvist, 1994; Polat and Mujumdar, 1995]. Compared to conventional drying with average SEC_hs of 3.4 to 5 MJ/kg evaporated water, air impingement drying can result in a 10-40% saving. A saving on the SEC_e is achieved because of the smaller drying section. However, extra electricity is required for air circulation. Overall an increase of 0-5% in SEC_e can be expected.

About eight air impingement modules are required to replace a conventional drying section [Sundqvist, 1994]. (Note: in Figure 4.3e only four modules are shown.) Although the complexity of the paper machine increases, the investment costs can be reduced due to the shortening of the drying section. We assume that the number of operators and the maintenance requirement will remain more or less the same.
This technology has already been applied in the production of sanitary paper, and is now being developed for the production of other paper types. Valmet has been doing pilot machine trials since 1994 [Sundqvist, 1994]. The results of the pilot plant tests for newsprint shows that air impingement drying has no harmful effects on paper quality [Kertulla, 1994]. The pilot plant has been run successfully at 2400 m/min [Kertulla, 1994]. Although air impingement drying is already being used in the drying of sanitary paper, the principle is new for other types of paper. The drying rate improves considerably. The paper-making process needs to be adapted substantially. Air impingement drying can therefore be categorized as a major change.

4.4.3 LATENT HEAT RECOVERY SYSTEMS

We will now discuss two concepts that make use of the latent heat of the evaporated water, namely steam impingement drying and airless drying. We have already discussed the use of a heat pump with condebelt drying to recover the latent heat from the evaporated water. Recovery and upgrading from the latent heat of evaporated water is a well-known process in some other industrial sectors. In the dairy industry, for instance, the vapor produced during the concentration of milk is upgraded by recompression and subsequently used as drying medium. This technology is known as mechanical vapor recompression.

<u>Steam impingement drying</u> (Figure 4.3f) is comparable to air impingement drying, with the difference being that superheated steam is used as the drying medium instead of hot air. A steam impingement dryer consists of a covered drying cylinder, a compressor or fan, a re-heater, and a heat exchanger to recover heat from the excess steam. The water is evaporated by superheated steam of about 300°C and 1.1 bar which is blown onto the sheet. The steam cools to about 150°C, releasing about 0.3 GJ of heat per mt of steam. It can be calculated that a minimum of about 10-11 mt of superheated steam is required to dry 1 mt of paper from a consistency of 43% ds to 95% ds. Additional steam is required to compensate for radiative losses and steam leakages, for heating the fibers and providing desorption heat, and for other inefficiencies. Reported design values are 15 to 20 mt of steam per mt of paper [Deventer, 1995]. Excess steam, equal to the amount of water evaporated, is purged. The latent heat can be recovered in a condenser. A compressor or fan is required to compensate for the pressure drop of about 0.05-0.1 bar. The drying rate can be considerably higher than with conventional drying [Kertulla, 1994; Deventer, 1995]. Drying rates of 100-200 kg/m^{2}h are reported [Polat and Mujumdar, 1995].
Several steam impingement dryers in a row can replace the entire conventional drying section [Kertulla, 1994]. Figure 4.3f shows three dryers, but the number of dryers can vary. If we assume that 15 mt of steam per mt of paper is necessary, the SEC_h is 4.5 GJ/mt of

paper. This value is more or less equal to that in conventional drying. Reduction of the heat requirement can be accomplished by recovering the latent heat from the purge steam. In a steady-state situation, all the evaporated water is condensed, and the latent heat recovered equals about 3 GJ/mt of paper. The maximum heat saving potential can be achieved when all the recovered heat can be used in the process. In practical situations the remaining heat requirement for processes other than drying is low, ranging from 0.5-1.0 GJ/mt of paper. The saving in SEC_h is therefore in the range of 10-15%. Theoretically, a larger saving can be achieved when a heat pump is installed which increases the temperature of the purge steam so that it can supply some of the heat required for superheating. However, heat pumps that can deliver heat at temperatures above 250°C are currently not available. Upgrading heat from 150°C to 300°C is therefore not possible, but this might change in the future [Smit *et al.*, 1994]. Another possibility is to work with lower temperatures, but this will have important consequences for the system lay-out and performance. One consequence would be that the volume of steam would increase. Then the equipment would need to be larger and the electricity requirement for circulating the steam would increase. In a situation without a heat pump, the additional electricity required to drive the fan is less than 1% of the original SEC_e. Because the drying section can be reduced considerably the electricity requirement for drive power will also be reduced. Contrary to air impingement drying, no electricity is required to drive air fans. Overall a 5-10% reduction in the SEC_e can be expected.

Steam impingement drying requires additional investment costs. However, the drying section can be much shorter, resulting in an overall reduction in the investment costs. The additional O&M-costs for the new equipment will be offset by lower O&M-costs for the shorter paper machine.

Superheated steam drying is already used for some industrial purposes, like the drying of sludge, coal and beet pulp [Mujumdar, 1995]. No commercial superheated steam dryer for paper exists at the moment. Laboratory scale tests have been done [Mujumdar, 1995], and research into the impact on paper quality has been performed [Poirier *et al.*, 1994]. The paper machinery manufacturer Valmet-Tampella and three research institutes are working together to develop the technology [Deventer, 1995]. A pilot plant of 1 meter width was become operational at VTT in Finland at the end of 1996 [Deventer, 1995]. Considerable RD&D effort is still required, for example to study the start-up and shut-down phase, the condensation of steam on the sheet, and sealing at high speed.

Steam impingement drying means drying according to a new principle. Implementation requires a major adaptation to the paper-making process. It affects the paper quality and the production parameters. Therefore, it can be categorized as a major change.

Airless drying (Figure 4.3g) also uses the latent heat of the evaporated moisture. It requires an airtight and well-insulated hood around the drying section of the paper machine. The paper is still dried by steam-heated cylinders, but the steam is produced by compressing the evaporated water from atmospheric pressure to 4 bar. Some of the superheated steam can also be recycled to the drying chamber to achieve additional drying by impingement. The problem of making the drying chamber airtight may be solved by making use of the large difference in the density of atmospheric steam inside the dryer and air outside the dryer. A very stable layer is formed between steam and air with a high resistance to cross-diffusion of both steam and air [Stubbing, 1993b].

When paper is dried airlessly from 43 to 95% consistency, 1.21 mt of water vapor is generated per mt of paper. This vapor can supply about 60-80% of the total amount of thermal energy needed. Recovery of the heat of vaporization can reduce the heat requirement of the paper mill by 5-10%. Overall the SEC_h can be reduced by 70-90%. Compressing the steam from 1 to 4 bar requires about 0.4 GJ_e, or 15-25% of the original SEC_e. However, since less ventilation is required, the resulting SEC_e will be approximately 15-20% higher than in conventional paper making.

An airless dryer involves constructing a closed hood around the drying section, restructuring the steam system and installing compressors. Direct investment costs will therefore be higher than for a conventional dryer. O&M-costs will increase also, mainly for the maintenance of the compressor [Stubbing, 1990].

The process is being developed by a British firm, Heat-Win, and is commercial for batch processes, like the drying of ceramics and laundry [Stubbing, 1993a]. Three continuous airless dryers have been built since 1993 to validate the theory that the large density difference between steam and air can be used to seal the airless dryer [Stubbing, 1993b; Stubbing, 1996]. Continuous airless drying will be applied first to dry ceramics, sludge and granular materials. Practical application to paper drying is not being considered at the moment [Stubbing, 1996]. Airless drying for paper making is still at the stage of applied research. It should be noted that airless drying is the only selected technology that is not being developed by a paper machinery manufacturer or a paper-related research institute. Airless drying still involves cylinder drying, but with considerable adaptations, although the old paper machine can be left in place. Effects on paper quality and production parameters should be small. We define airless drying as being on the borderline between evolutionary change and major change.

4.4.4 Comparison of the technologies

Table 4.6 summarizes the characteristics of the seven technologies. The change in SEC_h and SEC_e is converted to a SEC based on primary energy carriers, on the assumption that heat and steam will be generated with a 90% efficiency and electricity with a 50% efficiency. Figure 4.4 compares the technologies with regard to the degree of technical change and the stage of development.

Dry-sheet-forming techniques have the largest potential for improving the SEC_h: 50-100% of the current SEC_h. However, as the SEC_e increases considerably, the specific primary energy consumption increases by at least 55%. Dry-sheet-forming is already a commercial technology for some special papers and boards. However, currently no attempt is being made to develop this technology for other grades of paper. If this situation continues, it is highly unlikely that this technology will be commercialized for other grades of paper.

Paper-machine manufacturers and paper research institutes have done considerable research into high-temperature pressing technologies, i.e. press drying, impulse drying and

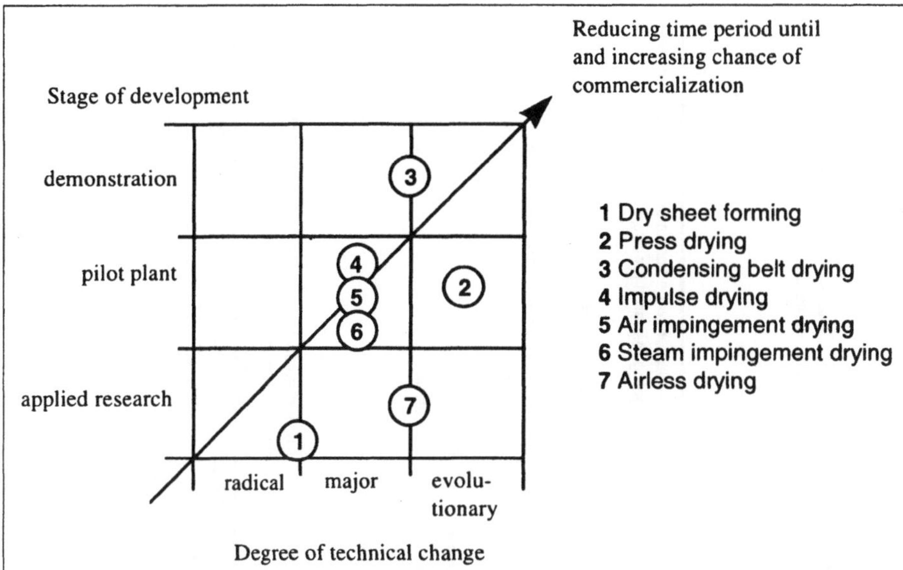

Figure 4.4: Comparison of the selected technologies on the degree of technical change required and the stage of development.

condensing belt drying. Of these technologies condensing belt drying is the one in the most advance stage of development. Impulse drying has a higher saving potential on SEC_h: 35-50% as compared to 10-20% for condensing belt drying. However, impulse drying requires more RD&D to solve technical problems. Press drying seems to be over the top of its research activity. The change in SEC_e of these technologies varies from -5% to +10% of the SEC_e of conventional paper making. Air impingement drying requires still further RD&D, but is being seriously developed by a paper machine manufacturer.

Finally, of the two technologies that make use of the fact that the latent heat of the evaporated moisture is lost and that low-quality heat is required, airless drying promises larger savings on SEC_h, 70-90%. The SEC_e increases by 15 to 20%. Although airless drying is commercial for batch processes, its application to continuous paper drying is still at the stage of applied research. Furthermore, it is the only technology that is not being developed by a firm related to the paper industry. Steam impingement drying, on the other hand, has a smaller saving potential, but has a higher chance of commercialization, because it is in a more advanced stage of development and it is being developed by a paper machinery manufacturer. It is not claimed that the two described technologies are the only ones to make use of the latent heat from the evaporated moisture. Other configurations are conceivable, and further research in this direction may result in additional solutions.

Table 4.6: Comparison of the selected technologies.

Technology	Change in SEC_h	Change in SEC_e	Change in SEC based on primary energy carriers[a]	Investment costs compared to current technology[b]	Stage of development	Degree of departure compared to current technology
Dry-sheet-forming	-50 to -100%	+150 to + 500%	+55 to +180%	smaller	Commercial for some specialties. No ongoing research for other types.	major/radical
Press drying	-5 to -30%	-5%	-5 to -20%	smaller	Pilot plant; over peak of research activity	evolutionary
Condensing belt drying (excl. heat pump)	-10 to -20%	0%	-5 to -10%	smaller	Demonstration unit at paper mill in Finland	evolutionary - major
Impulse drying	-35 to -50%	+5 to + 10%	-10 to -25%	smaller	Pilot plant	major
Air impingement drying	-10 to -40%	0 to +5%	-5 to -20%	smaller	Pilot plant	major
Steam impingement drying	-10 to -15%	-5 to -10%	-5 to -15%	smaller	Pilot plant	major
Airless drying	-70 to -90%	+15 to +20%	-20 to - 40%	larger	Applied research for paper drying; commer-cial for some batch drying processes;	evolutionary - major

[a] Assuming that in several decades heat will be generated in boilers with an ef ficiency of 90% and electricity will be generated in power plants with an efficiency of 50%. The reference SEC_h ranges from 2.5-6.0 GJ/mt of paper and the reference SEC_e from 1.5-3.0 GJ_e/mt of paper.

[b] Only the investments for the paper machine itself are taken into account. Lower investment cost for the steam boiler or CHP-unit due to a smaller capacity are not taken into account.

Combinations of short-term and long-term options to improve the energy efficiency are possible, resulting in lower SECs than those given in Table 4.6. For instance, an extended nip press and high-temperature pressing followed by impingement drying is a plausible configuration [Kertulla, 1994]. The SEC might be even further reduced by several evolutionary technologies, not mentioned in this study. We estimate that in the long term the SEC_h can be reduced by 50-75%, for example by a combination of impulse drying, a latent heat recovery system (e.g. steam impingement drying) and a number of small improvements. The SEC_e will remain equal or will increase slightly. Simultaneously, the production rate will increase substantially and the paper machine will become more compact. The direct investment costs will be lower than those for a conventional machine.

4.5 Discussion

We will discuss each step of the method on the basis of its application to the paper and board industry.

Process energy analysis We selected an energy service that restricted our study to paper making processes only. It should be realized that the range of technologies selected is affected by this choice. For instance, if the energy service was providing a medium for information exchange, distributing news via the Internet instead of newspapers would have to be taken into account. This may reduce the demand for paper, and hence the energy consumption for information exchange. The choice of the energy service we have used in this study also excluded more efficient pulping processes and recycling of waste paper from the analysis.
As an alternative approach, exergy analysis could have been used to analyze options for energy-efficiency improvements in more detail. However, as the temperature differences are small and as there are hardly any chemical modifications, this approach is not useful in this case. The level of analysis in our study was adequate to show where the major losses occurred, and how these losses can be reduced. Subsequently, the analysis of the water removal process pointed the way to the main improvement options.

Identification of technologies Given the restrictions specified in Section 6.1 concerning the definition of the energy service and the level of detail in analysis, we judge that the list of technologies covers all relevant fields of technologies that might be able to perform the energy service with a higher energy efficiency. It should be noted that, except for airless drying, all technologies selected are being developed to improve paper quality or production speed. We also conclude that improvement of energy efficiency is an important, but certainly not the main impetus for the development of new technologies for the paper industry. Consequently, technologies for improving energy efficiency only are not common. This is demonstrated by the fact that other technologies for recovering the latent

heat from the evaporated water are conceivable but are not being currently developed for paper making.
Most of the technologies described are already at a pilot plant stage, or are even being demonstrated in a paper mill. We did not come across any techniques that might enter the market after several decades. Since such technologies are still in early stages of development, it is not surprising that pertinent information is limited or not public.

Characterization of selected technologies A vital part was gathering and handling of data. Without access in all cases to reliable data on energy consumption, we used our own analyses, expert consultants, and the available literature to give a well-founded estimate of the expected SEC_h for all technologies. We indicated the uncertainty in the data by giving ranges of the expected reduction of the SEC.
The fact that investment costs for technologies developed by paper-machine manufacturers are estimated to be lower than for a conventional paper machine can be seen as a direct result of the tendency to shorten the production-rate limiting drying section. The only type of drying that might have higher investment costs is airless drying, that is being developed by a company that is not related to the paper and board industry.
Division into degrees of technical change is based on several characteristics of the technology, e.g. effect on performance and process parameters, the adaptations that will have to be made to the process. Information that allows these characteristics to be determined per technology is readily available. It was not always possible to place a new technology into one of the indicated categories, because the technology often had characteristics of more than one category. This underlines the notion that technical changes should be seen as a continuum rather than as discrete categories. Information about the stage of development was generally easy to obtain.

4.6 Conclusions and recommendations

A method for identifying and characterization of technologies that can improve the energy efficiency in the long term has been described and applied to the paper and board industry. By giving considerable attention to selection of the technologies we obtained a comprehensive picture of the saving potential. An overview is given of all technologies that can perform the energy service with less energy, irrespective of whether they have a high chance of commercialization or not. This helps to identify technologies that have a high potential for energy-efficiency improvement, but need additional RD&D effort. Technologies that work together to perform the same energy service, like condensing belt drying and air impingement drying, can also be assessed. Mutually exclusive technologies can be identified as well, like dry-sheet-forming and airless drying. The gathering and handling of data is transparent and reproducible. For some purposes a less-detailed set of data will be sufficient. More sector studies will be performed, to assess whether the method is generally applicable or needs adaptation.

The application of the method to the paper and board industry resulted in the characterization of seven technologies that can reduce the SEC_h for the paper-making process. Dry-sheet-forming technologies and airless drying both have the potential to significantly reduce the SEC_h by 50-100% and 70-90% respectively. However, the SEC_e will increase for both techniques. They are unlikely to be commercialized, unless RD&D-activity is intensified. This in contrast to condensing belt drying, impulse drying, air impingement drying and steam impingement drying, which are being developed by paper-related research institutes and paper-machinery manufacturers. Of these technologies, impulse drying has the highest saving potential, 35-50% of the SEC_h. The other technologies have potentials that range from 10-40% of the SEC_h. The change in SEC_e ranges from -10 to 20% of the SEC_e of conventional paper making. All the latter technologies have lower investment costs than conventional paper making and will bring benefits other than energy-efficiency improvement. They should lead to an improved paper quality or a higher production rate. It is concluded that in the long term the SEC_h can be reduced by 50-75% compared to the current average by the use of a combination of new pressing and drying technologies, latent heat recovery systems, and by the introduction of several smaller improvements. The SEC_e will remain about equal or slightly increase.

The method not only provides a comprehensive picture of the energy-efficiency improvement potential achievable with technologies currently under development, it also helps to identify fields where enhanced RD&D-activity might result in large energy-efficiency improvements. For improving energy efficiency in paper making beyond the potential given in this chapter it should be considered to do more research into the application of latent heat recovery systems in the paper and board industry, benefitting from the expertise in other sectors.

Acknowledgements - The authors would like to thank J. Lehtinen (Valmet Tampella), T. Stubbing (Heat Win), L. Nilsson (Lund University), R. Kemp (MERIT), H. van Deventer (TNO), E. Luiten and W. Turkenburg (Utrecht University) for providing information, suggestions and comments on our study. Financial support provided by the Netherlands Organization of Scientific Research is acknowledged.

CHAPTER 5

FUTURE TECHNOLOGIES FOR ENERGY-EFFICIENT IRON AND STEEL MAKING[1]

5.1 Introduction

The iron and steel industry is the largest energy-consuming manufacturing industry in the world. In 1990, its global energy consumption was estimated to be 18-19 EJ, or 10-15% of the annual industrial energy consumption [WEÇ, 1995]. Figure 5.1 shows that the annual world steel production has increased from about 100 million tonnes in 1945 to about 770 million tonnes in 1990 [IISI, 1992; IISI, 1996a]. The global steel production is expected to grow further by about 1.7% a year, mainly because of an increase in steel consumption in developing countries [Tilton, 1990; WEC, 1995]. At present the apparent steel consumption per capita in these countries is only one seventh of that in OECD-countries, but this situation is likely to change [WEC, 1995]. Whereas the crude steel production in OECD countries has remained fairly stable at 320-370 million tonnes per year since 1980, the production in developing countries is growing steadily at a rate of more than 6% annually and reached about 240 million tonnes in 1993 [WEC, 1995]. This growth is expected to continue. As a result, the global steel production might rise to 1280 million tonnes in the year 2020, assuming a business-as-usual scenario. In this scenario the global energy consumption of the iron and steel industry is projected to increase to more than 25 EJ in the year 2020 [WEC, 1995].

Improvement in the energy efficiency of steel production is one option to counteract the increasing demand for energy. There have been many studies of the potential for energy-efficiency improvement that can be realized in the short term, i.e. in less than 10-15 years from now, see e.g. [IISI, 1982; Maier and Angerer, 1986; Faure, 1993a; Worrell *et al.*, 1993; Chatterjee, 1996]. There have also been some estimates of the energy demand of the steel industry in the longer term. For instance, in a report of the World Energy Council it was estimated that on the basis of an advanced technology scenario the primary energy demand would grow to about 20 EJ in the year 2020 [WEC, 1995]. This amount would be a 20% reduction in the energy demand projected by the aforementioned business-as-usual scenario. Although scenario studies may give us some insight into possible developments, they usually give little information about the techniques required to bring about the energy-efficiency improvements. More information is needed on each technique, and the information needs to be collected and presented in a systematic way. Only then will it be

[1] Published in: Annual Review of Energy and Environment (23) 1998, p123-205. Co-authors: Ernst Worrell (Lawrence Berkeley National Laboratory, Berkeley CA, USA) and Kornelis Blok (Department of Science, Technology and Society, Utrecht University).

possible to assess the associated research and development (R&D) requirements, and to determine how much a specific technique will contribute to an improvement in energy efficiency in the longer term.

The objective of the present chapter is to identify and characterize techniques, through a systematic approach, that can contribute to an increase in the energy efficiency of steel making, in order to estimate the long-term potential for energy-efficiency improvement and to assess R&D-priorities. This approach has been described extensively in Chapter 2 of this book.

In this chapter first the historical perspective of iron and steel making processes is described. Next, an analysis is made of the theoretically lowest amount of energy required to produce iron and steel. In the following section of the chapter an exergy analysis is made of the currently prevailing steel production route, the blast furnace - basic oxygen furnace route. On the basis of the results of the energy and exergy analyses, a description is given of possible routes for energy efficiency improvement. Next, different techniques are described that can improve the energy efficiency of steel making. The potential impact and costs of each technique are evaluated. Finally, the methodology applied and the results are discussed and conclusions are drawn. In addition, recommendations for policy makers are given.

5.2 Past technological development of iron and steel production

In this section a brief history of the major iron and steel making processes is presented. The aim is to place these processes in a historical perspective and to describe energy-efficiency improvement in the past. We first discuss the main processes involved in the making of pig iron, which is reduced iron ore that still contains impurities, mainly carbon. Then we deal with the main processes used to improve the quality by removing impurities, with an emphasis on steel making processes.

5.2.1 History of iron making

The first record of the use of iron goes back to 2500-2000 BC [Chatterjee, 1996]. It is believed that in that period iron was not produced deliberately but was obtained from natural resources, e.g. meteorites [Kirk-Othmer, 1981]. Deliberate production of iron began in about 1300 BC with the use of charcoal as fuel and reducer, in small furnaces that made use of cold air. Evidence for the existence of such furnaces has been found in Africa, Asia and in Central Europe [Kirk-Othmer, 1981; Juleff, 1996]. The temperature that could be achieved in these furnaces was probably below the melting point of iron. The product had to be hammered for it to be freed from slag and to make wrought iron. When better blowing devices were introduced, the temperature could be raised, and liquid, high-carbon iron was formed. In 1300 AD the Stuckoven was introduced in Germany. Although the Stuckoven was only 3-5 m high and 1-1.2 m in diameter [Kirk-Othmer, 1981], its design was essentially the same as that of the modern blast furnace. Charcoal was used as fuel.

Based on data on the use of charcoal to produce pig iron and bar steel in the United Kingdom in the period 1540-1760 [Jeans, 1882; Hammersley, 1973], we can make an estimate of the reduction in the energy demand in this period. The charcoal consumption to make pig iron decreased from 5.5 to 2 loads[1] of charcoal per tonne of pig iron in this period. This is an improvement in energy efficiency of about 0.5% a year. Pig iron was converted to bar steel in the finery process. Between 1540 and 1760 the energy demand for the finery process decreased from 16 to 4 loads per tonne of bar steel, or a decrease of 0.6% a year [Hammersley, 1973]. Because both the demand for charcoal used for steel making and the amount of pig iron needed per tonne of steel decreased, the overall energy efficiency improvement is greater than 0.6% a year [Hammersley, 1973].
Because of the weak structure of charcoal, the height, thus the capacity of the blast furnaces was limited. This is because the coal in the mix forms the supporting structure of the furnace charge. Coke is much stronger and does not have this disadvantage. Coke was first used around 1718, but its application in the United Kingdom remained limited to one site until the 1750s [Hyde, 1977]. Before 1750 charcoal was cheaper than coke, but this situation changed in the period 1750-1790. In addition, the amount of coke required for pig iron reduction decreased markedly in this period [Hyde, 1977]. In 1750 coke pig iron made up 5% of the total UK pig iron production; by 1790 it made up 90% [Hyde, 1977]. The development of the average coke consumption for pig iron production from the time when coke-fired blast furnaces were introduced is shown in Figure 5.2 [Heal, 1975; IISI, 1996a]. Three main periods of energy efficiency improvement can be distinguished. First, in the period of the first diffusion of the process, between 1760 and 1800, a reduction in the coke demand of almost 2% a year was achieved, mainly by the introduction of steam engines, which permitted the use of higher blast pressures [Heal, 1975]. Second, in the 19th century the coke demand declined further, by an average of 1% a year. The use of regenerators to preheat the blast accounted for much of this reduction. Finally, in the period 1950-1990, reduction of demand for coke was 3.4% a year on average. This reduction in demand was achieved by, for instance, increasing the iron content of the ore, using ore agglomeration, raising the temperature of the hot blast, and the use of blast furnaces with a larger inner volume. On average an improvement in the energy efficiency of iron making of 1.4% a year was achieved in the period 1760-1990.

Up to the 1960s the blast furnace was the main process for reducing iron oxide. Direct reduction processes have been in use since ancient times, but gained renewed interest in the 1960s. Several direct reduction processes have been developed and are now in use. From a more recent date are the smelting reduction processes, which are still under development. These two processes are discussed later in this chapter.

---

[1] At that time charcoal was delivered in cartloads to the ironworks. A "load" did not have a standard measure. Hammersley [Hammersley, 1973] gives a range of 13.5-17.5 hundredweight (cwt) (1 cwt is about 50 kg) for a load of charcoal. Assuming an average lower heating value of 29.5 GJ/tonne [Rossillo-Calle *et al.*, 1996], 1 load of charcoal equals 20-26 GJ.

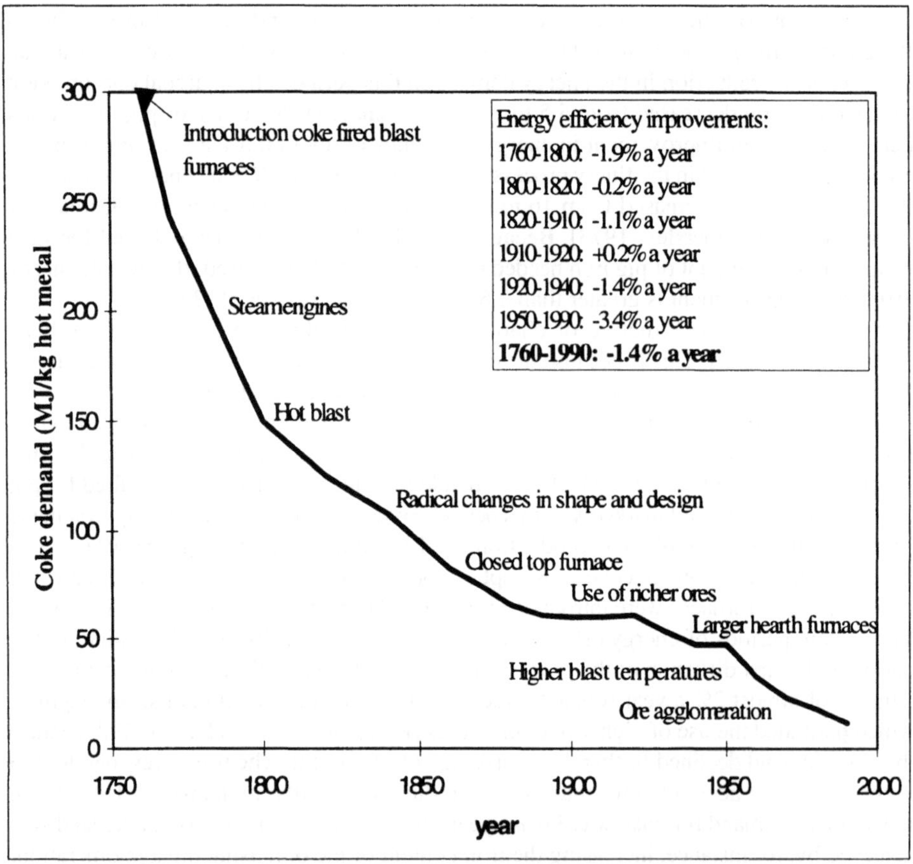

Figure 5.2: Development of the coke demand for pig iron making [Hammersley, 1973; IISI, 1996a].

5.2.2 HISTORY OF STEEL MAKING

Table 5.1 gives an overview of the history of processes to upgrade the quality of pig iron and steel making processes. Three lines of development are distinguished: (a)refining processes, (b) (re)melting processes, and (c) processes that both refine and melt. In refining processes, carbon and other impurities present in pig iron, like silicon and manganese, are removed. In this line of development we also consider processes that free the pig iron from slag. The product is not steel but is, for instance, wrought iron. Melting processes are processes whereby the steel is melted only to be cast. Because no refining takes place, the composition of the feed should equal the composition of the desired product.

The oldest process for refining iron is the inefficient, charcoal-fueled, finery process, which was widely used in the 11th and the 12th centuries [Kudrin, 1985]. The product was wrought iron rather than steel. At the end of the 18th century the puddle process was introduced. First it was used to make iron, but around 1850 it was converted to make steel by refining pig iron on the hearth of a reverberatory furnace. The product of a puddle process was a semifluid steel, which had to be forged. However, the technique had limited success.

Table 5.1: History of steel-making processes[a]. Approximate dates of first industrial application are shown in parenthesis.

Year	Refining processes	Refining and melting processes	Melting processes
	Finery (from 12th century)[b]		Crucible (18th century)[b]
	Puddling (1785)[b]		
1850	Bessemer (1860)	Open Hearth Furnace (1864)	
	Thomas (1878)		
1900			Electric Arc Furnace (1900)
1950		Basic oxygen furnace (1952)	
		Oxygen bottom blowing (1967) Combined blowing (1970)	UHP-EAF (1970) DC-EAF (1985)
2000			

[a] Based on [Faure, 1993b] and [Kudrin, 1985].
[b] Based on [Kudrin, 1985].

In 1855 Bessemer obtained a patent on a new process, at present known as the Bessemer converter. In 1860 the first Bessemer process went into operation [Rosenberg, 1982; Kudrin, 1985]. The principle of the Bessemer converter is still followed: The oxidation of carbon and other impurities provides enough heat to melt the metal. In the Bessemer converter cold air was blown from the bottom through a refractory-lined vessel. In theory, no additional fuel was required. In practice, about 1 tonne of coal per tonne of steel was consumed [Rosenberg, 1982], equal to about 30 GJ/tonne steel. Other advantages were the reduced refining time and investment costs. However, there were several disadvantages: It was impossible to remove sulphur and phosphor; the product became brittle after some time because of the large quantities of nitrogen dissolved in the steel; and the process of oxidation was so fast that is was very difficult to control the product quality [Rosenberg, 1982].

In 1878 an adapted version of the Bessemer process, the Thomas process, was introduced. This process allowed the production of low-phosphorus steel from high-phosphorus pig iron. The Thomas process used a basic refractory lining instead of the acid refractory lining of the Bessemer process; it is therefore also called the Basic Bessemer process [Kudrin, 1985].

In the meantime the Open Hearth Furnace (OHF) (or the Siemens-Martin furnace) was developed in France. In an OHF pig iron and scrap are melted on a hearth of a reverberatory furnace by the heat of a flame. The OHF resembles the puddle process; the difference is that in the OHF air and gaseous fuel are preheated by heat exchanging with the combustion gases in what was called a regenerative gas furnace [Siemens, 1873]. With the regenerative gas furnace it was possible to attain temperatures sufficiently high to melt steel. The process had two main advantages over the Bessemer process: (a) pig iron and scrap of any composition could be melted and (b) good control of the steel quality was possible. The price paid for this was higher investment, higher energy consumption and longer refining time.

In 1952 another new process for steel making was introduced: the Basic Oxygen Furnace (BOF). The process is also know as the Linz-Donawitz (LD) process, named after the two cities where the Austrian steel company VOEST built the first two BOFs. A BOF is an improved version of the Bessemer process. Oxygen is blown through a water-cooled lance from the top into the converter. The advantages of using pure oxygen instead of air are that the gas volume to be heated and compressed is smaller, no nitrogen can dissolve in the metal, and the heat generated by the oxidation of impurities is greater and adequate to melt 20-30% additional scrap. The BOF had a far better energy efficiency than the OHF and refining was ten times faster. The idea of using oxygen was already mentioned by Bessemer in 1856. Two factors impeded earlier implementation. First, industrial methods for producing large quantities of oxygen became available only around 1950. Second, experiments were initially directed at blowing oxygen from the bottom into the converter. This configuration generated so much heat that the tubes through which the oxygen was blown (tuyeres) could withstand only one single heat [Kudrin, 1985]. In the 1970s several processes were developed that used the concept of bottom-blowing. Currently, combined blowing processes, i.e. processes that combine the advantages of top- and bottom-blowing, are in operation at some sites (e.g. OBM and Q-BOP). The state-of-the-art process is a basic oxygen furnace that uses top-gas-recovery and additional scrap melting. Modern BOFs are net energy producers.

The oldest melting process is the crucible process. A closed pot with an average capacity of 25-35 kg – the crucible – was filled with solid wrought iron and heated in a shaft furnace [Kudrin, 1985]. The process required a charge with a composition close to that of the product [Kudrin, 1985] and about 7 tonnes of coke per tonne of steel [Rosenberg, 1982].

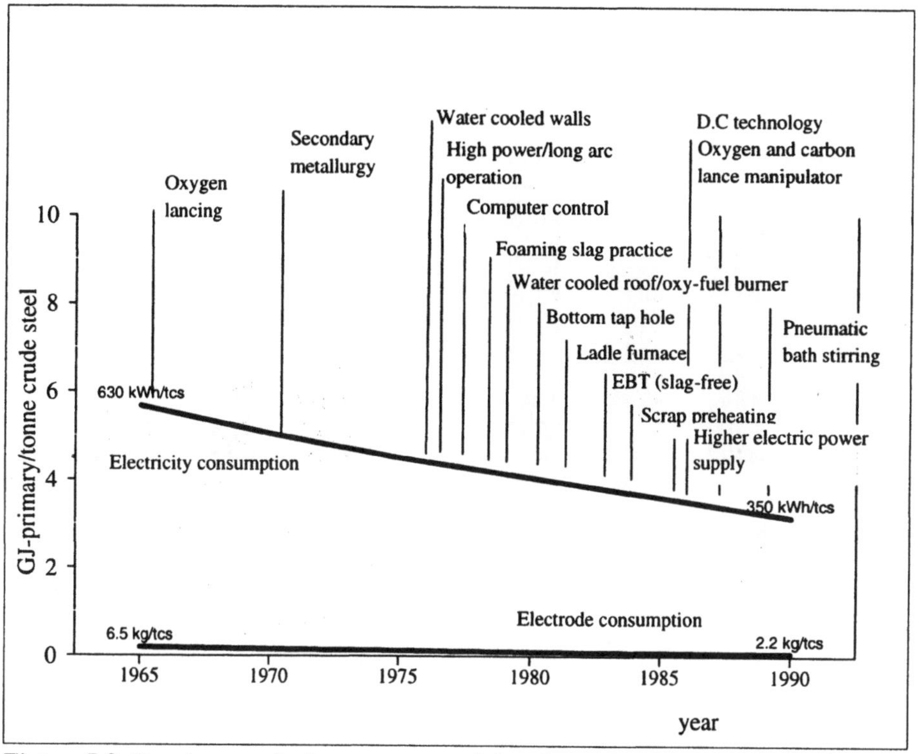

Figure 5.3: Development in the energy use of electric arc furnaces. Based on data from [Szekely, 1994] and [Bock, 1994].

A completely different route to steel is the melting of iron in a bath at a high temperature achieved with the help of electric arcs: the Electric Arc Furnace (EAF). First introduced in the late nineteenth century, its application was limited to specialty steels [Rosenberg, 1982]. At present EAFs are used to produce a whole range of products. EAF technology is very flexible with respect to inputs. All types of iron can be handled as can 100% scrap. Furthermore, it can be built separately from blast furnaces and coke ovens. Performance of EAFs has improved tremendously, as is illustrated in Figure 5.3. The figure shows that in the period 1965-1990 electricity demand declined from 630 to 350 kWh/tonne steel (2.3% a year on average) and electrode consumption declined from 6.5 to 2.2 kg/tonne (4.2% a year on average). So that a comparison can made, both consumptions are recalculated in terms of primary energy consumption (in GJ) per tonne of liquid steel[1]. Nowadays, refining also takes place in the EAF with the help of oxygen being blown into the furnace.

[1] It is assumed that the 1 kg electrode material equals 30 MJ and that electricity is generated with an efficiency of 40%.

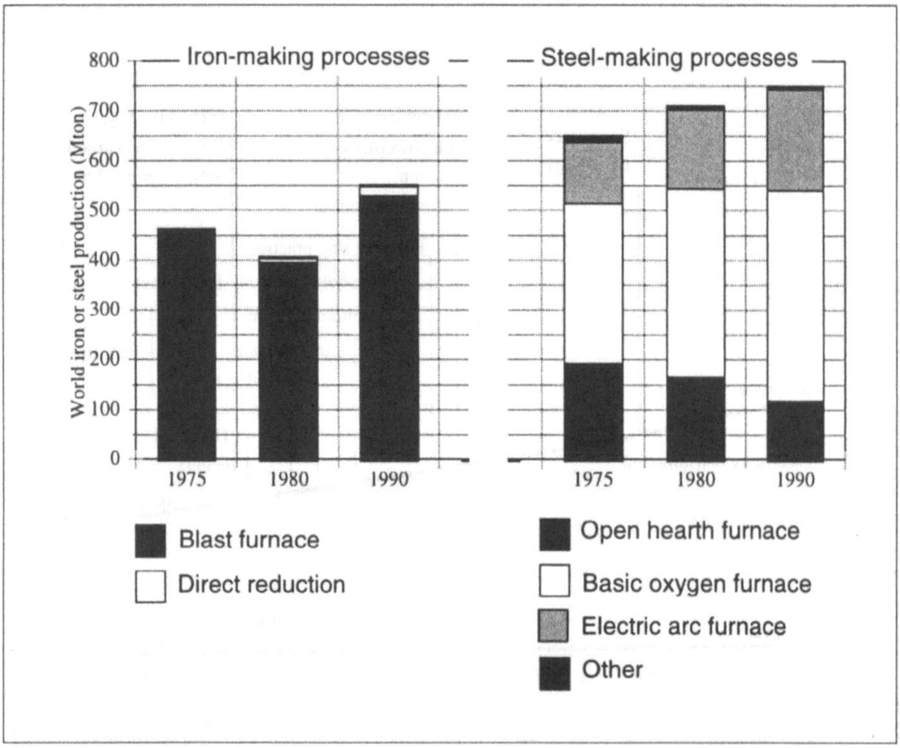

Figure 5.4: Production volumes of the main iron and steel production processes in 1975, 1980 and 1990. Data for 1975 and 1980, except for direct reduction, are taken from [Norris, 1996]; Data on direct reduction for 1975 and 1980 are taken from [Midrex, 1996]. All 1990 data are taken from [IISI, 1992].

5.2.3 THE CURRENT SITUATION

Figure 5.4 shows the proportion of different iron and steel production processes in the world production of iron and steel. The blast furnace is the most widely used production process for iron. The Basic Oxygen Furnace is still the main steel production technology, but the proportion of the Electric Arc Furnace is increasing steadily. Three steel production routes are illustrated schematically in Figure 5.5. This figure gives also specific energy consumptions (SECs), expressed in GJ primary energy per tonne of crude steel (GJ/tcs), for the different production routes [Worrell, 1995]. The SECs represent best-practice values, i.e the lowest values actually achieved in one plant. As can be seen from the figure , the SEC differs considerably depending on the process route. Even large differences in SEC occur with the same production method. The SEC for an integrated primary steel mill

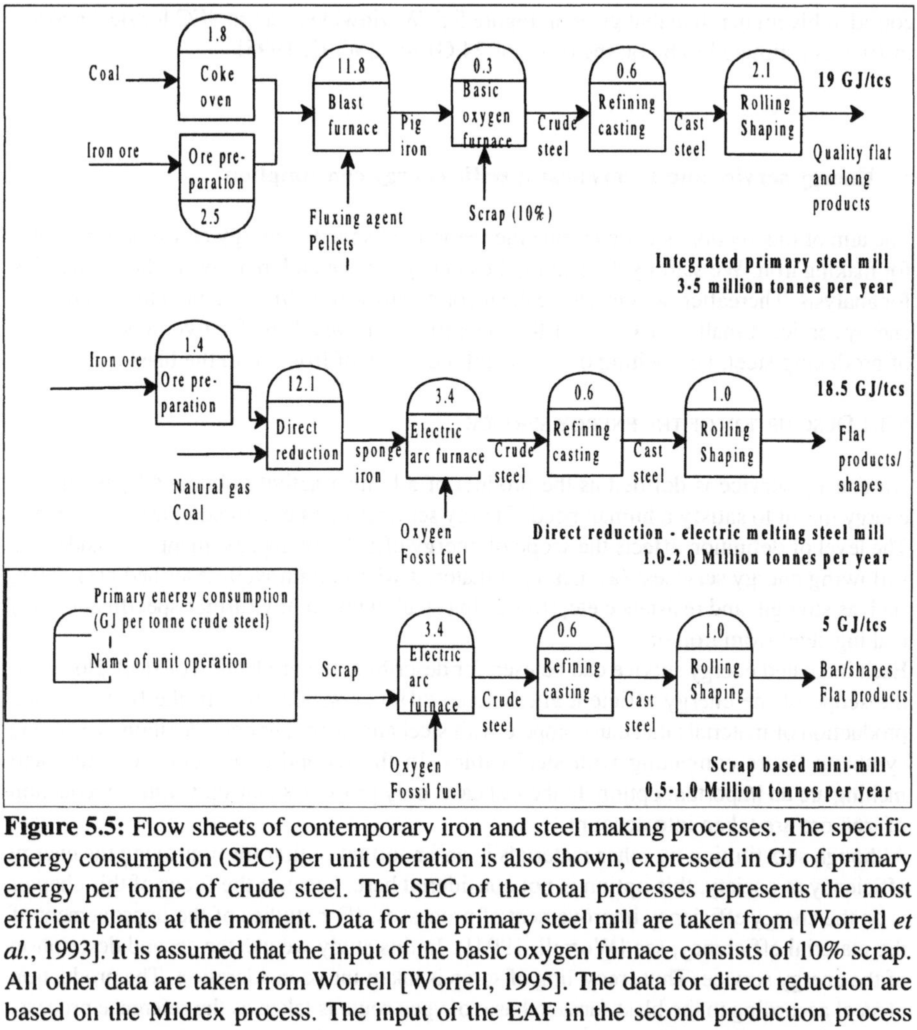

Figure 5.5: Flow sheets of contemporary iron and steel making processes. The specific energy consumption (SEC) per unit operation is also shown, expressed in GJ of primary energy per tonne of crude steel. The SEC of the total processes represents the most efficient plants at the moment. Data for the primary steel mill are taken from [Worrell *et al.*, 1993]. It is assumed that the input of the basic oxygen furnace consists of 10% scrap. All other data are taken from Worrell [Worrell, 1995]. The data for direct reduction are based on the Midrex process. The input of the EAF in the second production process consists solely of direct reduced iron. All processes end with a hot rolling mill. A cold rolling mill and other finishing operations are not taken into consideration, because of the large variation per product. The typical annual capacity is also given.

varies from 19 to 40 GJ/tcs [WEC, 1995]. The direct reduction-electric arc furnace (DR-EAF) production route shows less variation in SEC, because the technology is newer and there are far fewer plants in operation. EAF steel making itself has become far more efficient over the past 25 years, as we have shown. Old EAF plants have a SEC that is considerably higher than that given in Figure 5.5. Worldwide average SEC for steel making in 1990 is estimated to be on the order of 24 GJ/tcs[1] [WEC, 1995].

5.3 Energy service and theoretical specific energy consumption

The aim of this section is to determine the theoretical specific energy consumption (SEC) for making iron. We start by describing the energy service and thereby set the boundaries for analysis. Thereafter, we determine the theoretically lowest SEC required to perform this energy service. Finally, we consider the theoretically lowest SEC for two important ways of producing steel, i.e. melting of scrap and reduction of iron ore in the blast furnace.

5.3.1 Description of the energy service

An energy service is defined as the product of a human activity obtained by the use of energy meant to satisfy a human need. Energy services can be defined at different levels. The level of definition affects the scope of energy efficiency improvements. Consider the following energy services: (a) making a material with certain well-described properties, such as strength and resistance etc; (b) making steel, without any further specification; (c) making steel from iron ore.

Each indicated energy service can be used for describing the production of steel. However, the scope of the energy efficient alternatives differs considerably. In the first case, the production of materials that can compete with steel are taken into consideration, e.g. strong synthetic fibers competing with steel cables. In the second case, scrap recycling and melting are an important option. In the last case, only processes that start with the reduction of iron ore are taken into account.

Although substitution by other materials is an important option for improving the energy efficiency of society, this option is not considered here, because the focus of this chapter is on the energy efficiency improvement of processes. (For studies of the improvement of the material efficiency, see [Worrell, 1994]). In this study we use the second description of the energy service. Thus, recycling of scrap is taken into consideration. The production of steel according to the blast furnace-basic oxygen route is taken as the reference process, because this process is the main production route for steel.

5.3.2 Calculation of the theoretically lowest energy demand

The theoretically lowest energy demand is the amount of energy required to perform the selected energy service without taking into account practical processes. The theoretical steps required for the production of steel from ore are (a) separation of iron oxide from other compounds in the ore, (b) reduction of iron ore, (c) adjusting the composition to

[1] World crude steel production for 1990 is estimated at 771 million tonne and the primary energy demand at 18.6 EJ [WEC, 1995]

make the desired steel, and (d) shaping the steel in the form of the product. When steel is made from scrap, the theoretical steps are (a) upgrading the scrap and (b) shaping the steel in the form of the desired product. The energy required for mining and transporting the ore and for recycling the scrap are not taken into consideration. We give a brief explanation of each step.

Production of steel from iron ore

Many minerals in the earth crust contain iron. Besides iron, these minerals, or ores, can contain many other compounds, mainly other oxides, e.g. SiO_2, Al_2O_3 and MnO_2. The iron content of iron ore can be as low as 30%, but it is usually in the 60-70% range [Considine and Maxwell, 1974]. Oxides are the most important iron ores. There are three types of iron oxides: hematite (Fe_2O_3), magnetite (Fe_3O_4) and wustite ($Fe_{1-y}O$, with y=0.045-0.135) [Considine and Maxwell, 1974]. Hematite is the most abundant oxide of iron.

<u>a. Separation of iron ore from other compounds in the ore</u> In homogeneous oxides there is a three-dimensional network of covalent bonds. There is strong ionic and covalent bonding. Breaking this bonding requires a high energy input, which is reflected by the relatively high melting points [Porterfield, 1984]. In multicomponent solids, such as ores, the entropy of mixing should also be taken into account. The entropy of two components that are mixed is smaller than the entropy of the separate components together. Mixing the compounds results in a decrease in the entropy and thus an increase in Gibbs free energy by an amount that is equal to the temperature times the entropy of mixing. For iron ore, consisting of 77% haematite, 15 % SiO_2, 5% magnetite and 3% other compounds [Tierney and Linehan, 1994], the difference in Gibbs free energy compared to the separate components is calculated to be -0.04 to -0.08 GJ/tonne Fe. It is assumed that all iron in the ore can be recovered. To separate the mixture into the individual compounds, at least the same amount of energy has to be supplied to the mixture. There are two reasons for regarding this amount of energy as an upper limit for the minimum energy required to recover iron compounds from the ore. First, energy demand is based on the separation of the mixture into its pure compounds, whereas we are interested only in iron oxide. Second, ideal mixing is assumed, while in practice compounds will appear in clusters in the ore. In these clusters no mixing, or less mixing occurs, between different compounds; thus the entropy of mixing is smaller.

<u>b. Reduction of iron oxide</u> The pure oxides can be decomposed into elements, according to the following reactions:

				$\Delta G°$(GJ/tonne Fe)[Weast, 1983]	
hematite:	Fe_2O_3 (s)	$\rightleftharpoons$	2 Fe (s) + 3/2 O_2 (g)	6.6	(1)
magnetite:	Fe_3O_4 (s)	$\rightleftharpoons$	3 Fe (s) + 2 O_2 (g)	6.1	(2)
wustite:	$Fe_{0.947}O$ (s)	$\rightleftharpoons$	0.947 Fe + ½ O_2 (g)	~4.7	(3)

We base the theoretically lowest SEC for the reduction of iron ore on the Gibbs free energy for reaction (1), because hematite is the most abundant iron oxide.

c. Adjusting the composition of the iron to make the desired steel Steel consists mainly of iron. Besides the elements derived from the ore and coke, mainly C and some Si, Mn, P, and S, other elements are added to make alloy steels, including Zn, Cr, Cu, Ni, and Mo. [Considine and Maxwell, 1974]. Although adding these elements does not require energy, the production of these additives may require a significant amount of energy. Because this amount varies, depending on the type and amount of additive, we do not take it into account. New compounds can be formed by a reaction of the elements and compounds present during cooling or heat treatment, e.g. iron carbide and ferromanganese. As an indication for the energy required for these reactions, we consider the formation of iron carbide. The Gibbs free energy of formation of iron carbide (Fe_3C) is 0.11 GJ/tonne Fe_3C [Weast, 1983]. Because the average value for the carbon content in steel is less than 0.5% by weight, the maximum theoretical energy demand for iron carbide formation is 0.002 GJ/tcs.

d. Shaping the steel into the form of the desired product Finally, the steel is shaped into the desired form and the surface can be adjusted to give the steel certain properties. The difference in the energy content of shaped and nonshaped steel is small. Also, in theory, the changes in the surface properties require hardly any energy.

We can conclude that the theoretical SEC for making steel from ore equals that of one step: the reduction of iron ore. The theoretical energy demand for the other steps is less than 1% of that for the reduction of iron ore. In fact, the energy for iron ore reduction is liberated when iron returns to the more stable iron oxide, a process known as rusting. Unfortunately, this energy is hard to recover. In practice the energy demand for crushing and grinding iron ore, pelletizing or sintering, and shaping, may constitute a considerable part of the energy demand for making a steel product.

Production of steel from recycled scrap

Steel scrap is recycled from many sources. The quality of the scrap depends on the source. One of the largest sources for recycled steel is from automobiles bodies and frames. This scrap contains large amounts of zinc, which was used as a surface layer. If not removed, the zinc negatively affects the quality of the steel.

a. Upgrading scrap The quality of scrap is not uniform. It is possible to recycle homogeneous, relatively pure scrap. However, as steel is increasingly being used in combination with other materials, or is being coated, a major part of the scrap resource will be contaminated with other metals like zinc, nickel, copper and tin and with polymers and other materials. If we assume that there are no covalent or ionic bonds, the minimum amount of energy that has to be supplied to the mixture to obtain the pure components equals the entropy of mixing times the temperature. We assume that this amount of energy is on the same order of magnitude as that required for the separation of ore into its components, thus less than a maximum of 0.1 GJ/tonne iron.

<u>b. Shaping of the steel into the form of the desired product</u> For shaping steel the same conclusion can be drawn as for shaping of primary steel: The theoretically lowest energy demand for this process is negligible.

We can conclude that, in theory, making steel out of scrap requires hardly any energy. However, practical processes require more energy than in the theoretical cases discussed above. Particularly, ore preparation and shaping of steel, of which we neglected the energy demand in this theoretical discussion, will contribute to a higher energy demand. In the next sections we discuss the theoretically lowest energy demand for melting iron and the chemical conversion that take place in a blast furnace.

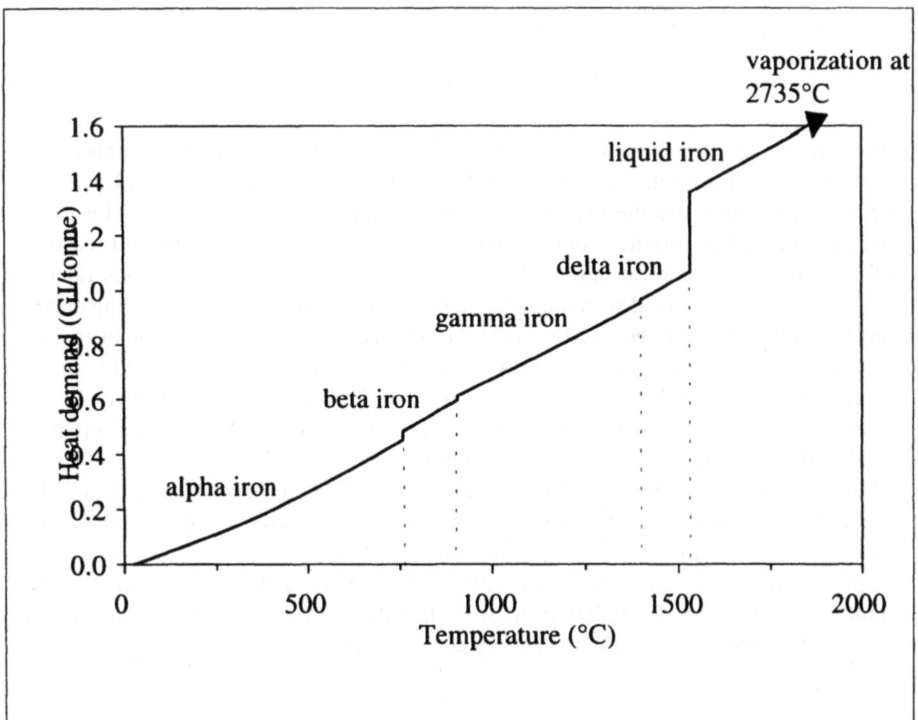

Figure 5.6: Heat demand for heating and melting pure iron (data taken from [Weast, 1983].

Box 1: Reactions that take place in the blast furnace (based on [Evans, 1991]).

Iron ore, coke and limestone are added to the blast furnace from the top. At the bottom hot compressed air is blown into the furnace. The reduction of iron ore takes place in two stages, the gasification of carbon, and the reduction of ore by carbon monoxide. The main reactions that occur in a blast furnace are:

150-600°C	$3\,Fe_2O_3 + CO \rightleftharpoons 2\,Fe_3O_4 + CO_2$ (5)
	$2\,CO \rightleftharpoons C + CO_2$ (6)
600-1000°C	$Fe_3O_4 + CO \rightleftharpoons 3\,FeO + CO_2$ (7)
	$FeO + CO \rightleftharpoons Fe + CO_2$ (8)
1000-1400°C	$FeO + C \rightleftharpoons Fe + CO$ (9)
	$CO_2 + C \rightleftharpoons 2\,CO$ (10)
1400-2000°C	$C + O_2 \rightleftharpoons 2\,CO_2$ (11)
	$2\,C + O_2 \rightleftharpoons 2\,CO$ (12)

The temperature zones indicate the zones that can be found in a blast furnace, the hottest zone is at the bottom. Coke is gasified at the bottom (e.g. reactions 11 and 12), providing the heat and the high temperature required for some reactions. Hot gases ascend, and carbon dioxide can react with coke according to the Boudouard reaction (10) to form more carbon monoxide. The temperature of the gas decreases rapidly because heat is exchanged with the coke bed and with molten materials coming down, and because of the endothermic Boudouard reaction and the direct reduction of molten iron oxide. Direct reduction of FeO with carbon (9) occurs only when FeO is in the liquid phase. The melting point of FeO is 1370°C. Carbon monoxide rises in the furnace, reacting with wustite (8), magnetite (7) and hematite (5). At lower temperatures the Boudouard reaction proceeds in the opposite direction (6).

Molten iron trickles down and collects in a well at the base of the furnace. Although the melting point of iron is 1530°C, a pasty, porous mass is already formed at 1200°C; this is related to the fact that carbon is dissolved.

Impurities are removed by reaction with calcium oxide, and a slag is formed. The molten slag floats on the molten iron. Silica that does not react with calcium oxide is reduced by carbon, increasing the energy consumption.

5.3.3 Heating and melting of iron

Figure 5.6 shows the heat demand for heating and melting pure iron. When pure iron is heated the lattice structure changes three times. Each change requires the input of transition energy. There are four forms of pure iron, known as α, β, γ, and δ, with transition points at 760, 907, and 1400°C. Each transition has its own transition enthalpy. Fe-δ melts at a temperature of 1535°C. Heating iron from 25 to 1535°C and subsequent melting requires 1.36 GJ/tonne Fe. Of this amount, the total enthalpy demand for all transitions is 0.35 GJ/tonne Fe. The melting requires the largest part of this: 0.29 GJ/tonne. The melting point of iron is lowered when carbon is dissolved in the iron. When the carbon content is 4.3%, a typical value for pig iron, the melting point is lowered to 1150°C. The enthalpy demand for heating iron from 25°C to the melting point is reduced by about 0.3 GJ/tonne by this temperature decrease. Heating and melting of pig iron theoretically requires 1.05 GJ/tonne; melting of steel, which has a low carbon content, is close to 1.36 GJ/tonne. When the iron cools to environmental temperature, this energy is released again.

5.3.4 Iron ore reduction in the blast furnace

Iron in oxides has a positive oxidation state, and therefore must gain electrons to become free iron. This result can be achieved in several ways, for instance chemically – a chemical reductant provides electrons –, or electrochemically – a direct current provides the electrons. In many metallurgical processes high temperatures are used to promote reactions kinetics and to shift thermodynamic equilibria. A combination is also possible. Aluminum production using the Hall-Herault process, for instance, is a combination of both routes, performed at high temperature.
Iron ore is reduced in the blast furnace with a chemical reductant, carbon (actually carbon monoxide) at high temperatures. Theoretically, this reaction can be described as follows:

$$2\,Fe_2O_3 + 3\,C \rightleftharpoons 4\,Fe + 3\,CO_2 \qquad \Delta G^\circ = 1.5 \text{ GJ/tonne Fe} \qquad (4)$$

The calorific value of pure carbon is 32.8 MJ/kg [Weast, 1983]. According to reaction (4), 161 kg of carbon is required to produce 1 tonne of iron, equal to 5.3 GJ. The total minimum energy demand for the reaction is therefore 6.8 GJ/tonne Fe. Note that this amount is only slightly higher than the theoretical minimum energy demand of reaction (3). A temperature of more than 900°C is required to let reaction (4) proceed thermodynamically. In practice, the set of reactions that take place in a blast furnace is far more complex. In box 1 a short description of these reactions is given. Carbon is largely converted to CO, which is the main reducing agent in the blast furnace.

5.3.5 Comparison with practical processes

When the theoretically lowest SEC is compared with the SEC of practical processes the following conclusions can be drawn:

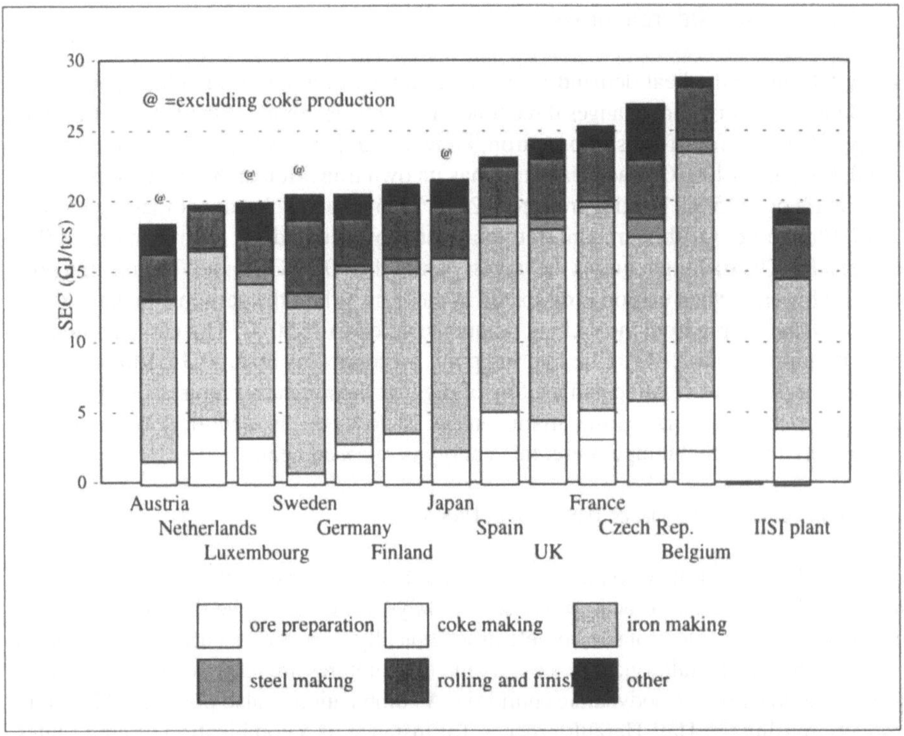

Figure 5.7: Countries' average SEC (1994) for primary steel making, and the SEC of the IISI reference plant. Data are derived from [IISI, 1996 #190]. Note that coke making is not included in the data for Austria, Japan, Luxembourg and Sweden (marked with @) Consequently, in these cases the SEC should be adjusted upwards by about 2-3 GJ/tcs. The data for Austria are for 1993

1. The minimum SEC for making a steel product from iron ore equals the energy demand for iron ore reduction that is 6.6 GJ/tonne steel. In modern blast furnaces carbon is supplied in the form of coke, coal and sometimes fuel oil. The total carbon demand is in the range of 350-400 kg of carbon/kg pig iron [IISI, 1996a]. Additional energy is supplied by the hot blast. Furthermore, energy is recovered with the blast furnace gas. The net SEC of a modern blast furnace is in the range of 12.5-15 GJ/tonne pig iron [IISI, 1996a]. This figure is about twice the theoretically lowest SEC of reaction 4 and also of reaction 1. In theory the SEC for pig iron reduction can be reduced by about 50%. The SEC of modern integrated steel plants, including all other processes, is three times as high. Consequently, the theoretical potential for improvement of the SEC for steel making from iron ore is 65%.
2. The minimum SEC for making a steel product from scrap is negligible. Scrap is melted in modern EAFs with a final energy input of about 1.5 GJ/tonne (3.5 GJ/tonne on a primary energy basis). In theory, the potential for reduction of the SEC is 100%, when

the minimum SEC for making steel from scrap is used as the reference. Note that the value of 1.5 GJ/tonne is about 10% above the energy required for heating and melting steel (the composition of scrap is almost similar to that of steel).

5.4 Exergy analysis of an integrated steel plant

In this section we perform an exergy analysis of an integrated primary steel plant (see Figure 5.5 for the unit operations of such a plant) to locate the main exergy losses in the process and evaluate their cause. Exergy is the amount of work obtainable when some matter is brought to a state of thermodynamic equilibrium with the common components of the natural surroundings by means of a reversible process [Szargut *et al.*, 1988]. It is comparable to Gibbs free energy; the difference is that the common compounds in the environment are taken as reference, rather than the elements.
Exergy analyses of selected processes in an integrated steel mill have been described in the literature. For example, Bisio and Poggi present exergy analyses of the sinter plant [Bisio and Poggi, 1991], thermal energy recovery from semifinished products and by-products [Bisio and Poggi, 1990], and blast furnace top gas pressure recovery [Bisio and Poggi, 1990]. Bisio performed exergy analyses of scrap remelting [Bisio, 1993] and to investigate the opportunities for recovery of heat from molten slag [Bisio, 1997]. Szargut et al give exergy analyses of the blast furnace, the BOF plant, walking beam furnaces and an Open Hearth Furnace [Szargut *et al.*, 1988]. Stepanov gives an analysis of a complete integrated steel mill [Stepanov, 1993]. However, Stepanov's mill differs considerably from modern steel mills. For example, it includes an Open Hearth Furnace, instead of a BOF plant. To our knowledge there has not yet been published an exergy analysis of a complete modern integrated steel mill.
Because we needed information on the location and the cause of exergy losses, we conducted such an exergy analysis ourselves. Before we present the results of our analysis, we describe the reference plant that we used for the analysis and discuss whether this plant is representative of other integrated primary steel plants.

5.4.1 THE REFERENCE PLANT

The exergy analysis is based on a hypothetical reference plant described by the International Iron and Steel Institute (IISI) [IISI, 1982]. The data for this plant were compiled by a group of international experts on energy use in the iron and steel industry and are based on actual operation data from plants in many countries. The plant is made up of components that were considered to be the most energy efficient techniques at that time (early 1980s); they were technically proven and commercially viable. The specifications of the main unit operations of this reference plant are given in Table 5.2. To assess whether use of this plant as the reference for our analysis is justified, we compare the reference plant with modern integrated steel mills.

- The IISI reference plant does not use fuel injection into the blast furnace. Nowadays, coke is partially replaced by fuel (coal or oil) injected through the tuyeres into the blast furnace. Fuel injection varies from plant to plant. At present the maximum is about 40% of the coke rate achieved at Hoogovens in the Netherlands [IISI, 1996a]. Coal injection reduces the energy demand for coke making. On the other hand, the energy demand for the blast furnace increases because the coal/coke substitution ratio is on the order of 1.04 [IISI, 1982] to 1.25 [Worrell *et al.*, 1993] and more oxygen is required. The overall SEC will decline by 0.2 GJ/tcs [Worrell *et al.*, 1993] to 0.5 GJ/tcs [IISI, 1982].
- The IISI reference plant has a net primary energy consumption of 19.2 GJ/tcs [IISI, 1982]. Figure 5.7 compares this value with average SECs for primary steel making in several countries. The IISI reference plant, although designed in the early 1980s, is still fairly efficient compared with the practices in most countries.
- The ore input into the blast furnace of the reference plant is 70% sinter and 30% pellets. This ratio differs greatly among countries. For instance, in the United States the sinter input is 20% of the raw material input whereas in Luxembourg it is more than 90% [IISI, 1996b]. The 70:30 ratio seems to be a reasonable choice. Production of pellets is not included in the reference plant. Several integrated mills also have a pellet plant. The SEC for pelletizing in a modern facility is about 1.0 GJ/tonne pellet on a primary energy basis [Worrell *et al.*, 1993]. The SEC of the reference plant would increase by 0.45 GJ/tonne rolled steel if pelletizing were to be included.
- On other points there is no noteworthy difference in operational practice between contemporary mills and what is described for the IISI plant. The differences remain within the range of differences between mills.

On the basis of these observations we conclude that we are justified in using the plant described by IISI as the reference plant for our exergy analysis.

For clarity we simplify the flow sheet of the IISI reference plant at two main points:

1. In the IISI reference plant, 58% of the crude steel is continuously cast into slabs, 29% is cast into ingots, and 11% is continuously cast into blooms. In our analysis we consider only continuous slab casting. Since the 1980s, continuous casting has become a well-accepted technique; in 1994, 72% of the world crude steel production was continuously cast [IISI, 1996b]. In modern integrated mills, ingot casting is rarely used. Continuous bloom casting resembles continuous slab casting and is therefore not treated separately.
2. The IISI reference plant considers finishing operations at a high level of detail. Several products are taken into account. Finishing operations are exxkuded from the analysis for two reasons. First, the energy consumption of these operations is small compared with that of the the front end of the plant. Second, the configuration and capacities of finishing operations vary from mill to mill. The IISI configuration is not necessarily representative of other integrated mills.

Table 5.2: Specifications of the reference plant [IISI, 1982].

Unit operation	Basic specification	Underfiring by	Others
coke oven	4 batteries of 100 ovens	enriched BF-gas	- wet charging and water quenching
sinter plant	2 strands of 400 m^2	coke breeze and CO-gas	- combustion air pre-heated in sinter cooler by heat exchange with hot sinter
hot blast stoves	blast heated to 1100°C and compressed to 4.0 bar	37% CO-gas and 63% BF-gas (enthalpy basis)	- compressor driven by condensing steam turbine; - cold blast air pre-heated by heat exchange with exhaust to 200°C.
blast furnace	2 furnaces of 4,400 m^3	coke rate is 470 kg/tonne pig iron	- equipped with top gas pressure turbines - no coal injection
basic oxygen furnace	2 out of 3 converters of 360 tonne		- hot metal ratio is 75%; - BOF gas recovery
continuous casting	2 twin strand slab casters		
reheating furnace	3 multi zone walking beam furnaces	87% CO gas and 13% BF gas	- air preheat to 500°C
hot strip mill	fully continuous with 5 roughing and 7 finishing stands		- waste heat boilers installed
power plant	steam boiler, back-pressure steam turbine and condensing steam turbine	CO gas, BF gas and BOF gas	- plants' electricity demand is exactly satisfied; - MP and HP steam is produced

These assumptions reduce the specific energy consumption. From the energy balance of the reference plant in Table 5.3, it can be seen that the SEC is 18.0 GJ/ton of hot rolled steel.

The capacity of the IISI reference plant is 8 million tonne of crude steel per year. All figures are presented in relation to the production of 1 tonne of hot rolled steel (trs).
The composition of several flows was not given or was only partially given by IISI. We used information from literature to complete the data. An average composition of iron ore was taken from Tierney and Linehan [1994]. The composition of coke oven gas was taken from Szargut and Morris [1987], and adjusted slightly to match the lower heating value given by IISI. The same procedure was used for coal [Ghamarin and Cambel, 1982] and coke [Szargut *et al.*, 1988]. Finally, the composition of coal tar was taken from Spielmann [Spielmann]. The lower heating values of coal and the in-house generated fuels are given in Table 5.4.

The exergy analysis is performed with the software package Enerpack [Nieuwlaar, 1996]. Exergy values of flows are related to an environmental reference system (ERS). Enerpack uses the environmental temperature (in our analysis, 298.15 K) and pressure (101.325 kPa) as reference for hot and cold flows and flows with an elevated or reduced pressure. It also uses the most stable compound that occurs in the natural environment as reference for chemical elements. The chemical elements nitrogen, oxygen, and carbon and the noble gases have their reference compounds in the atmosphere. Nitrogen, oxygen, and the noble gases are themselves reference compounds. Carbon dioxide is the reference compound for the element carbon. For the remaining elements, the reference compounds are taken from the lithosphere, the hydrosphere, or a combination of both. Liquid water is taken as the reference compound for the element hydrogen. Table 5.5 gives an overview of the ERS values of the chemical elements that are of importance in this study. Once the exergy of the elements is determined, the exergy of all other compounds can be calculated.

Table 5.3: Energy balance of the reference plant.

INPUT				OUTPUT			
flow	volume	unit	GJ/trs	flow	volume	unit	GJ/trs
coal	658.33	kg	20.80	BOF gas	15.71	Nm^3	0.14
oxygen	48.51	Nm^3	0.17	CO gas	43.87	Nm^3	0.85
				coal tar	24.23	kg	0.90
				benzole	8.56	kg	0.36
				coke breeze	23.60	kg	0.75
		Total	20.97			Total	3.00

Table 5.4: Lower heating values of energy carriers

energy carrier	LHV	energy carrier	LHV
coal	31.6 MJ/kg	coke oven (CO) gas	19.3 MJ/Nm^3
coke	29.8 MJ/kg	blast furnace (BF) gas	2.8 MJ/Nm^3
		basic oxygen furnace (BOF) gas	9.0 MJ/Nm^3

Table 5.5: ERS-values of selected chemical elements [Nieuwlaar, 1996].

Element	Reference compound	Enthalpy (kJ/mole)	Entropy (J/K/mole)	Exergy (kJ/mole)
H_2	H_2O	285.69	169.34	235.20
C	CO_2	393.52	-56.82	410.46
N_2	N_2	0.00	-2.31	0.69
O_2	O_2	0.00	-13.26	3.96
Si	SiO_2	910.94	195.29	852.71
Ca	$CaCO_3$	813.40	338.53	712.47
Fe	Fe_2O_3	412.10	147.31	368.18

Process description

Figure 5.8 is a simplified flow sheet for the plant. The process starts with the preparation of the raw materials. The blast furnace requires an open structure to allow gases to ascend and liquid material to descend. For this purpose coal is converted to coke in the coke ovens and iron ore is agglomerated in the sinter plant. An additional objective of sintering is to increase the surface reactivity of the ore. The temperatures in the coke ovens and the sinter plant are 700 and 1000°C respectively. Both coke and sinter cool to the environmental temperature before being fed to the blast furnace, together with pellets and lime. Here iron ore is reduced to pig iron, which leaves the furnace at about 1400°C. The reactions that take place in the blast furnace have already been described in box 1. During transport to the BOF plant the pig iron cools by about 140°C. In the BOF, carbon from pig iron reacts with oxygen, injected through a lance. In the reference plant 25%, of the charge to the BOF is cold scrap, which is heated and melted in the BOF. The liquid steel with a temperature of about 1650°C is transported to the continuous caster, where slabs are cast. Three quarters of the cast steel cools to the environmental temperature. The remaining 25% cools to 500°C. All steel is reheated to 1200°C in the reheating furnace and then rolled in the hot strip mill. Finally, the hot rolled products cool to environmental temperature.

5.4.2 RESULTS OF THE EXERGY ANALYSIS

Table 5.6 presents the results of the exergy analysis. Of the 22.6 GJ per tonne of hot rolled steel (trs) that goes into the process, mainly in the form of coal, 10.9 GJ/trs is inherited by useful products. The exergy of 1 tonne rolled steel almost equals the minimum amount of energy needed to produce iron from hematite according to reaction (4). This is due to the fact that hematite is the reference substance for iron in the ERS. The difference between exergy input and useful output, equal to 11.7 GJ/trs, is considered to be lost. External losses, i.e. losses associated with flows that are not recovered for utilization purposes, account for 5.5 GJ/trs of this loss. The remainder, 6.2 GJ/trs, is caused by exergy losses that occur within the system boundaries of the plant, the internal losses. The external and internal losses are specified in Table 5.7.

From Table 5.7 it can be seen that radiation and convection losses are the largest source of external losses. This type of loss accounts for about 3.6 GJ/trs, or 30% of the total exergy loss of the steel mill. Of this, about 2.5 GJ/trs is due to exergy lost by cooling materials. The large share of convection and radiation losses will probably come as no surprise considering the large temperature differences that occur. This is illustrated in Figure 5.9, which shows the change in temperature and enthalpy of the solid flows. Four different material flows are distinguished.

Table 5.6: Exergy balance of the reference integrated steel plant. Numbers in brackets refer to the flows in Figure 5.8. The external and internal exergy losses are specified in Table 5.7.

INPUT	GJ/trs	OUTPUT	GJ/trs
Coal (1)	20.2	Rolled steel (9)	6.62
Scrap (2)	1.87	Coal tar (10)	0.92
Iron ore (3)	0.22	CO export gas (11)	0.84
Fluxes (4)	0.2	Recollected steel (12)	0.76
LPG (5)	0.05	Coke breeze (13)	0.72
Air - various flows (6)	0.03	BF slag (14)	0.56
Pellets (7)	0.03	Benzole (15)	0.25
Oxygen (8)	0.01	BOF export gas (16)	0.14
		MP steam (17)	0.1
		Total useful products	*10.9*
		External losses	5.47
		Internal losses	6.15
Total	22.6	Total	22.6

About 1.6 GJ/trs of exergy is lost as chemical and physical[1] exergy of waste gaseous streams. In this category the exhaust of combustion reactions forms an important group. Because the chemical exergy of these waste streams is low, almost all exergy of these streams is physical exergy loss, mainly resulting from elevated temperature.

The last category of external losses is the loss of material. Coal is lost as dust that is removed from the coke oven gas. Blast furnace gas also contains dust (coal, iron, ore), which has to be removed before the gas can be used. The dust is considered lost. In total 0.5% by weight of the pig iron is lost with dust and with blast furnace slag. Steel is lost in several operations, e.g. tapping the steel from the BOF, casting, and rolling. It is assumed that these steel losses are collected and can be reprocessed. Therefore, they are considered to be useful products (flow 12 in Figure 5.8).

Table 5.7: Specification of external and internal exergy losses (GJ/trs).

	EXTERNAL EXERGY LOSSES (GJ/trs)				Total internal exergy losses	Total exergy losses
	Radiation and convection losses	Chemical or physical exergy of waste streams	Material losses	Total		
Coke oven	0.28	0.47	0.24	0.99	0.87	1.86
Sinter plant	0.29	0.39		0.68	0.98	1.66
Hot blast stoves	0.25	0.11		0.36	0.41	0.77
Blast furnace	0.44	0.18	0.04	0.66	1.35	2.01
BOF plant	0.12	0.06		0.18	0.34	0.52
Continuous caster	1.05			1.05	0.06	1.11
Reheating furnace	0.04	0.2		0.24	0.5	0.74
Hot strip mill	0.62			0.62	0.12	0.74
Power plant	0.2	0.21		0.41	1.51	1.92
Others	0.28			0.28	0.03	0.31
Total	3.57	1.62	0.28	5.47	6.15	11.62

[1] A flow has positive physical exergy when the pressure and/or the temperature differs from the reference pressure and temperatures, respectively.

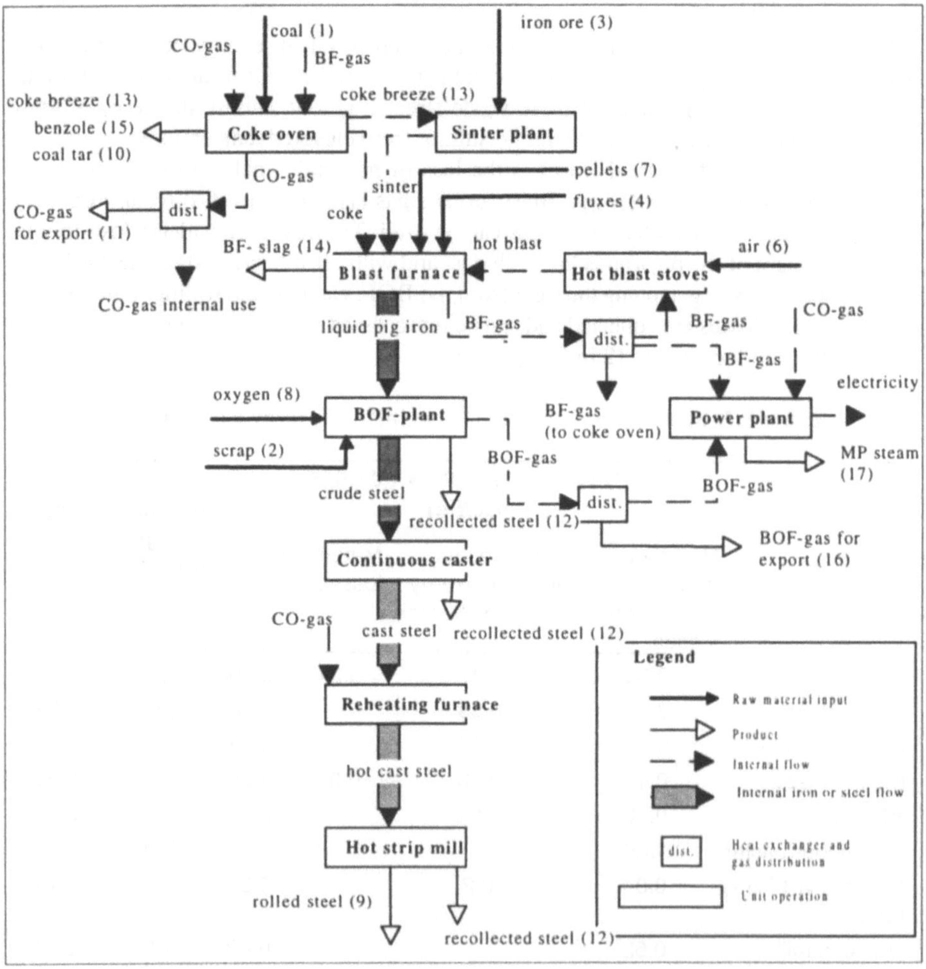

Figure 5.8: Flow sheet of the integrated steel mill used in the exergy analysis. Numbers (in parenthesis) refer to the flows given in Table 5.6. Raw material and product flows are shown, as are the internal flows of product gases and steel. In total, 141 flows were taken into account.

The main sources of <u>internal losses</u> are combustion reactions, other chemical reactions, heat transfer, and compression and expansion. Calculating internal losses is difficult because the underlying processes are often complex, e.g. the set of chemical reactions in the blast furnace (see box 1), and because different types of internal losses are interlinked, e.g. combustion and heat transfer. Here the discussion is restricted to an estimation of the size of four different categories.

The largest amount of internal loss is related to combustion reactions. The exergy loss resulting from the irreversibility of the reaction is equal to the difference in the exergy of the reactants at input temperature and pressure and of the combustion products at the combustion temperature [Szargut *et al.*, 1988; Nieuwlaar, 1996]. Without preheating of the reactants, this loss amounts to about 30% of the exergy content of fuel [Nieuwlaar, 1996]. In the integrated steel plant, combustion reactions take place in the coke oven, sinter plant, hot blast stoves, reheating furnace, and power plant. The exergy of all fuels, both in-house generated gases and bought gases, is about 7 GJ/trs. The exergy loss of combustion reactions depends on the conditions in which the combustion takes place. Assuming that about 30-35% of the exergy is lost during combustion, we can estimate the exergy losses from combustion reactions to be 2.0-2.5 GJ/trs.

A second type of internal exergy loss is caused by irreversibilities in chemical reactions other than combustion, for instance the conversion of coal to carbon monoxide. In this category we also group exergy losses that occur as a result of (a) the friction of gases with solids when they flow through a reactor (mainly in the blast furnace), (b) mixing of flows, and (c) pressure drops over reactors. The losses in this category occur mainly in the coke oven, the blast furnace and the BOF. If we assume that all losses in these processes that are not due to combustion reactions and heat transfer are, in one way or another, related to chemical reactions other than combustion, the losses total 1.5-2.0 GJ/trs.

A third type of internal loss is caused by heat transfer, which takes place in heat exchangers and can occur simultaneously with combustion reactions. Heat transfer can take place through surfaces or by mixing of the flows. The driving force behind heat transfer is temperature difference. The larger the temperature difference, the larger the exergy loss. However, exergy loss is smaller at high temperatures. In heat exchangers, exergy losses are usually minimized by using a countercurrent operation that ensures the smallest temperature difference. On the basis of calculations of exergy loss of the main heat transfer processes (in the power plant, in hot blast stoves, in the blast furnace and the sinter plant, and in the coke oven) we estimate the internal losses resultingfrom heat transfer at 1.0-1.5 GJ/trs.

Finally, a category of other internal losses can be distinguished that contains mainly losses resulting from irreversible compression and expansion. These losses occur, for example, in the compressing of the blast, in the expansion of the blast furnace top gas and in the power plant. These losses are estimated to be about 0.5-1.0 GJ/trs.

5.4.3 CONCLUSIONS

We can conclude that exergy losses are due mainly to the application of high temperatures and the need for several cooling and reheating steps. Radiation and convection losses, physical exergy lost with gaseous streams, losses resulting from the conversion of chemical energy to gases with high temperature, irreversibilities in heat transfer, and even irreversibilities in some undesired chemical reactions that occur only at higher temperatures all contribute to these exergy losses. Reducing the exergy loss should therefore be directed

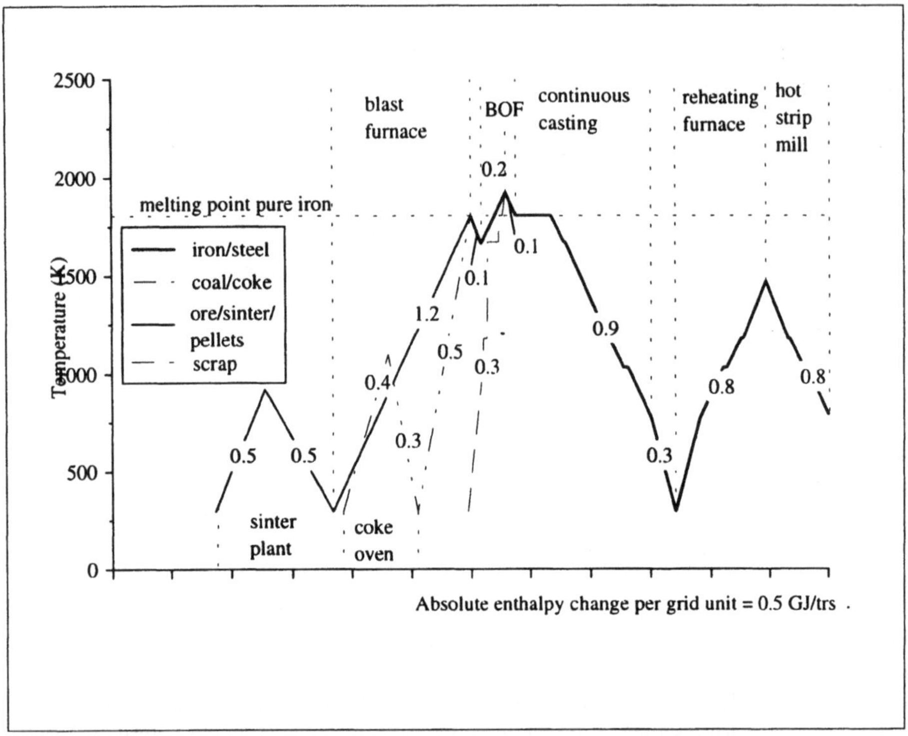

Figure 5.9: Enthalpy and temperature changes of the iron and steel flows in an integrated steel plant. Data on heat capacity are taken from [Weast, 1983], except data for coal. These are based on [Gomez, 1965], where it should be noted that the relation between heat capacity and temperature is only valid in the temperature range between 273 and 473 K. It is assumed that above 473 K the heat capacity of coal is temperature-independent. Different slopes for the same material flow indicate that the volumes of the flows differ. Positive slopes indicate that energy has to be added, negative slopes indicate that energy is released. The numbers indicate the absolute change in enthalpy in GJ/trs

at reducing the temperature or decreasing the number of temperature changes. In the next section we investigate whether techniques under development can achieve these objectives.

5.5 Identification and selection of long-term energy efficient techniques

In this section we discuss the way information on long-term energy-efficient techniques was gathered and how we selected the techniques that may reduce the SEC of steel making in the long run.

5.5.1 Gathering of information

The identification of new techniques started with a search for relevant literature, performed in two ways. First, the following literature databases were searched: Applied Science and Technology Index, Environline/Energyline, Metadex, and Compendex. These databases were searched in two steps. At the start of the research a general search was performed. Later, when more specific key words were known (e.g. names of techniques), the searches were repeated using these keywords. The second method of literature search was scanning volumes of journals specific to the iron and steel industry to identify emerging techniques. Of the following journals, the volumes from 1988 to 1995 were scanned: *Journal of the Iron and Steel Institute of Japan; Stahl und Eisen* and *Steel Times*. We expanded our database of literature by checking the references of the collected literature.
The next step in the gathering of information was contacting the developers of the techniques to obtain the most recent data.
We checked all data for accuracy and reliability by consulting experts, and by making our own calculations and judgments, or by obtaining evidence from other sources.

5.5.2 Selection of energy-efficient techniques

In the previous section we concluded that the main exergy losses are due to the application of high temperatures and the need for several cooling and heating steps. In current steel making, high temperatures are necessary to achieve several goals, e.g. to change the structure of the ore and coal so that they can be processed in the blast furnace, to overcome kinetic and thermodynamic limits to chemical reactions in the reduction of iron oxide, and to provide steel in a liquid form so that it can be shaped.
Techniques that reduce exergy losses resulting from high-temperature applications can be divided into three groups according to the degree to which the need for high temperatures is avoided or reduced.

1. Techniques that avoid at least one heating and cooling step The avoidance of one heating and cooling step can be achieved by techniques that combine two or more processes. The two major groups of techniques are smelting reduction processes and near-net-shape casting techniques. Smelting reduction processes make direct use of coal and usually also of iron ore, without having to convert coal to coke and ore to sinter or pellets. Near-net-shape casting techniques reduce or eliminate the reheating demand in the shaping of products. A completely different route involves avoiding the iron ore reduction by processing recycled scrap and subsequent melting, casting, and shaping.

2. Techniques that reduce the temperatures required in different process steps Reduction of iron ore below the melting point is already commercially feasible in direct reduction processes. Coke making at lower temperatures is a topic of research. Casting and shaping without melting can be accomplished by powder metallurgy, a process that is already used commercially for the production of some speciality products.

3. Technologies that recover and apply heat at high temperatures Technologies that recover and apply heat at high temperatures do not alter the need for high temperatures. In an integrated steel mill waste heat recovery from clean gaseous flows like combustion gases is normal practice. Recovery of the heat from gaseous flows that are contaminated with, for example, organic compounds and small solid particles, runs into technical problems, such as the fouling of heat exchangers. At present, recovery of the sensible heat from solid flows is not an important point of research interest; therefore information on this issue is not available.

5.6 Characterization of long-term energy-efficient techniques

In this section we characterize the selected techniques. The focus is first on techniques that avoid at least one heating and cooling step. Smelting reduction processes are dealt with in section 6.1, and near-net-shape casting techniques are dealt with in section 6.2. Both sections start with a general description including the formulation of a general basis for comparison, i.e. the way the SEC and the costs are determined, and a description of the main production parameters. Then, separate techniques are described. Both sections conclude with a comparison of the techniques.
With regard to the techniques described in sections 6.1 and 6.2 we assess the degree of technical change that is required to replace the current technique by the new technique. We distinguish three categories of required technical change. First, techniques that require an *evolutionary change* imply a continuation of the trend in technological development. No changes in the way the energy service is performed are expected, and the effects on the following aspects are small or negligible: performance, process parameters, quality and nature of the products, the purchasing and supply industry and the plant layout. Second, a *major change* is required when at least part of the energy service is performed according to a new principle, the performance of the process increases more than one can expect by trend extrapolation, and there are considerable effects on the other aspects. Finally, a *radical change* is required when a new energy service arises or all the aspects change to a large extent. (For a more extensive description of this categorization see [Beer *et al.*, 1997] and chapter 2 of this book.)
In section 6.3 the state of the art and the developments in making steel from scrap are discussed. Section 6.4 deals with steel making at lower temperatures and section 6.5 with waste heat recovery techniques. Finally, in section 6.6 future process routes for steel making are sketched and the potential for the reduction of the SEC is determined. In this concluding section we evaluate to what extent exergy losses can be reduced by the techniques described, and we suggest what needs to be done to achieve further reduction.

5.6.1 SMELTING REDUCTION PROCESSES

Smelting reduction (SR) processes involve reduction of iron ore without the need for coke and – in most cases – agglomerated ore. The driving forces behind the development of SR

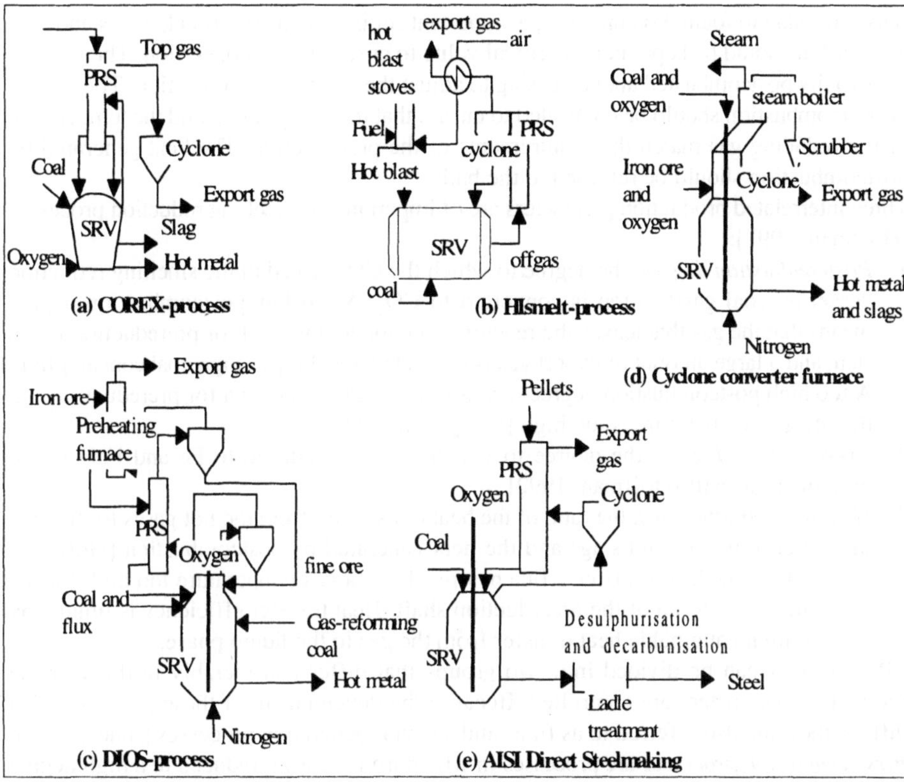

Figure 5.10: Schematic representations of smelting reduction processes (not on the same scale). PRS = Prereduction shaft; SRV = Smelting reduction vessel.

processes are the reduction of capital and operation costs and the smaller environmental impact, both of which can be achieved by eliminating coke ovens and ore agglomeration. The principle behind SR is that iron oxide is reduced in the liquid state by carbon or carbon monoxide. Liquid state reactions are much faster that solid state reactions. Because the reduction in a blast furnace is a solid state reaction, the reduction time can be reduced. In principle, an SR process can consist of a single reactor in which unprepared iron ore and coal react to form a product similar to steel; that is decarburization of the iron takes place in the same reactor. In practice, SR processes consist of at least two reactors and the product resembles pig iron, which has to be refined in a separate reactor for steel to be obtained. Figure 5.10 gives some schematic representations of SR processes.

In SR processes iron ore is prereduced in the solid state in a prereduction shaft by a reducing gas generated in a smelting reduction vessel. Melting and final reduction generally take place in this smelting reduction vessel as well. In many SR processes, the reaction site is the slag floating on the bath of liquid iron. Coal reacts with oxygen or iron ore in the liquid state to form a gas that consists mainly of carbon monoxide. The gas causes the slag to foam. Foaming slag is important for improving reaction kinetics and heat transfer but should be kept under a critical value to ensure normal operation. The gas can be partially postcombusted above the slag to adjust the chemical composition. The degree of postcombustion should be controlled to ensure that the composition and the temperature of the reducing gas match the requirements of the prereduction. The heat generated by postcombustion should be returned to the bath.

Three interrelated production parameters are of importance in smelting reduction processes [Hoffman, 1992]:

1. *Postcombustion degree*: the degree to which the CO formed in the smelting reduction vessel by coal gasification is converted to CO_2. A too low postcombustion degree means that the gas that leaves the reactor at the top and is used for prereduction is too rich, and a large amount of export gas is generated, resulting in high coal consumption. A too high postcombustion degree means that the gas is too lean for prereduction and the off-gas temperature is too high [Hoogovens, 1995].
2. *Prereduction degree*: the degree to which Fe_2O_3 is reduced to Fe and FeO in the prereduction shaft [Hoffman, 1992].
3. *Heat transfer efficiency*: the ratio of the heat transferred from the hot gases to the bath of molten iron, ore and slags and the heat generated by postcombustion [Hoffman, 1992]. Too low heat transfer efficiency results in a gas temperature too high for the constructing material of the prereduction shaft. Heat transfer efficiency is limited by the maximum attainable heat transfer from the gas to the liquid phase.

SR processes can be divided into two groups that differ considerably in the way the production parameters are controlled. Because the development of these processes also differs, they are also referred to as first- and second- generation processes [Innes, 1995]. *First-generation processes* are processes in which iron ore is prereduced to a high degree (up to 90%) before being fed to a shaftlike SR furnace. Prereduced ore and coal are present as solids in the reactor, in either a fluidized or a permeable bed. No postcombustion takes place in this furnace. The only commercial SR process, COREX, has a shaft-type furnace. Kawasaki (Japan) developed a smelting reduction process with a shaft-type furnace before it became a partner in the joint Japanese effort to develop the Direct Iron Ore Reduction-process. Hoogovens (in the Netherlands) initially studied a shaft-type process (Converted Blast Furnace), before it focused on the Cyclone Converter Furnace.

In *Second-generation processes* the SR reactor is derived from the converter process for steel making. The final reduction takes place in a bath of molten iron and prereduced ore, with a molten slag floating on it. These processes are characterized by rapid reduction of iron ore in the molten slag layer, high heat transfer efficiency, and postcombustion of the process gases above the molten slag layer. No commercial second generation SR process is available, although many processes have been studied or are under development. The most important of these processes are: Direct Iron-ore Smelting (Japan), High Intensity smelting (Australia), American Iron and Steel Institute Direct Steel Making (USA), Converter Cyclone Furnace (The Netherlands), Jupiter (France, Germany), and Romelt (former-USSR, USA).

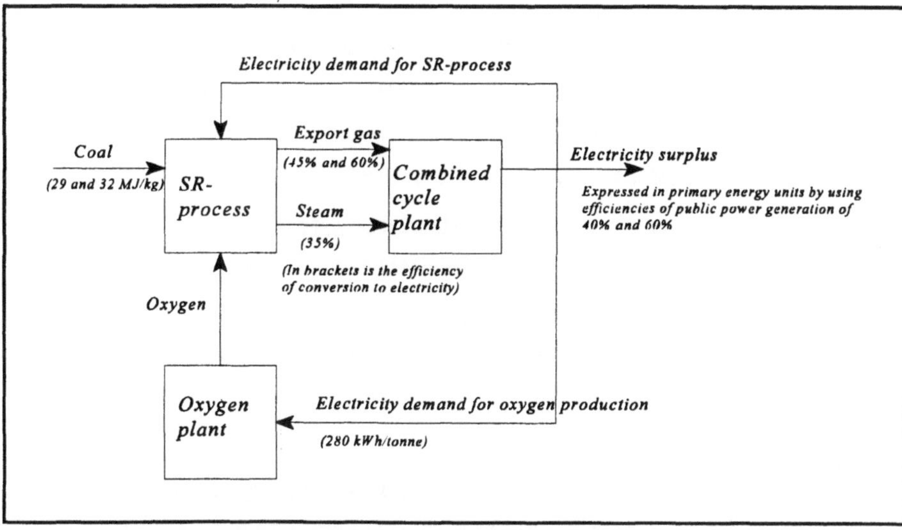

Figure 5.11: Model that summarizes the assumptions that were made to calculate and compare the SECs of Smelting Reduction processes.

A group of smelting reduction processes, which can be both first and second generation, uses coal for reduction and electricity for melting, e.g. by electric arcs, plasma, or flash smelting. Examples are INRED, ELRED, and Plasmamelt. For the INRED process it is claimed that all electricity required can be generated by using the heat of the off gas [Chatterjee, 1996]. However, 620 kg of coal per tonne of hot metal is required, which is about 30% more than for modern blast furnaces. The Plasmamelt process requires 275 kg of coal and coke and 1120 kWh electricity per of tonne hot metal [Chatterjee, 1996]. From an energy point of view, this is competitive with modern blast furnaces only when the electricity is generated with an efficiency of almost 100%. Because these processes do not have the potential to reduce the SEC of iron making, we do not consider them further.

Assumptions for comparison of SR processes
Before the techniques can be compared assumptions has to be made concerning the way in which the SEC and the investment and operation costs are determined.
The specific energy consumption (SEC) of smelting reduction processes is presented on a primary energy basis. A breakdown into energy carriers is given. So that final energy can be converted to primary energy, several assumptions have been made. To take the sensitivity for these assumptions into account, we consider a low and a high case. The energy input of smelting reduction processes is often expressed in tonnes of coal. The lower heating value (LHV) of coal varies considerably with the composition[1]. Van Gool

[1] Carbon, hydrogen and sulphur contribute to an increase in heating value, while nitrogen, oxygen, ash and water have a reducing effect [Ghamarin and Cambel, 1982].

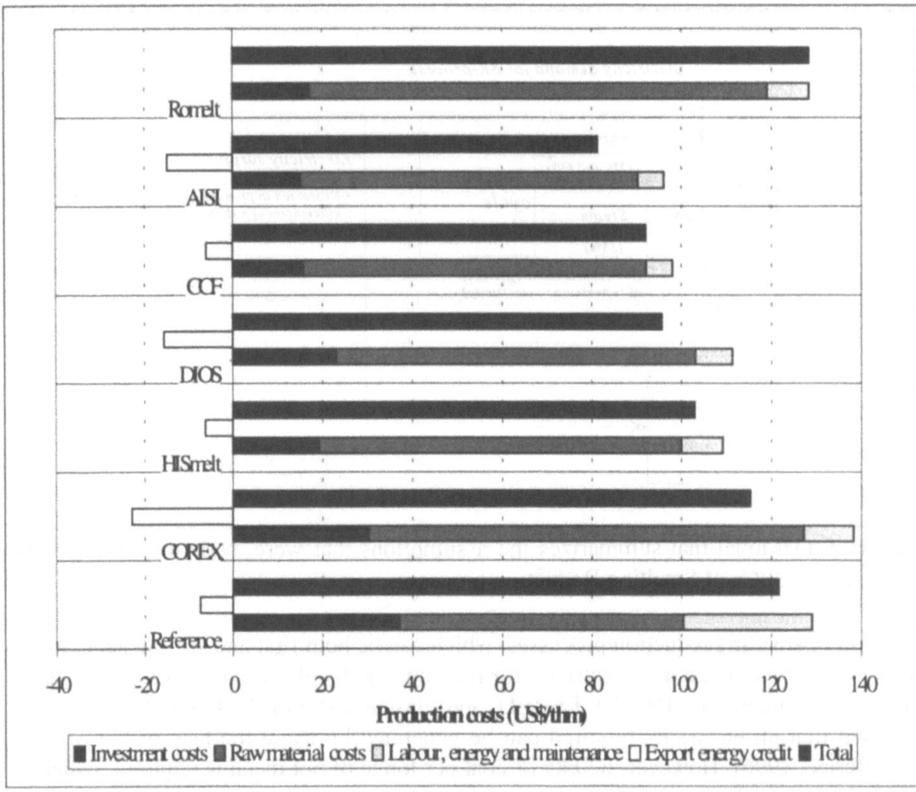

Figure 5.12: Production costs of hot metal (in US dolar per tonne of hot metal) of smelting reduction processes and a reference process consisting of coke ovens, sinter plant and blast furnace. Basic assumptions are given in Table 5.9. The production costs represent the low case.

reports values varying from 20 to 38 GJ/tonne coal [Gool, 1986]. IISI uses 30 GJ/tonne as an average value [IISI, 1990]. Here we use a lower limit of 29 GJ/tonne and an upper limit of 32 GJ/tonne. The composition of the export gas, thus its LHV, depends on the type of coal used. However, because the composition is often not reported, we do not consider this variation here. Another input to most smelting reduction processes is oxygen. We assume that oxygen is produced in an air liquefaction plant, with an electricity use of 280 kWh/tonne oxygen (0.4 kWh/m^3) [Hendriks, 1994]. To credit for the export gas and steam produced in the SR process, we use a simple model, represented in Figure 5.11. We assume that electricity is generated in house from the gas produced in the SR process in a combined cycle plant. Firing low-calorific gas in a gas turbine is already a commercial technology. In Japan a combined cycle plant, fired by gas with an LHV of 3.7-4.6 MJ/Nm^3, can achieve an electric efficiency of 46% [Takano *et al.*, 1988]. In September 1997 a combined cycle plant went into operation at the Hoogovens site in the Netherlands. The projected efficiency is also 46%, using gas with an LHV of about 4 GJ/Nm^3 [Verweij, 1997]. Both combined cycle plants use a gas turbine developed by Mitsubishi Heavy Industries. The LHV of export gas produced in SR processes varies from about 2 MJ/Nm^3 to 7.5 MJ/Nm^3. The efficiency of about 45% can also be achieved with other low-calorific gases provided that an adiabatic flame temperature of about 1100°C can be achieved [Verweij, 1997]. To credit for the gas produced in SR processes we assume that the chemical energy of the gas is converted to electricity in a combined cycle plant with a 45% (current) and 60% (estimate for the future) efficiency. We also assume that steam produced in the SR process is fed to a steam turbine and converted to additional electricity with an efficiency of 35% (based on [Hoogovens, 1995]). It is assumed that the electricity is first used for in-plant demand and oxygen generation. The surplus is expressed in primary energy units using generation efficiencies of both 40% (as a current average for electricity generation in public power plants) and 60% (future).

Because the SEC depends strongly on the method of crediting for the export gas, we perform a sensitivity analysis by varying the credit factor for the gas.

The SEC of smelting reduction processes is compared with the SEC for the production of 1 tonne molten pig iron with the most efficient process in operation, including coke ovens, sinter and pellet plant, and blast furnace. The plant of Hoogovens is used as the reference. Figure 5.7 shows that Hoogovens (the only integrated steel plant in the Netherlands) is one of the most efficient plants in the world. Using the conversion factors described above, we can derive an SEC for pig iron production at Hoogovens of 16-17.5 GJ/tonne pig iron (tpi) on a primary energy basis [Worrell *et al.*, 1993]. Table 5.8 shows a breakdown of the SEC of this reference plant into energy carriers.

A good basis for comparing the economics of the processes is the production cost for one tonne of hot metal (thm). Because hot metal is an intermediary product, exact costs are not always known. Fruehan reports a typical selling price of US$120-140/thm [Fruehan, 1994]. Another source reports production costs on the order of US$165/thm for a 2 million tonnes per year blast furnace [Bosley *et al.*, 1987]. We made our own estimate of the production costs, presented in Table 5.9. According to our calculations, the production costs range from US$120 to 160/thm. Variable costs make up 70% of the total production costs and investment costs about 30%.

Only direct investment costs are considered. Annual investment costs are determined by allowing for depreciation of the investment on an annuity basis over 15 years, using real interest rates of 5% and 10%. Variable costs consist of costs for raw material, labor, energy and maintenance. Credits are considered for the production of gas and steam.

All investment costs of SR processes are based on estimates found in literature and expressed in US dollars per tonne hot metal. The variable costs of SR processes are determined in the same way as for the reference process. Regarding raw material costs, there are two major differences between the reference process and SR processes. First, steam coal is used instead of coking coal. Steam coal is on average US$ 5/tonne cheaper than coking coal [IEA, 1993]. The costs of steam coal are taken at US$40-45/tonne. Second, oxygen is usually required instead of air. For a 1500 tonnes per day oxygen plant, a typical capacity for SR processes, investment costs are estimated to range from US$30 million [Hendriks, 1994] to US$55 million [Dijk *et al.*, 1994]. Reported production costs are US$0.015/Nm^3 [Weston and Thompson, 1996b] and US$0.04/Nm^3 [Stelco, 1993]. In this study the first value is used as the lower and the second as the upper limit of the costs.

The resulting production costs estimates are presented in Figure 5.12. The values are based on design values for raw material and energy consumption, and developers' estimates of investment costs. We evaluate the effect of higher investment costs on the production costs of hot metal. Furthermore, our determination of the production costs ignores most indirect investment and manufacturing costs. Therefore, these production costs should be treated as rough estimates and used for comparison only.

Table 5.8: Specific energy consumption (SEC) of smelting reduction processes and of a reference process expressed in GJ primary energy per tonne hot metal[a].

Process	energy carrier	Input	low case (GJ/thm)	high case (GJ/thm)
Reference process[b]	coal	0.59 t	17.1	18.9
	other fuel		2.1	2.1
	oxygen			
	electricity demand	69 kWh	0.4	0.6
	export gas	-3.7 GJ	-3.7	-4.2
	export steam			
	SEC		15.9	17.4
COREX	coal	0.88 t	25.5	28.2
	other fuel			
	oxygen	0.71 t	1.2	2.8
	electricity demand	60 kWh	0.4	0.5
	export gas	-11.5 GJ	-11.5	-12.9
	export steam			
	SEC		15.6	17.5
HIsmelt	coal		19.3	19.3
	other fuel		0.7	0.7
	oxygen			
	electricity demand			
	export gas	-3.1 GJ	-3.1	-3.5
	export steam	-2.7 GJ	-1.6	-2.4
	SEC		15.3	14.2
DIOS	coal	0.81 t	23.6	26
	other fuel			

Process	energy carrier	Input	low case (GJ/thm)	high case (GJ/thm)
	oxygen	0.66 t	1.1	1.7
	electricity demand			
	export gas	-7.8 GJ	-7.8	-8.8
	export steam			
	SEC		16.9	18.9
CCF	coal	0.64 t	18.6	20.5
	other fuel			
	oxygen	0.67 t	1.1	1.7
	electricity demand			
	export gas	-3 GJ	-3	-3.4
	export steam	-5.78 GJ	-3.4	-5.1
	SEC		13.3	13.7
AISI	coal	0.7 t	20.3	22.4
	other fuel	1.02 GJ	1	1
	oxygen	0.56 t	0.9	1.4
	electricity demand	62 kWh	0.4	0.6
	export gas	-7.4 GJ	-4.4	-3.3
	export steam			
	SEC		15.2	17.1
Romelt	coal	1.35 t	39.2	43.2
	other fuel			
	oxygen	1.35 t	2.3	3.4
	electricity demand			
	export gas		p.m	p.m.
	export steam			
	SEC			
Jupiter	coal	0.57 t	16.5	18.2
	other fuel			
	oxygen	0.57 t	1	1.4
	electricity demand			
	export gas	-2.6 GJ	-2.6	-2.9
	export steam			
	SEC		14.9	16.7

[a] It is assumed that export gas and steam are converted in a combined cycle to electricity that is used for oxygen production and other electricity demand. The surplus is expressed in primary energy carriers using the efficiency of public electricity generation. Low and high cases are distinguished. Low case: electricity generation efficiency in the combined cycle plant and of the public grid of 60% and LHV of coal of 29 GJ/tonne. High case: electricity generation efficiency of 45% in the combined cycle and 40% of the public grid and LHV of coal of 32 GJ/tonne. Figure 5.11 illustrates these assumptions.

[b] The reference process consists of coke ovens, sinter plant, pellet plant, blast furnace and hot blast stoves. source: [Worrell *et al.*, 1993].

Table 5.9: Calculation of the costs of the production of one tonne hot metal in the reference plant.

Investment costs	unit/thm	unit costs (US$)		Input data (unit/thm)	annual costs (US$/thm) 5%[x]	10%[x]
blast furnace	US$			195[u]	18.8	25.6
coke oven	US$			145[u]	14	19.1
sinter plant	US$			45[u]	4.3	5.9
TOTAL	US$			385[u]	37.1	50.6
Variable costs		low	high		low	high
Raw materials						
coking coal	tonne	45[a]	50[b]	0.59[v]	26.6	29.5
steam coal	tonne	40[c]	45[d]			
iron ore	tonne	25	30[f]	1.27[v]	31.8	38.1
fluxes	tonne	50[g]	60[h]	0.1[v]	5	6
oxygen	Nm^3	0.015[i]	0.04[j]			
Labor	man hours	10[k]	30[l]	0.4[w]	4	12
Energy						
natural gas	Nm^3	0.1[m]	0.15[n]	107[v]	10.7	16.1
electricity	kWh	0.035[o]	0.05[p]	70[v]	2.5	3.5
Maintenance	% of invesment	3[q]	4[r]		11.6	15.4
Export gas credit	GJ	-2[s]	-3[t]	3.7[v]	-7.4	-11.1
Total					84.6	109.5
TOTAL COSTS					121.7	160.1

Sources and assumptions: [a-d] [IEA, 1993];[c] [Weston and Thompson, 1996b]; [f] [Gielen and Dril, 1997]; [g] Assumption based on [h][Weston and Thompson, 1996b]; [i] [Weston and Thompson, 1996b]; [j] [Stelco, 1993]; [k] [Stelco, 1993]; [l] [Weston and Thompson, 1996b]; [m-n] [EZ, 1994]; [o] [Weston and Thompson, 1996b]; [p][EZ, 1994]; [q] Assumption based on [r] [Stelco, 1993]; [s-t] [Faure, 1993b; Stelco, 1993]; [u][COREX, 1996] [v] [Worrell and Beer, 1991];[w] [Stelco, 1993]; [x] Investments are depreciated over 15 years using the annuity method with real interest rates of 5% and 10%.

Characterization of smelting reduction processes

A characterization of the following smelting reduction processes is presented: COREX, HIsmelt, DIOS, CCF, and AISI Direct Steel making. The Romelt and the Jupiter processes are also discussed briefly. Each characterization starts with a short history of the development, followed by a description of the technique and assessment of the main process parameters. Next the SEC is determined, and finally an estimate is given of the production costs of one tonne hot metal.

COREX (Germany/Austria) Development of the COREX process dates back to the 1980s and was performed by Voest-Alpine (Germany/Austria). A 300,000-ktpa commercial plant has been in operation in South Africa since 1989. This plant is named the C-1000 plant, referring to the daily capacity. Since 1995 a COREX plant has been in operation in the Republic of Korea, with twice the capacity of the South African plant (C-2000) [COREX, 1996]. Several more orders have been placed for C-2000 plants in India, South Korea, and South Africa [Cheeley *et al.*, 1996; COREX, 1996].
Figure 5.10.a gives a simplified flow sheet of the COREX process. In the dome of the smelting reduction vessel, coal is gasified in a fluidized bed, maintained by an upstream of oxygen injected halfway down the reactor. The coal gas reduces iron ore by 90-95% in the prereduction shaft. Prereduced iron is introduced into the fluidized bed, further reduced, and melted. The COREX process is characterized by a high prereduction degree and no postcombustion. The gas that leaves the prereduction shaft has therefore a relatively high LHV (7.5 MJ/Nm^3).
Coal use in the C-1000 plant is about 900 kg/thm [Puehringer *et al.*, 1991]. The oxygen requirement is 540 Nm^3/thm. The production of export gas is around 11.5 GJ/thm [Puehringer *et al.*, 1991]. A small amount of electricity is required, 60 kWh/tonne. On the basis of these data, the SEC of the C-1000 plant can be calculated to be 15.5-17.5 GJ/thm. First operational results of the C-2000 plant in South Korea indicate a similar SEC [Eberle *et al.*, 1996]. Besides generating electricity, several other ways of using the large volumes of relative high calorific gas have been investigated, e.g. production of Direct Reduced Iron[1] and synthesis gas generation [COREX, 1996].
Direct investment costs of a COREX plant with an annual capacity of 1.5 million thm plant are US$250/thm [COREX, 1996]. COREX reports the production costs per tonne hot metal to be 20% lower than that of the blast furnace process. Our calculations, presented in Figure 5.12, show that the investment cost are indeed about 20% lower but that the variable costs are more or less equal. As a result, production costs are estimated to be US$115-150/thm, about 5% lower than for pig iron from the blast furnace.

HIsmelt (Australia/Germany) The development of the High Intensity smelting reduction (HIsmelt) process originates from cooperation between CRA (Australia) and Klöckner Werke (Germany) in the early 1980s [Millbank, 1995]. A small-scale pilot plant (10-12 ktpa) was built in Germany in 1984 and operated until mid-1980 [Prideaux, 1996]. In 1989 CRA formed a joint venture with MIDREX Corporation (USA) to develop the HIsmelt process [Cusack *et al.*, 1995]. Since 1994 a 100-ktpa pilot plant has been in operation in Kwinana in Australia [Innes, 1995; Prideaux, 1996]. Hismelt scheduled to have the first

[1] Two COREX/MIDREX combination plants are under construction, in South Korea (Hanbo Steel site at Asian Bay), and in South Africa (Saldanha Steel) [Cheeley *et al.*, 1996] [Millbank, 1995; COREX, 1996].

commercial plant in operation in 1999 [Cusack *et al.*, 1995]; the capacity goal is 500-1000 ktpa [Prideaux, 1996]. CRA's interest in the HIsmelt process is related to the availability of noncoking coals in Australia and the desire to use local low-grade ores. Furthermore, the process should be able to operate in an environment where infrastructure is lacking [Meijer, 1996]. This is the reason that the process is designed around the use of hot blast instead of oxygen and that the export gas production is kept to a minimum.

Figure 5.10.b is a simplified flow sheet of the HIsmelt process. The HIsmelt process uses a horizontal cylindrical smelting reduction vessel. Noncoking coal is injected from the bottom, while prereduced ore and hot blast are injected from the top. Carbon dissolves in the bath and reacts with oxygen from iron ore. The CO that is formed in this way reacts above the bath with oxygen from the blast air. The generated heat is returned to the bath via a fountain of droplets of molten iron, giving a high heat transfer efficiency. The prereduction shaft is of the fluidized bed type, facilitating the processing of fine ores. The hot gases leaving the gas cleaning cyclone are used to preheat the incoming blast air [Prideaux, 1996]. The advantage of using a hot blast instead of oxygen is that the costs of oxygen production are avoided. Furthermore, the nitrogen in the air is believed to promote heat transfer and to control postcombustion temperature. A disadvantage, however, is that the size of the equipment has to be adjusted to handle the large gas volumes. The gas cleaning equipment can be especially expensive [Stelco, 1993]. Smaller gas volumes can be achieved by using air enriched with oxygen up to a maximum of 30% [Cusack *et al.*, 1995].

It is claimed that a postcombustion degree of more than 60% and a heat transfer efficiency of 90% have been achieved in the pilot plant [Prideaux, 1996]. The high postcombustion degree is achieved by using a shallow bath with a high surface area [Stelco, 1993]. The prereduction degree is in the 20-25% range [Chatterjee, 1996].

A heat balance of the HIsmelt process – with all values expressed in GJ/thm – is given by Chatterjee [Chatterjee, 1996]. The input is 19.3 GJ/thm of coke and 0.7 GJ/thm of natural gas to fire the hot blast stoves. Then 4.0 GJ/thm of steam is generated, of which 1.3 GJ/thm is used in the turbo blowers to compress the blast air. The calorific value of the export gas is about 1.5 MJ/m^3. In total 3.1 GJ/thm is produced [Kreulitsch *et al.*, 1993]. On the basis of these data a SEC of 14-15 GJ/thm is determined.

The capital costs of the pilot plant in Kwinana amounted to more than US$100 million [Cusack *et al.*, 1995], or US$1000/thm. The target capital cost is US$200/thm [Prideaux, 1996]. The annual variable costs are estimated to be US$80-95/thm, see Figure 5.12. Total production costs are US$100-120/thm, about 20-25% lower than for hot metal produced in a blast furnace.

<u>DIOS process (Japan)</u> The Japanese Direct Iron-ore Smelting (DIOS) reduction process is being developed by cooperation between the Japanese Iron and Steel Federation, the Center for Coal Utilization, and eight Japanese integrated steel manufacturers. This national program started in 1988 and aims to have a first commercial unit in 2000 [Furukawa, 1994]. The process should be available around 2010-2015 at the latest, when most of the Japanese coke ovens will have completed their 40 years of service . A 180,000

tonnes per year pilot plant operated in the period 1993-1995. The capacity of commercial plants that is aimed for is 1-1.8 million tonnes per year [Furukawa, 1994].
The DIOS process has three fluidized bed furnaces, one placed above the other: On the top is a preheating shaft, in the middle the prereduction shaft, and at the bottom the smelting reduction vessel [Furukawa, 1994] (see Figure 5. 10.c). Ascending gas fluidizes materials in the furnaces. In the prereduction shaft, iron ore, preheated in the preheating furnace, is prereduced by means of the gas generated in the smelting reduction vessel. The bath in the smelting reduction vessel consists of three layers. The top layer is a mixture of coal and prereduced iron ore. Final reduction takes place in the middle layer, which is molten slag and coal combustion products, e.g. chars. Molten iron sinks to the bottom layer. Oxygen is blown through a lance into the first two layers, but partial side-blowing can enhance the postcombustion degree [Chatterjee, 1996]. The reducing gas, generated by coal gasification in the top layer, can be reformed by additional coal injection so as to increase its reduction capacity and reduce the temperature of the reducing gas [Stelco, 1993].
The prereduction degree is 25%, of which 5% is achieved in the preheating furnace. The postcombustion degree is 40-60% [Chatterjee, 1996]. The heat transfer efficiency is not known.
The SEC has not been published so far, but it is said to be lower than 16.5 GJ/thm [Stelco, 1993]. However, a SEC of 17-19 GJ/thm is determined using values given by Stelco [Stelco, 1993]. Inputs are 814 kg/thm coal and 504 Nm^3/thm oxygen. The production of export gas amounts to 7.8 GJ/thm [Stelco, 1993].
Investment costs have not been published. The costs of the pilot plant were US$70 million [Furukawa, 1994]. Investment costs of a commercial green field DIOS plant, including oxygen plant, are 80-85% of a green field coke oven/sinter plant/blast furnace. The costs of the latter are US$385/thm [Faure, 1993b]. A DIOS plant with a capacity of 1.5 million tpa requires an oxygen plant with a capacity of 2700 tpd. The investment costs of such a plant are about US$80 million [Dijk *et al.*, 1994]. The investment costs of a DIOS plant can be estimated to be US$230-250/thm. Variable costs are estimated to be US$75-95/thm, see Figure 5.12. Total production costs are determined to be in the range of US$95-125/thm, which means that they are about 20% lower than for hot metal from a blast furnace.

<u>Cyclone Converter Furnace (Hoogovens, The Netherlands)</u> The development of the Cyclone Converter Furnace (CCF) started in 1986. From 1986 to 1992, Hoogovens worked with British Steel and Ilva (Italy) (since 1988) on the development. Initially, a shaft type SR process was being studied. However, because ore agglomeration was still required a shift was made to the development of a converter type of process in 1989 [Langen *et al.*, 1992]. In 1992 British Steel decided to focus on direct coal injection and stopped their involvement in the development of CCF. Since 1994 Ilva and Hoogovens have continued their research separately [Meijer, 1996]. Ilva operated a 5-tph pilot plant [Millbank, 1995]. Tests by Ilva appeared to be less successful, and plans for a new series of experiments were stalled because of the high financial risk [Meijer, 1997]. Hoogovens still continues its research and development connected with the CCF.

The CCF being developed by Hoogovens consists of a melter/prereduction cyclone mounted on a converter type vessel (see Figure 5.10.d). Granular coal and oxygen are injected through a lance into the bath of molten metal. The gases produced rise and mix with the fine ore and oxygen that are injected tangentially into the cyclone. Here, the ore is not only prereduced but also molten. The molten prereduced ore trickles down to the bath, where the final reduction takes place. Postcombustion takes place not only directly above the bath but also in the cyclone, which allows for efficient heat transfer.

In 1994 a 20-tph test facility for the melting cyclone was built and successfully operated at the Hoogovens site. The converter has not yet been tested on a pilot-plant scale. In 1995 Hoogovens acquired the rights of the AISI direct iron making technology, which is directed primarily at in-bath smelting. Hoogovens aims to achieve plant capacities of 500 and 1,000 ktpa [Meijer, 1995]. In 1997 the Dutch government decided to support the development of the demonstration plant with 60 million Dutch guilders (US$32 million) [Soldaat, 1997]. The demonstration plant with a final capacity of 700 ktpa should be on stream in 2000 [Meijer, 1997].

A prereduction degree of 20% on average was obtained in the pilot plant experiments with the cyclone alone. The postcombustion degree is expected to be 25% and heat transfer efficiency 80% in the bath, but this has not been tested. A lower postcombustion degree can be maintained than in other SR processes, because the iron ore is not only reduced but also melted in the prereduction shaft. The direct connection between the smelting reduction vessel and the prereduction cyclone permits optimum use of the heat of the postcombustion gas, since no hot de-dusting and cooling are necessary. Since postcombustion also takes place in the cyclone the final postcombustion degree is more than 75% [Meijer, 1996].

Design values for the energy input are 640 kg/thm coal and 510 Nm^3/thm oxygen. Energy output is 3.0 GJ/thm export gas and 5.7 GJ/thm steam. A SEC of 13-14 GJ/thm is estimated on the basis of these data [Meijer *et al.*, 1995].

The investment costs of the CCF process are estimated at US$150-180/thm [Meijer *et al.*, 1995]. These low investment costs seem reasonable considering that the CCF process is simple compared to other smelting reduction processes, mainly because the prereduction shaft and the smelting reduction vessel are combined. The variable costs are estimated to be US$70-90/thm. Total production costs are estimated to be US$90-115/tonne, which means they are about 30% lower than for producing hot metal in a blast furnace (see Figure 5.12).

<u>AISI Direct Steel making process (USA)</u> The program of the American Iron and Steel Institute (AISI) and the US Department of Energy (DOE) for developping the Direct Steel Making process ran from 1988 to 1994 . A number of universities and industries participated in the research [Farley and Koros, 1992]. Steel manufacturers were also closely involved. The aim of the program was to prove the technical and economic feasibility of a direct steel-making process, including continuous desulphurization and decarbonization and ladle treatment. However, early in the program the focus shifted to direct iron making, and particularly to in-bath smelting. Tests in a 15 tpd pilot plant showed that smelting of wustite in a high-intensity bath process is a manageable process [Farley

and Koros, 1992]. The productivity and the fuel rate, however, were far behind the goals formulated at the start of the program [Badra, 1995]. Nevertheless, several steel companies believed that the technical problems could be overcome and considered a 350 ktpa-demonstration plant. This plant has not yet been built. Instead, AISI and DOE launched a new program to determine the feasibility of converting steel plant waste to pig iron with the in-bath smelting technology to be used in steel making and foundries [Badra, 1995; Millbank, 1995]. The development of the direct iron making from ore process stalled [Badra, 1995]. However, as mentioned in the discussion of CCF, Hoogovens is considering the use of the AISI technology in connection with its cyclone reduction and melting technology.

The AISI process involves the melting of partially prereduced iron ore pellets. Figure 5.10.e is a simplified flow sheet. Final reduction takes place in foaming slag above the molten iron bath. Coal is top-fed into the slag. Oxygen is injected through a lance. The carbon monoxide that is formed is postcombusted above the slag layer. The aim of the AISI technology was to prereduce pellets to wustite [Farley and Koros, 1992], implying a prereduction degree of about 50%. To match the reducing potential of the off-gas with this prereduction degree, a postcombustion ratio of 40% is required. In the pilot-plant tests, it has been proven that this ratio is attainable [Farley and Koros, 1992]. The heat transfer efficiency has not been reported. The AISI process is distinct from other smelting reduction processes in that pellets are used rather than ore and that the export of gas is minimized – at least in intention. The background to the use of pellets is that in the United States a modernization of the pellet plant capacity had just been completed. Minimizing export gas production should increase hot metal production per unit of working volume per day.

Design values for the coal input of 700 kg/thm and oxygen use of 430 Nm^3/thm have been reported [Farley and Koros, 1992; Stelco, 1993]. The production of export gas amounts to 7.4 GJ/thm. We can estimate the SEC to be 15-17 GJ/thm on the basis of these data. In the SEC the energy demand for pelletizing is included; this amounts to about 1.5 GJ/thm [Worrell and Beer, 1991].

Faure reports investment costs of US$160/thm [Faure, 1993a]. Variable costs are estimated to be US$65-85/thm (see Figure 5.12). Total production costs are therefore about US$80-105/thm, or 35% lower than for hot metal produced in the blast furnace.

Other developments Besides the above-mentioned techniques, R&D has been done on several other SR processes. Because little information about these processes is available, the discussion here is limited.

<u>Romelt process (former-USSR, USA)</u> The Romelt process is a single-stage SR process. Coal and ore are fed to a horizontal large-volume vessel. Oxygen is injected through side tuyeres at two levels, enabling a highly agitated bath and a high postcombustion degree of 70%. Reduction takes place in a foamy slag layer. This process is based on the Vanyukov process for copper smelting [Millbank, 1995; Chatterjee, 1996].

An 18-tph pilot plant operated at the Novolipeski Metallurgical Kombinat (NLMK) in Lipetsk from 1985 to 1988. A 1000-ktpa demonstration plant was approved but never realized [Millbank, 1995]. Because of the collapse of the USSR, Soviet R&D stopped. In 1995 the engineering and constructing company ICF Kaiser International (USA) and Nippon Steel Corp (J) were granted licences to market and commercialize the process [Weston and Thompson, 1996a]. ICF Kaiser believes that the Romelt process is ready for commercialization, especially for processing steel plant wastes [Weston and Thompson, 1996a]. The energy input is relatively high [Weston and Thompson, 1996b]. Pilot-plant results showed a minimum coal input of 1350 kg/thm and an oxygen demand of 1900 Nm^3/thm (78% O_2 in blast). Although no figure for export gas production is given, the SEC of the Romelt process is probably considerably higher than the SEC of other SR processes and the SEC of the blast furnace route. An economic analysis of a 400-ktpa Romelt plant yielded investment costs of US$180/thm [Weston and Thompson, 1996b]. Variable costs are in the range of US$110-155/thm, or 30-40% higher than that for pig iron produced in a blast furnace. Total production costs are estimated to be US$130-180 /thm, or 5-10% higher than that for pig iron produced in a blast furnace. When steel plant waste oxides are processed, as ICF Kaiser aims to do, the production costs are reduced to US$85-130/thm, if no costs are assumed for the waste oxides.

<u>Jupiter (France, Germany)</u> The Jupiter process was developed by IRSID, the research center of Usinor-Sacilor (France), in cooperation with Thyssen Stahl and Lurgi (both Germany) [Lassat de Pressigny, 1993]. However, R&D seems to have stopped [Badra, 1995] [Meijer, 1996]. The aim was to develop a process to supply virgin metal to EAF plants [Lassat de Pressigny, 1993]. A coal input of 570 kg/thm and an oxygen demand of 435 Nm^3/thm were reported. Export gas production was 2.6 GJ/thm [Faure, 1993b]. The SEC can be estimated to be in the range of 15-17 GJ/thm. No costs figures are available for the Jupiter process.

<u>Other processes</u> Numerous other smelting reduction processes – not considered here – have been or are being developed. For instance, besides being involved in the DIOS program, Kawasaki Steel in Japan is developing the Kawasaki XR process. The Kawasaki XR process has a shaft-type furnace and is now being developed to process BOF dust [Chatterjee, 1996] [Stelco, 1993]. Another example is the Chinese effort to develop a smelting reduction process without postcombustion and foaming slag,[Yemin and Zhihong, 1996]. Iron ore concentrate is prereduced and deposited along with fine carbon in a separate reactor. In the next step, the deposited carbon is combusted quickly, resulting in flash smelting of the ore. The melted ore falls on a hot coal surface. This process is still at the stage of applied research [Yemin and Zhihong, 1996].

Conclusions concerning smelting reduction Now that several SR processes have been characterized, a founded estimate can be made for the potential of energy-efficiency improvement when these processes are implemented and a comparison can be made of the production costs. We also evaluate the chance of successful commercialization of the SR

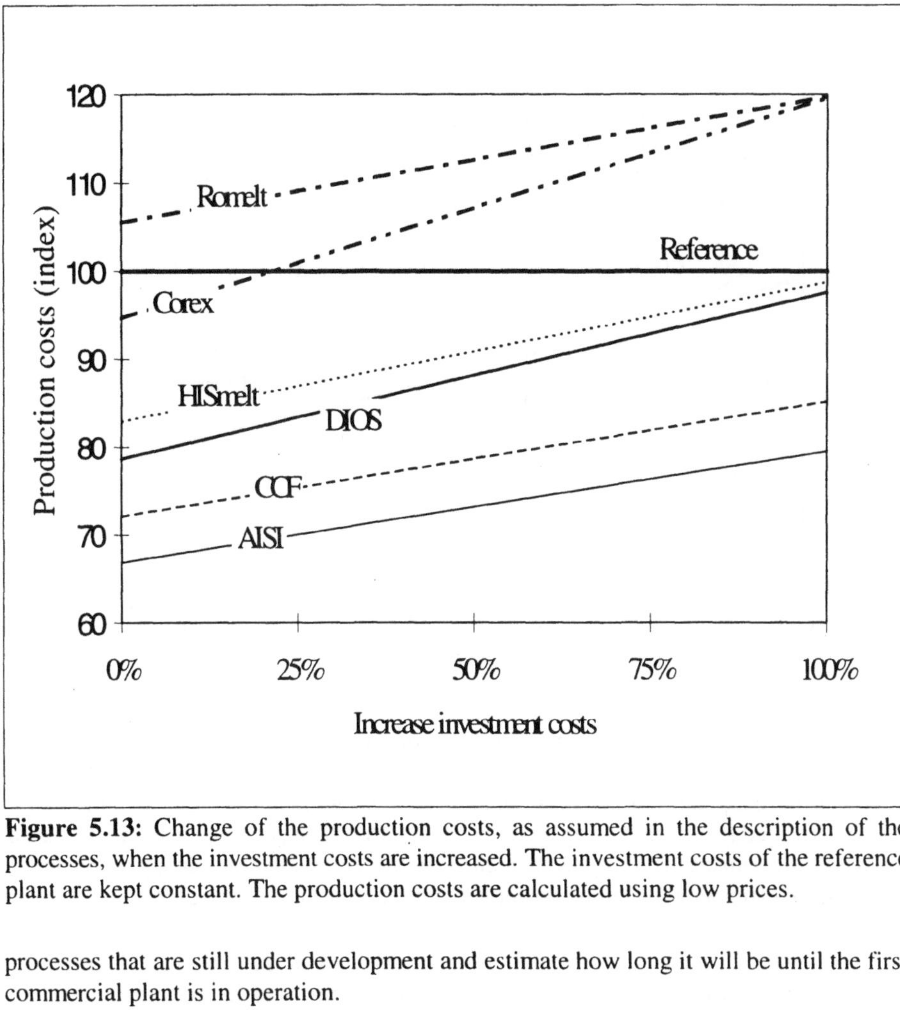

Figure 5.13: Change of the production costs, as assumed in the description of the processes, when the investment costs are increased. The investment costs of the reference plant are kept constant. The production costs are calculated using low prices.

processes that are still under development and estimate how long it will be until the first commercial plant is in operation.

Table 5.8 presents an overview of the SECs of the SR processes and the reference process. The SEC of smelting reduction processes – except the Romelt process – varies from 13 to 19 GJ/thm. The CCF process seems to be the most energy efficient; the SEC is about 20% lower than the SEC of the reference process. The HIsmelt and Jupiter processes have SECs that are 5-10% lower than those of the reference, whereas the AISI, COREX, and DIOS processes have SECs equal to or even higher than the reference.

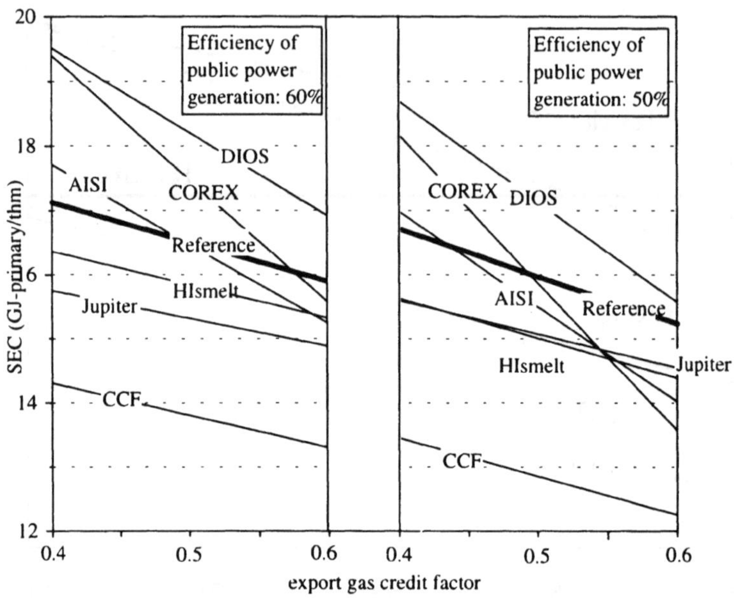

Figure 5.14: Dependency of the SEC of SR-processes on the credit that is given to the export gas. The export gas credit factor is the efficiency of the export gas to electricity in a combined cycle plant. For comparison, at present gas with an LHV of about 4 MJ/Nm^3 can be converted to electricity with an efficiency of 46%. The generated electricity is used for the plant's demand. The surplus is credited for by expressing it in primary energy units, assuming an efficiency for public power generation of both 60% and 50%.

The SEC depends strongly on the production and utilization of the export gas. Figure 5.13 shows how the SEC varies when the export gas credit factor is varied from 0.4 to 0.6[1]. This is done for two efficiencies of public power generation: 60% is an estimate of the maximum achievable efficiency for a gas-fired power plant, and 50% is an estimate of the maximum achievable efficiency for a coal-fired power plant. The figure shows that the SR processes can be divided into two groups according to the sensitivity of the SEC to the export gas credit factor. Processes in the first group – COREX, DIOS and AISI – have

[1] The export gas credit factor is the efficiency at which the gas is converted to electricity in a combined cycle. Therefore, credit factors higher than 0.6 are not considered. This is not to say that higher credits cannot be achieved, e.g. by substitution of fuel oil in heating or as reducing gas in DRI production. However, in these cases another way of crediting should be considered. This may provide new insights, but it is beyond the illustrative scope of this sensitivity analysis.

relatively large export gas production (more than 7 GJ/thm), because of a low degree of postcombustion. Therefore, the SEC depends on the export gas credit factor. The use of the export gas should be carefully considered. For instance, Figure 5.13 shows that a credit factor of 0.5 is required for the AISI process to be more efficient than the reference. This is on condition that electricity can be generated in central power stations with an efficiency of 60%. When we use a 50% efficiency of power generation, AISI is already more efficient with an export credit gas factor of 0.45. Processes in the second group – HIsmelt, Jupiter, and CCF – are in all cases more efficient than the reference.

The conclusion that emerges is that SR processes are not necessarily more efficient than conventional iron making. The energy requirement for coke making and, in most cases, ore agglomeration is avoided. Energy consumption of the iron ore reduction itself increases, as a result of the higher coal consumption and the need for pure oxygen. The energy consumption can be minimized by selecting optimum values for the process parameters. Careful attention should be paid to the utilization of the export gas, both in the reactor and outside. The maximum energy-efficiency improvement appears to be about 20% compared to the current best-practice iron-making process. However, that all data on energy requirement are still based on design values or on pilot-plant results. But SR technology is still in an early stage of development, and further work on these technologies might well lead to even more energy-efficient designs.

Figure 5.12 gives an overview of the production costs of one tonne hot metal of the SR processes compared with the reference process. All second generation processes – except the energy-intensive Romelt process – have production costs that are 20-35% lower than those of the reference. This reduction is achieved predominantly by a 35-55% reduction in investment costs. The variable costs are 5-25% lower. The expected cost reduction by using steam coal instead of the more expensive metallurgical coal is partially offset by the larger coal demand. To get an idea of the sensitivity of the production costs of hot metal to changes in the investment costs, we increased the investment costs of the SR processes by up to 100%. We kept the investment costs of the reference process at the same level. The results are shown in Figure 5.14. It can be seen that the production costs of hot metal in the COREX process become higher than those of the reference process, if the investment increases by 25%. However, since the COREX process is already commercial, the investment costs are based on actual figures. An increase of 25% is not to be expected. The production costs of all other SR processes (except the Romelt process) remain below those of the reference up to a 100% increase of the investment costs. The sensitivity to changes in variable costs is taken into account by calculating with low and high costs (see Table 5.9). The high cost figures resulted in a 0.5% increase in the production costs for the DIOS process and a 7% increase for the HIsmelt process.

It can be safely concluded that the production costs of hot metal in a second-generation SR process will be lower than in conventional iron making.

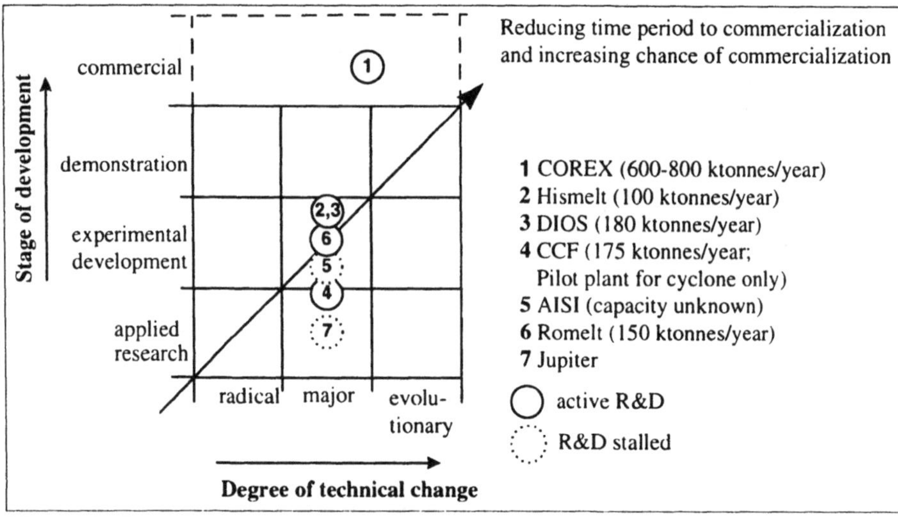

Figure 5.15: Comparison of smelting reduction processes with respect to the degree of technical change compared with the current technology and the stage of development.

What chance is there that a SR process will be commercialized, and when can we expect the first commercial plant? To answer this question we assess the technical change required to bring the technology to commercialization and give a résumé of the stage of development.

As far as the technical change is concerned, a distinction can be made between first- and second-generation SR processes. Second-generation processes require a new type of reactor, whereas for first-generation processes the blast furnace can in principle be converted to serve as a smelting reduction vessel. Industry has less technical experience with the converter-type reactor than with the shaft-type reactor. On the other hand, the technology for smelting reduction differs considerably from that needed in blast furnaces, although the principle of iron ore reduction is more or less the same. The front end of the plant, i.e. all processes up to the blast furnace, has to be replaced completely. The processes after the blast furnace can remain essentially the same.

Considering the large process adaptations, we can say that changing from a blast furnace to a smelting reduction process is a major technical change and that first generation smelting reduction technology is a smaller technical change than second generation technology. In Figure 5.15 this information is combined with the stage of development.

There are large differences in the stages of development. The only first-generation process, COREX, is already commercial. Of the six second-generation processes, only three have ongoing research into reduction of iron ore, namely DIOS, HIsmelt, and CCF. HIsmelt and DIOS seem to be in the most advanced stage of development. Both processes have been fully tested on a pilot-plant scale. The developers expect the first commercial unit to be in

operation in 1999 (HIsmelt) and 2000 (DIOS). The CCF has been only partially tested on a pilot-plant scale, namely only the smelting cyclone. A demonstration plant is planned to be in operation in 2000. The Romelt and AISI processes are now being developed to process steel plant wastes. However, there may still be a future for the AISI technology in ore reduction, as Hoogovens has acquired the rights and intends to combine it with the CCF. The Jupiter process has never been tested on a pilot-plant scale.
From Figure 5.15 it can be concluded that at least three second-generation smelting reduction processes are on the verge of being demonstrated. It can reasonably be expected that one or more of these processes will result in a commercial plant within the next decade.

5.6.2 NEAR-NET-SHAPE CASTING

The second group of techniques that avoid at least one heating and cooling step is concerned with casting and shaping of steel. Traditionally, steel is cast into ingots of different shapes and weights ranging from several tonnes to about 300 tonnes. Nowadays, more than 60% of the crude steel is cast directly into blooms (square blocks with an outline of 0.15-1m) and billets (small bars with an outline of less than 0.15m) or slabs of 0.15-0.2 m thick using a continuous caster [IISI, 1996b]. Blooms and billets are further processed in hot rolling mills to long products to change the shape into e.g. beams, profiles, and rails. Slabs are converted to flat products in a hot strip mill or hot plate mill to reduce the thickness to 1-10 mm for strips and 10-25 mm for plate. The thickness of flat products may then be further reduced to about 0.1-3 mm in a cold rolling mill. The casting and shaping process is characterized by its discontinuity, requiring intermediate storage and putting high demands on logistics.
Near-net-shape casting processes use techniques that can attain the final shape with fewer operations, or even in one step. The main advantages of near-net-shape casting are: (a) reduction in investment and operation costs; (b) reduction in processing time between casting and final product; (c) reduction in intermediate heating and cooling and storage; and (d) improved (surface) properties resulting from a finer, more homogeneous microstructure.
The state of the art in near-net-shape casting is thin slab casting: Slabs are cast with a thickness of 40-90 mm. Thin slab casting has been applied successfully since 1985 in connection with EAF steel plants. With a combination of an EAF and thin slab caster, flat products can be produced at costs that are competitive with the costs of flat products made in an integrated steel plant. This opened the market for flat products for EAF steel, a market that had been restricted to integrated steel producers. For a few years now, the technique has been applied in integrated steel plants as well. It is estimated that 10% of the world hot strip is produced using a thin slab caster . The direct casting of beams is also a commercial technique [VAI, 1994]. This section focuses on the near-net-shape casting of flat products.
Four categories of near-net-shape casting techniques can distinguished for flat products:
1. Thin slab casting: thickness range 40-80 mm;

2. Thin slab casting with liquid core reduction: thickness range 10-25 mm;
3. Strip casting: thickness range 1-10 mm;
4. Spray casting: thickness range 5-20 mm.

The first two techniques resemble the continuous caster and still require a reheating furnace, albeit with a smaller heating capacity. The third technique makes the hot strip mill redundant, and is therefore interesting from the point of view of energy conservation. Spray casting produces semifinished products of different geometry by spraying and rapid solidification of small metal particles onto a substrate surface [Parodi, 1993].

Specific energy consumption The integrated analyzed in section 4 is used here as the reference process. In this process casting requires only a small amount of heat for preheating of the ladle and electricity for, e.g., drive power and crane handling. Shaping requires the most energy, particularly the reheating furnace (1.82 GJ/trs). The electricity required for driving the machines is 0.37 GJ of electricity/trs. Steam (0.15 GJ/trs) is generated in the cooling section of the hot strip mill. Presented in primary energy units this is 2.4-2.7 GJ/trs, or about 15% of the total primary energy demand of the integrated mill. We assume that electricity is produced in a central power plant with efficiencies of 40% and 60%, respectively. We credit the generated steam with a factor 0.35, representing the ratio between exergy and enthalpy of the steam.

Economics Direct investment costs of conventional casting and shaping of hot strip can be divided as follows: 71% to the hot rolling mill and reheating furnace, 22% to the caster, and 7% to finishing [Flemming *et al.*, 1988]. Absolute investment costs are not available. Reported production costs for conversion from crude steel to hot rolled strip for an integrated mill (at the Great Lakes, USA) are US$67/tonne hot rolled strip [Szekely and Trapaga, 1994]. About 30% of these costs are for maintenance, 20% for energy and 15% for labor. The remainder consists of costs for replacements of rolls, refractories, supplies and of credit for yield losses.

Both investment costs and operation costs of near-net-shape casting techniques are presented with reference to conventional casting and shaping per tonne of hot rolled strip.

Thin slab casting Although thin slab casting (TSC) is already a commercial technique, this technique is discussed here because experience with TSC in integrated steel mills is still small. Compact Strip Production of SMS Schloemann-Siemag, Inline Strip Production of Mannessmann Demag, and Continuous Thin Slab Casting and Rolling Technology (Conroll) of VAI are the major TSC techniques, but others are available or are being developed [Parodi, 1993; Stelco, 1993]. In 1995 about 15 thin slab casters had already been installed worldwide or had been ordered, and several pilot plants were in operation [Parodi, 1993; Flemming, 1995]. Most of these installation are combined with an EAF. However, construction of the first facility to combine a BOF with a Compact Strip Production plant started operation in 1994 in the USA . Recently, two European integrated steel manufacturers, Hoogovens (Netherlands) and Thyssen (Germany), announced that

they will install thin slab casters with annual capacities of 1.2 and 1.5 million tonnes strip respectively [Stam, 1997].
A TSC facility resembles a continuous caster. The casting muld, which gives shape to the cast steel, is adapted to cast slabs with thicknesses of 40-50 mm (Compact Strip Production), 30-60 mm (Inline Strip Production), and 70-90 mm (Conroll). TSC facilities combine the caster and the rolling mill in one plant by tight control and the use of a reheating furnace that brings the steel to a uniform temperature of about 1100-1200°C [Rohde and Flemming, 1995]. The type of furnace differs depending on the configuration: SMS and Conroll apply a hearth furnace and Inline Strip Production uses an induction furnace. After the rolling mill, a thickness of less than 2 mm can be achieved. Industrial plants are designed with slab widths of 900-1700 mm [Flemming, 1995]. Thin slab casters can produce different types of steel, including stainless and carbon steel.
The capital costs (per tonne of product) are 30% [SMS, 1995] to 55% [Hendricks, 1995] of the costs for a conventional caster with hot strip mill. The largest reduction is achieved when a two-strand TSC plant is used with an annual capacity of about 1.5 million tonnes of rolled steel. At that capacity the hot rolling mill is used optimally [Flemming *et al.*, 1988]. Total investment costs for the TSC plants that will be installed at Hoogovens and Thyssen are US$200 and US$300/thm, respectively. Operation costs vary from 80 to 110% of the reference, depending on capacity utilization [Flemming *et al.*, 1988; Hendricks, 1995]. TSC is already competitive at half of the capacity at which a hot strip mill operates most cost-effectively (i.e. 4 million tpy) [Hendricks, 1995].
The SEC of the Inline Strip Production process is 196 kWh/tonne, including electricity for the induction furnace [Tomassetti, 1995]. The fuel requirement for reheating is eliminated in this case. In terms of primary energy, a SEC of 1.2 to 1.8 can be achieved or a reduction of 35 to 50%. For comparison, savings of 50% on the primary SEC for the Compact Strip Production are reported [SMS, 1995].

TSC is a continuation of the developments in continuous casting. The quality and nature of the products are comparable to those of flat products from a continuous caster. The products can be further processed in the existing cold rolling mill and finishing operations. TSC can be integrated into existing steel plants, including integrated steel plants and plants with an EAF. On the basis of these considerations, TSC can be seen as an evolutionary change.

<u>Thin slab casting with liquid core reduction (TSC with LCR)</u> Slabs with a thickness of less than 25 mm can be cast by compressing the cast steel shortly after it leaves the mold, i.e. while the edges are already solid and the core is still liquid. To roll strips of 1-10 mm a reheating furnace and rolling mill are still required. Several techniques are under development. Inline Strip Production also uses liquid core reduction, but the thickness is reduced from only 60 mm to 45 mm after casting [Hendricks, 1995].
Thyssen Stahl (Germany) and SMS (Germany), and partially Usinor-Sacilor (France), are working together to develop the Casting Pressing Rolling (CPR) [Kruger, 1995]. A pilot plant was built in Germany and has been in operation since 1993 [Kruger, 1995]. After use

of the pressing rolls that squeeze the just cast steel, the strip thickness is 10-15 mm [Hendricks, 1995].
The capital costs (per tonne output) of CPR plants are said to range from 100 to 120% of the reference [Hendricks, 1995]. Kruger reports capital costs of about US$260 /trs for a CPR plant with a capacity of 450 ktpa [Kruger, 1995]. At this capacity a CPR plant is competitive with a conventional caster and rolling mill with a capacity of 4 million tpa [Hendricks, 1995]. Operation and maintenance costs are comparable to the reference [Hendricks, 1995].
No data on energy demand are available. As the thin slabs have hardly any hot core, the heat demand for homogenizing the temperature is expected to be more or less the same as that for thin slab casting without liquid core reduction. The temperature increase is about 20% of that needed in a conventional reheating furnace. On the other hand, since less rolling is required, savings on electricity are expected to be higher than with TSC without LCR. We estimate the heat demand to be 20% of the reference and the electricity demand to be 80%, resulting in a SEC of 0.9-1.1 GJ/trs based on primary energy units. This is a reduction of 60-65% compared to the reference.

The step from TSC to TSC with LCR is not large as far as degree of technical change is concerned. Therefore, we use the same categorization as for TSC: evolutionary change.

Strip casting Contrary to continuous casting (and thin slab casting) strip casting does not require a casting muld. Instead the liquid steel is cast directly on a belt or on rolls. Separate rolling is no longer required. Figure 5.16 gives an illustration of some possible configurations. The configuration with the two casting rolls (Figure 5.16.b) closely resembles the strip casting machine that Bessemer patented around 1860.
Many R&D efforts have been made with regard to the development of strip casting. R&D is generally performed by a joint venture between a casting machine manufacturer and a steel manufacturer [Fruehan, 1994]. Examples are Allegheny Ludlum and Voest Alpine, CSM and Ilva. More than 30 R&D projects have been reported; most of them are still in the pilot-plant stage with ladle capacities of less than 3 tonnes [Stelco, 1993]. However, at least six strip casting development projects are already in the demonstration stage; i.e. a ladle capacity of more than 10 tonnes can be handled [Parodi, 1993]. Commercial strip casters should have a ladle capacity of 20-25 tonnes. So far, no commercial strip casters have been developed. Problems with the geometry of the strip appear to be the main bottleneck for development, although surface quality and mechanical-technical properties of cold-rolled cast strip also require serious R&D. Furthermore, production of carbon steel, the most important type of steel, is difficult. Differences in steel microstructure and resistance to oxidation are characteristics that make stainless steel favorable to carbon steel [Stelco, 1993]. However, some companies are developing strip casters to produce carbon steels [Parodi, 1993; Fruehan, 1994]. Parodi (Ilva) states that eventually it will be possible to cast carbon steels [Parodi, 1993].
Strip casting can achieve considerable energy savings because hot rolling, thus intermediate reheating of the steel, is no longer required. The potential for reduction of the

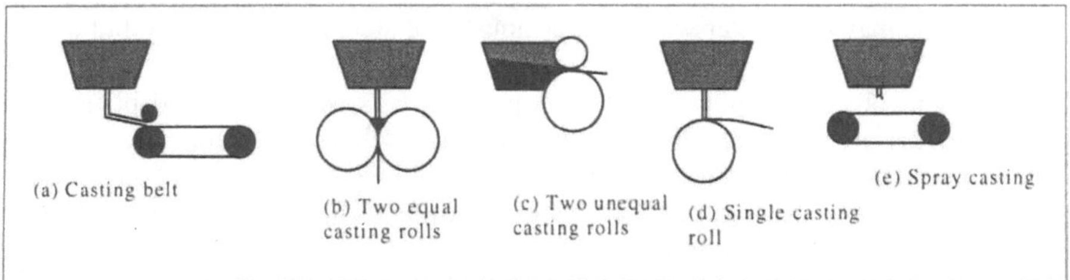

Figure 5.16: Four possible configurations for strip casting (a-d) and one for spray casting (e) (sketches adapted from [Hendricks, 1995]).

heat requirement for rolling is therefore 100%. Some electricity is still required for various small operations: e.g. ladle handling. We estimate the electricity requirement of strip casters to be 25% of the reference (no data available, own estimate). The SEC based on primary energy units will then be 0.15-0.25 GJ/trs, which means that it is 90-95% lower than that of the SEC of the reference.
Reduction of the capital costs compared with a conventional continuous caster with a hot strip mill is estimated to be 55-65% (per tonne of product) [Hendricks, 1995]. O&M costs of a strip caster are estimated to be 50% higher [Hendricks, 1995]. Strip-casting plants are already commercially attractive with a capacity of 0.5-0.7 million tonnes per year [Hendricks, 1995].

Strip casting implies the application of a new principle of casting. Major adaptations have to be made to the casting and shaping processes. The product characteristics may also change and open new markets. Strip casting is therefore considered to be a major change.

Spray casting Spray casting involves atomization of the liquid metal and depositing of the formed droplets on a substrate. The droplets are cooled by a gas stream while being deposited. With regard to quality and mechanical-technical properties spray cast steel has some advantages over strip cast steel. A disadvantage is the low yield. The technique is being applied to some nonflat products [Parodi, 1993; Stelco, 1993].
A spray-casting process is being developed by Mannesmann Demag (Germany). A pilot plant has been in operation in Germany, producing flat products of 12-20 mm thickness [Stelco, 1993]. Other companies that have worked on the development of spray casting are Sandvik Steel (Sweden), Sumitomo Heavy Industries (Japan), and General Electric (USA). Spray casting for the production of large quantities of steel is still in an early stage of development. Realization of the process for certain specific products, e.g. super alloys, is more probable [Parodi, 1993].
As far as the SEC is concerned, no savings over strip casting are. No heat is required, as the metal is already molten. We expect a higher electricity demand than for strip casting because electricity is required for spraying the droplets, for maintaining the cooling gas stream and for some drive power. We estimate the electricity demand to be 50% of the reference. The SEC is then 0.31-0.46 GJ/trs, or about 85% lower than the reference. The

overall savings may be smaller when a large amount of steel is lost as a result of a low yield.

Spray casting involves the application of a new principle for casting. Major adaptations to existing process are required. Product characteristics will be different, affecting the finishing operations. The technique might be used only for speciality steels. Although spray casting can be considered a radical change in casting technique, it is a major change if steel making as a whole is considered.

Conclusions concerning near-net-shape casting Table 5.10 gives a breakdown of the SEC of the near-net-shape casting techniques and a reference process, namely continuous casting and a hot strip mill.

Table 5.10: Breakdown into energy carriers of the SEC for near-net-shape casting techniques and a reference process.

	Final energy	Primary energy	
		low case	high case
	(GJ/trs)	(GJ/trs)	(GJ/trs)
Reference (based on [IISI, 1982])			
fuel	1.82	1.82	1.82
electricity	0.37	0.62	0.93
steam	-0.15	-0.05	-0.05
SEC		2.38	2.69
Thin slab casting (ISP)[Tomassetti, 1995]			
fuel	0.00	0.00	0.00
electricity	0.71	1.18	1.76
SEC		1.18	1.76
Thin slab casting with liquid core reduction (own estimate)			
fuel	0.36	0.36	0.36
electricity	0.30	0.49	0.74
SEC		0.86	1.10
Strip casting (own estimate)			
fuel	0.00	0.00	0.00
electricity	0.09	0.15	0.23
SEC		0.15	0.23
Spray casting (own estimate)			
fuel	0.00	0.00	0.00
electricity	0.19	0.31	0.46
SEC		0.31	0.46

The maximum saving can be achieved with strip casting. This technology might reduce the SEC of casting and shaping by 90-95%. If this is done, the SEC of an integrated primary steel mill can be reduced from 19 to 16.5 GJ/tcs, or 13%.
Table 5.11 gives an overview of the other characteristics of the near-net-shape casting techniques. The degree of technical change and the stage of development of all techniques are also shown in Figure 5.17. TSC is already commercial and is therefore placed outside the framework. The techniques with the largest saving potential, strip casting and spray casting, are still not available. The degree to which the techniques differ from the current technique is characterized as major, implying that implementation of these techniques requires major adaptations to the process.

Table 5.11: Overview of the characteristics of near-net-shape casting techniques.

Technique	Stage of development	Casting thickness (mm)	Capital costs per tonne (index)	O&M (index)	Degree of technical change
Continuous casting and hot strip mill	state of the art	>150	100	100	-
Thin slab casting	commercial	40-80	30-55	80-110	evolutionary
Thin slab casting with liquid core reduction	commercial/ demonstration	<25	100-120	90-110	evolutionary
Strip casting	demonstration/pilot plant	1-10	55-65	135-165	major
Spray casting	pilot plant	5-20	?	?	major

5.6.3 SCRAP-BASED PROCESS

In primary steel production, most energy is required to prepare the raw materials and to reduce iron ore. In modern energy-efficient steel mills, the proportion of these processes (including heating and melting of the iron) can be as high as 90% of the total primary energy demand. It is obvious that large energy savings can be achieved if these processes are avoided. Recycling and reprocessing of steel scrap offers this possibility. To adjust shape and properties, however, melting is still required. Several options for melting scrap are available or under development:

1. in a Basic Oxygen Furnace;
2. in an Electric Arc Furnace;

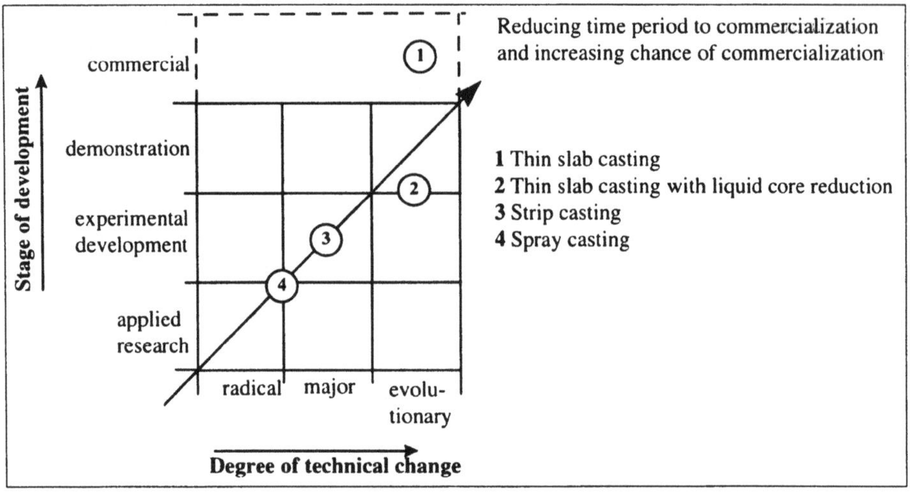

Figure 5.17: Comparison of near-net-shape casting techniques with regard to the degree of technical change compared to the current technique and the stage of development.

3. in a scrap melter using both electricity and fossil fuel;

4. in an all-fossil fuel melter;

The quality of the steel depends largely on the quality of the scrap. Because high-quality scrap is expensive, virgin iron-containing materials can be added to upgrade the quality of the product. Direct reduced iron is frequently used for this purpose. Pig iron can also be used, and recently experiments have been performed in which an EAF is charged with iron carbide [Smith, 1995]. Obviously, the use of these virgin materials increases the overall energy consumption, as their production consumes a considerable amount of energy. The quality of scrap can also be upgraded by chemical and mechanical separation processes. The additional energy demand for these processes can be estimated to be 0.5-2 GJ/tonne scrap [Stelco, 1993].

Below we discuss the four options for melting scrap and then assess the potential for energy-efficiency improvement of scrap-based processes.

<u>Scrap melting in the BOF</u> Scrap melting in the BOF is common practice. The heat for melting is generated by the oxidation of carbon in the pig iron. Without additional fuel injection the maximum scrap input is limited to about 25-30% of the charge. A higher scrap ratio can be processed when additional fuel is injected. In fact, Klöckner developed a process for melting a 100% scrap charge in a BOF-like converter. This process is explored further in the discussion scrap melting with fossil fuel only.

<u>Electric Arc Furnaces</u> Electric arc furnaces (EAFs) can operate on a 100% scrap charge, but they can also process mixed charges including DRI, iron carbide, and cold or hot pig

iron. EAF capacity is growing rapidly, from about 18% of the world steel production in 1975 to 27% in 1990 (see Figure 5. 4). EAFs have some advantages over integrated steel mills, e.g. lower capital costs, the possibility of economic operation at low capacity, and smaller environmental impact [Smith, 1995]. A drawback is that not all products can be produced. Traditionally, EAFs produce low-quality long products. The low quality of the products is due to the fact that scrap is usually more contaminated than pig iron. Hot rolling mills for flat products can be operated competitively only at large capacities, and are therefore usually not installed at EAF facilities. This picture is changing. Near-net-shape casting techniques make the production of flat products from EAF steel competitive with primary steel products. In fact, Nucor in the United States has already entered the flat product market with EAF steel [Mott *et al.*, 1994].

The principle of EAFs is that steel is melted via electric arcs between cathode and one (for DC) or three (for AC) anodes. The anodes can be placed just above the baths or be submerged in the bath. Oxygen can be injected to promote metallurgical reactions, coal powder can be added to promote slag foaming through CO formation, and oxy-fuel burners may be directed at cold spots.
The major energy input to EAFs is electricity. A reduction of 35% in electricity consumption has been achieved in the past 30 years, as is shown in Figure 5.3. Even lower electricity consumption levels have been attained by partially replacing electricity by fossil fuel and scrap preheating. At present the most energy-efficient EAF is the 'Finger Shaft Furnace' from VAI and FUCHS [Hofer, 1996; Hofer, 1997b]. The Finger Shaft Furnace makes optimum use of the energy available in the process gases by preheating scrap in a shaft placed above the furnace. Operational results of a 100% scrap charged Finger Shaft Furnace at Von Roll/SWG in Switzerland show that a primary energy demand of 3.7 GJ/tcs can be achieved, using a 40% efficiency to convert electricity to primary energy carriers [Rong *et al.*, 1996; Hofer, 1997b]. Table 5.12 gives a breakdown of the SEC into energy carriers. Virgin iron materials, like pig iron and DRI, cannot be preheated in the shaft, because they would reoxidize. However, these materials can be charged directly into the furnace. With a 55% DRI and 45% scrap charge an SEC of 4.6 GJ/tcs (primary energy) has been achieved in a DC Finger Shaft Furnace at HYSLA in Mexico (see Table 5.12) [Rong *et al.*, 1996]. The higher SEC compared to the 100% scrap charged furnace is due to the fact that (a) the DRI is not preheated and (b) the DRI has slags, which have to be melted as well. The production of DRI requires about 10 GJ/tonne, which has not been included in this SEC. A third Finger Shaft Furnace at Cockerill Sambre in Belgium is charged with up to 35% hot metal. Because the hot metal does not have to be melted, the electricity requirement can be as low as 230 kWh/tcs; the primary SEC is 2.4 GJ/tcs [Hofer, 1997b]. The losses in modern EAFs are due to cooling water losses, slag heat loss, and chemical and thermal energy loss with the off-gas [McIntyre and Landry, 1992].

The energy required for heating scrap from 25°C to melting point and subsequent melting is between 1.05 to 1.36 GJ/tonne; the lower limit is for pig iron and the upper limit is for pure iron. However, the theoretically lowest energy consumption for making steel from scrap is low. When we compare the SEC of modern EAFs with the energy required for melting we can conclude that EAFs are efficient melters. The recovery of the thermal energy from the hot molten steel, on the other hand, is far less efficient. This needs to be improved to reduce the SEC of steel making from scrap still further. We discuss this option in section 6.5.

Large losses occur in the conversion of primary energy to electricity. If the efficiency of electricity generation can be raised from 40% to 60%, the SEC of the Finger Shaft Furnace (100% scrap charge) will reduce from 3.7 to 2.7 GJ/tcs, without improvements in the EAF itself. Melting of scrap with fossil fuel only, thus avoiding the use of only electricity, is discussed later.

Developments in EAF technology are directed towards (a) increasing the productivity by decreasing the tap-to-tap time and increasing the capacity and (b) reducing the operation costs by reducing power and electrode consumption [Stelco, 1993]. The Double Shaft Furnace (VAI/FUCHS), which has two separate shaft furnaces that are served by one set of electrodes and one transformer, can achieve a production of 1.2 million tpa, as compared to 0.7-0.8 million tpa for a Finger Shaft Furnace [Rong *et al.*, 1996]. The SEC is on the same order as that for the Finger Shaft Furnace. One step further is the Combination Shaft Furnace (VAI/FUCHS), which has an efficient scrap preheating shaft that can be rotated from one furnace to the other. The scrap is preheated to about 1000°C. From Figure 5.6 it can be seen that preheating to this temperature requires approximately 0.7 GJ/tonne. Thus, half of the theoretical energy demand for heating and melting is provided in the preheating shaft. According to the developer, this furnace can produce crude steel with a power consumption of 180 kWh/tonne, 65 Nm^3 of oxygen, 35 kg of carbon and 0.8 kg of electrode [Stelco, 1993]. A SEC of 3.0 GJ/tonne (primary energy; 40% efficiency of electricity generation) results (see Table 5.12). When electricity is generated with 60% efficiency, the SEC will come down to 2.3 GJ/tonne.

Table 5.12: Breakdown of the SEC of scrap smelting processes, expressed in GJ primary energy per tonne of crude steel[a].

Process energy carrier	**unit** **per tcs**	**low case** **(GJ/tcs)**	**high case** **(GJ/tcs)**
Von Roll Finger shaft furnace (100 % scrap) [Hofer, 1997b]			
power	315 kWh	1.9	2.8
oxygen	27.5 Nm^3	0.1	0.1
natural gas	7 Nm^3	0.2	0.2
electrode	1.4 kg	0.0	0.0
carbon powder	15 kg	0.4	0.5
SEC		2.7	3.7

Process energy carrier	unit per tcs	low case (GJ/tcs)	high case (GJ/tcs)
HYLSA Finger shaft furnace (55% DRI; 45% scrap) [Hofer, 1996]			
power	432 kWh	2.6	3.9
oxygen	31 Nm^3	0.1	0.1
natural gas	4 Nm^3	0.1	0.1
electrode	1.2 kg	0.0	0.0
carbon powder	15 kg	0.4	0.5
SEC		3.3	4.6
Combination shaft furnace (100% scrap) (design value) [Stelco, 1993]			
power	180 kWh	1.1	1.6
oxygen	65 Nm^3	0.2	0.2
natural gas	0 Nm^3	0.0	0.0
electrode	0.8 kg	0.0	0.0
carbon powder	35 kg	1.0	1.1
SEC		2.3	3.0
K-ES (100% scrap) [Teoh, 1989; Patuzzi *et al.*, 1990]			
power	300 kWh	1.8	2.7
oxygen	50 Nm^3	0.1	0.2
natural gas	4 Nm^3	0.1	0.1
electrode	3 kg	0.1	0.1
carbon powder	22 kg	0.6	0.7
SEC		2.8	3.8
KS (100% scrap) [Patuzzi *et al.*, 1990]			
power	0 kWh	0.0	0.0
oxygen	300 Nm^3	0.7	1.1
natural gas	1.5 Nm^3	0.0	0.0
electrode	0 kg	0.0	0.0
coal	255 kg	7.4	8.2
SEC		8.2	9.3
KVA (100% scrap) [Patuzzi *et al.*, 1990]			
power	0 kWh	0.0	0.0
oxygen	70 Nm^3	0.2	0.3
natural gas	40 Nm^3	1.3	1.3
electrode	0 kg	0.0	0.0
FeSi	15 kg	2.2	2.2
SEC		3.6	3.7

[a] The following conversion factors are used: from electricity to primary energy carriers 0.6/0.4 (low/high); electrode 30.95 GJ/tonne; oxygen 280 kWh/tonne; natural gas 31.65 MJ/Nm^3; coal/carbon powder 29/32 GJ/tonne (low/high); FeSi 0.14 GJ/kg.

Scrap melting using both electricity and fossil fuel Scrap melting using both electricity and fossil fuel is a technique already used in EAFs. The Klöckner Electric Steel making (K-ES) process was developed to partially replace electricity by coal or coke [Teoh, 1989; Patuzzi *et al.*, 1990]. Enhanced bath stirring is achieved by bottom injection of inert gases. With a coal injection of 22-30 kg/tcs, the electricity consumption is 300 kWh/tcs [Teoh, 1989; Patuzzi *et al.*, 1990]. The SEC on a primary energy basis is 2.8-3.8 GJ/tcs (with an efficiency of electricity generation of 40%) (see Table 5.12). Three companies (two in Italy and one in Japan) operated the K-ES process in 1993. Mannesmann-Demag has been studying the Melting Machine, which also used bottom fuel injection and electricity for scrap melting. A design electricity use of less than 300 kWh/tcs has been reported [Stelco, 1993]. However, other data are not available.

Scrap melting with fossil fuel only At least two processes that melt scrap with fossil fuel only have been proposed, the Klöckner Steel making (KS) process and the Klöckner Voest Alpine (KVA) process [Teoh, 1989; Patuzzi *et al.*, 1990]. The KVA is a continuous scrap melting process using oxygen and natural gas burners to melt the scrap, applying postcombustion [Stelco, 1993]. Refining of the molten steel has to be done in a separate reactor. R&D concerning this process stopped because of environmental problems and a high scrap price [Hofer, 1997a]. The KS process melts scrap using bottom coal injection, oxygen and postcombustion. After melting the converter can switch to the refining operation. The SEC of the KVA process is about 3.6 GJ/tcs; the SEC of the KS process is 8.6 GJ/tcs (see Table 5.12) [Patuzzi *et al.*, 1990].
It can be concluded that at the moment the SEC of the KVA process is more or less equal to the most efficient EAF in terms of primary energy consumption. In terms of final energy consumption, a modern EAF requires about 1.4 GJ of electricity to melt scrap, whereas the KVA process requires 3.6 GJ of fossil fuel. The transfer of the heat of hot gases to a bulk of metal is apparently less efficient than heating by electricity. The result is a relatively large volume of waste gas that is produced in the furnace. When electric melting is applied, this volume of gas is produced in a power plant, which is designed to optimally recover the thermal and chemical energy of the gas. Utilization of the waste gas of the melting furnace is hampered by environmental problems caused by impurities in the scrap. To reduce the SEC of fossil fuel-fired melting, R&D should be directed towards improving the heat transfer of the hot gases to the scrap, production of clean waste gas, and optimal use of the thermal and chemical energy of the waste gas.

Conclusions concerning scrap-based processes The most efficient process for melting a 100% scrap charge is the Single Shaft Furnace with a SEC of 3.5 GJ/tcs (assuming a 40% conversion efficiency for power plants). The most efficient EAF process under development is the Combination Shaft Furnace, which has a design SEC of 2.9 (40%) to 2.3 (60%) GJ/tcs. This process uses both fossil fuel and electricity to melt scrap. The most efficient fossil fuel-based process, which is not yet commercial, has a SEC that is comparable to the currently most efficient EAF in terms of primary energy. To decrease the SEC of fossil fuel-based processes, the large volume of gas needs to be used efficiently.

Although the SEC of the future EAF approaches the minimum energy requirement for heating and melting iron, a further reduction of the SEC may be achieved when ways are found of recovering the thermal energy that is now lost with the waste gases, the hot metal, and the slag. Heat recovery from the waste gases to preheat scrap is already commercial. Slag waste heat recovery and heat recovery from the molten steel have not yet been demonstrated. We deal with these options in section 6.4.

The technical changes needed for the implementation of new melting processes are considered to be evolutionary. Many improvements have been realized in the past few decades, but the principle of electric melting has not changed.

5.6.4 STEEL MAKING AT LOWER TEMPERATURES

The ultimate technique for reducing the need for high temperatures would be steel making at room temperature, without any temperature rise. Since the reduction of iron ore at room temperature is thermodynamically and kinetically unfavorable, such a process is hard to conceive. The various unit operations, however, can be operated at lower temperatures than in present processes, although these temperatures are usually still far above room temperature.

<u>Coke making</u> Coke can be produced at a lower temperature (800°C instead of 1100°C) by completing the heating of the coke while it descends into the blast furnace [Stelco, 1993]. This process has been tested on a small scale by Kobe Steel in Japan. A saving of about 15% on the fuel consumption of coke making can be achieved [Stelco, 1993]. This process partially integrates coke making and iron making. However, cooling of the hot coke still occurs.

<u>Iron oxide reduction</u> Iron oxide can be reduced without melting the iron. In fact, Direct Reduction (DR) processes do this. DR technology uses a synthesis gas (or solid fuel directly) to reduce the iron oxides at temperatures below the melting point (900-1000°C, compared with 2000°C at maximum in a blast furnace). Direct Reduced Iron (DRI) is physically similar to the ore feedstock and contains the minerals that were originally present in the ore. Many DR processes have been developed [Zervas *et al.*, 1996]. At present, the MIDREX DR process has the largest share. It is used to produce 65% of the global DRI [Midrex, 1996]. Only 5% of the iron produced globally is DRI, the remainder being pig iron produced in blast furnaces [IISI, 1996b].

The MIDREX process consists of three reactors. In the first reactor, natural gas is preheated by heat exchanging with off-gases. In the second reactor the preheated natural gas is reformed to a mixture of CO and H_2. The reformed gas is fed to a shaft furnace where iron ore, in the form of pellets or lump ore, is reduced. An improvement to the MIDREX process is the Arex process, which has been applied in Venezuela [Zervas *et al.*, 1996]. The improvement lies in the fact that hot DRI in the shaft reactor is used as catalyst

for the natural gas reforming. The gas reforming and the iron ore reduction can therefore be performed in the same reactor, resulting in lower investment costs and a lower SEC. Not only does the MIDREX process (thus, the Arex process) work at a lower temperature than a blast furnace, but coke making and sintering are no longer required. Thus, the high temperature processes in the raw material processing are eliminated as well. However, one new high-temperature process is added, namely the production of reducing gas. Ore can be fed as lumps and pellets. Pelletizing still requires a high temperature. The DRI is either cooled in the lower part of the reduction furnace, giving off its heat to a cooling gas, or pressed while hot in briquets, which are then cooled. In both cases the heat, amounting to about 0.4 GJ/tonne DRI, is lost [Midrex, 1991].
Two values of the energy requirement for an Arex plant are reported: (a) 8.7 GJ of natural gas and 70 kWh of electricity per tonne DRI [Faure, 1993a] and (b) 9.7 GJ/tonne of DRI [Zervas *et al.*, 1996]. We use an average of 9.5 GJ/tonne[1]. Assuming that all iron ore is charged as pellets, an additional 1.3 GJ of primary energy per tonne is required for pelletizing (assuming a 40% efficiency for electricity generation). The total SEC of 10.8 GJ/tonne is about 25% lower than the SEC of making pig iron in a blast furnace, including coke making and ore agglomeration. However, because of its low carbon content DRI cannot be charged as the sole raw material in a BOF plant. DRI has to be melted and refined in an EAF plant, or can be charged with pig iron to a BOF. The energy use for melting in an EAF plant should therefore be accounted for when different steel-making processes are being compared, resulting in an SEC more or less equal to that of the blast furnace route. A comparison of SECs is presented in the last section of this chapter.

Shaping Powder metallurgy can be used to mix iron with other components and to shape iron into specific forms without melting [Thümmler and Oberacker, 1993; Salak, 1995]. Powder metallurgy is used commercially to make special products; it is characterized by small production volumes and high demands on precise shape and well-defined properties. Worldwide iron powder consumption in 1992 is estimated to have been about 570,000 tonnes in 1992 [Thümmler and Oberacker, 1993]. Powder metallurgy involves shaping directly from a ferrous (or nonferrous) powder by pressing it into a mold of the desired shape and subsequently heating it in a furnace to bond the fibers together. Thus, shaping does not eliminate the application of high temperatures. The production of powder requires high temperatures as well. More than 50% of the iron powder is made by direct reduction using the Högenäs-process – especially designed for powder production – or the FIOR process [Thümmler and Oberacker, 1993]. Other common methods of producing iron powders start with pure iron, e.g. atomization or the carbonyl reaction [Thümmler and Oberacker, 1993]. As pure iron is required for these processes, they do not offer possibilities for energy-efficiency improvement.
It is unlikely that bulk steel products will ever be made at a commercial rate with this technique. Furthermore, making steel products with powder production requires an amount

[1] When the electricity demand is converted to primary energy units using a 40% efficiency of public power generation, the energy figures reported by the first source are 9.3 GJ/tonne DRI on a primary energy basis.

of energy that is at least equal to the amount required in conventional processes. Powder metallurgy is therefore not considered further in this study.

Conclusion concerning steel making at lower temperatures No technology avoids the melting of steel; melting remains necessary for shaping steel. Although direct reduction of iron ore is performed at a temperature that lies far below the melting temperature, the DRI still has to be heated further to be melted. When we compare the DRI-EAF route to the blast furnace-BOF route, we see that both routes have comparable SECs. It can be concluded that as long as melting is required to shape the products, a decline in the temperature of iron ore reduction will not result in a significant decrease of the SEC.

5.6.5 WASTE HEAT RECOVERY AT HIGH TEMPERATURES

The techniques discussed in the previous sections involved a reduction in the application of high temperatures. In this section, we explore techniques under development that can recover heat at high temperatures and make it available as a high-quality energy carrier. First, we discuss techniques that can be applied in the conventional integrated steel mill. Then we look at possible ways of recovering high-temperature heat from streams from future processes.

<u>Existing integrated steel plants</u> Table 5.13 gives an overview of the main hot flows in an integrated steel mill. The sensible heat and the exergy of the flows are based on the reference plant described in section 4. The total amount of energy lost as a result of heat leaking to the environment is about 5.5 GJ/trs. The exergy is about 3.1 GJ/trs. Table 5.13 also presents techniques for heat recovery and the present stage of development.
Waste heat recovery is already applied in integrated steel mills for some clean gaseous flows, like exhaust gases from combustion processes. Most process gases, however, contain dust and have to be cleaned before they can be redistributed. The feasibility of heat recovery at higher temperatures depends on the development of a high-temperature gas-cleaning system. Several systems are being developed or are already commercial, e.g those based on electrostatic precipitation, ceramic filters and high-efficiency cyclone separators [Ree *et al.*, 1995].

The main process gases are blast furnace gas, coke oven gas, and basic oxygen furnace gas. Blast furnace gas is usually wet-cleaned before being expanded in top gas recovery turbines. The gas enters the turbine at environmental temperature. To recover the sensible heat of the gas, dry cleaning techniques should be applied. The entry temperature to the expansion turbine can be raised to about 120°C, which increases the power output by 30-35% [Stelco, 1993]. Coke oven gas is processed in the by-product plant to recover tars and benzol. A waste heat boiler can be installed at the ammonia incinerator, recovering about 0.01 GJ/trs of steam [Stelco, 1993]. Some experiments with high-temperature heat recovery from the coke oven gas failed, because of fouling of the heat exchanger [Worrell *et al.*, 1993]. The sensible heat of blast furnace gas can be partially recovered using a closed BOF gas system or a combined boiler/suppressed combustion gas recovery system [IISI, 1982; Stelco, 1993]. About 0.1 GJ/trs of steam can be produced [IISI, 1982].

The sensible heat from solid streams is yet not being recovered at a large scale, although some techniques are commercial or under development. Dry coke quenching, for instance, is a commercial technique for recovering the sensible heat from hot cokes. The heat is recovered by blowing an inert gas over the incandescent cokes and generating steam by heat exchange with the hot gas [Stelco, 1993; Dungs and Tschirner, 1994]. However, this technique is expensive [Worrell *et al.*, 1993] and is therefore seldom installed.
Blast furnace slag, and to a lesser extent BOF slag, are also sources of sensible heat losses. In the 1980s several systems were considered for blast furnace- and BOF-slag heat recovery. They have never been realized in practice, however, mainly because of the high investment costs [Stelco, 1993; Bisio, 1997]. In most designs the slag is granulated while heat is recovered in the temperature range of 1500-1000°C in a steam boiler. Subsequently, heat is recovered by convection to air in the temperature range of 1000-150°C. One concept used a low-boiling, organic fluid, for the recovery of the sensitive heat. Granulation can be achieved by using an air-jet, rotating drums, or a stirrer. These processes can recover about 65-75% of the enthalpy of the slag [Bisio, 1997]. Developments of these processes have taken place in Japan, Germany and Great Britain. Using an exergy analysis, Bisio showed that the recovered heat can be best used for preheating combustion air used in the hot blast stoves [Bisio, 1997].
Another flow of solid hot material is sinter. The sensible heat of hot sinter is usually partially recovered in the sinter cooler. In our reference plant, for instance, it is assumed that the combustion air is preheated by heat exchanging with sinter cooler air. However, more heat can be recovered by advanced sintering or emission-optimized sintering [Stelco, 1993]. Advanced sintering involves cooling and sintering in one machine. The waste gas temperature is much higher than in conventional sintering. In emission-optimized sintering, the exhaust of the sinter process is partially recycled to this process. The sinter bed functions as a filter that catches small particles present in the exhaust. The part that is not recycled is water quenched. Emission-optimized sintering is a commercial technology [Hoogovens, 1996].

Finally, there are two flows of hot steel that give off heat to the environment: in the continuous caster and after the hot rolling mill. The initial temperature of the first flow is 1400-1500°C, and the second flow is about 800-900°C. The sensible heat of the first flow is usually lost to the environment. This loss can be reduced by charging the slabs to the reheating furnace while they are hot, by transporting the slabs through an insulated tunnel and by charging them directly to the reheating furnace. This technology – called hot charging – is commercial, and implemented at several steel mills around the world. Hot charging is usually not possible for the whole production for logistic reasons, e.g. the capacities of the caster and the rolling mill do not match, or part of the slabs are sold. Furthermore, there is always a temperature drop of about 500-600°C between casting mold and the entry of the reheating mill.

Table 5.13: Waste heat recovery techniques for process gases and solid flows in an integrated steel plant. Based on [Stelco, 1993] and [IISI, 1982].

Unit operation	Hot flow (gas/solid)	Sensible heat (GJ/trs)	Exergy (GJ/trs)	Maximum temperature (°C)	Technique	Stage of development
Coke making	Hot cokes (s)	0.24	0.14	1100	Dry coke quenching	commercial
	CO gas (g)	0.24	0.12	700	Waste heat recovery	stopped
Sintering	Sinter cooler gas (g)	0.97	0.28	350	Advanced sintering machine or	commercial
	Sinter exhaust gas (g)	0.23	0.12	350	Emission Optimized Sintering	demonstration
Blast furnace	BF gas (g)	0.82	0.33	500	TRT using dry-cleaning	commercial
	BF slag (s)	0.39	0.26	1300	Radiant heat boiler	prototype, R&D stopped since end 1980's
Basic oxygen furnace	BOF gas (g)	0.19	0.12	1200	BOF gas recovery	commercial
					combined boiler /suppressed combustion	commercial
	BOF slag (s)	0.02	0.01	1500	Radiant heat boiler	prototype, R&D stopped since end 1980's
Casting	Cast steel (s)	1.39	1.06	1600	Radiant heat boilers with heat pipes	commercial
					Slab cooler boiler	commercial
Hot strip mill	Hot rolled steel (s)	1.04	0.62	900	Water spraying and heat pumps	commercial
Total		5.53	3.06			

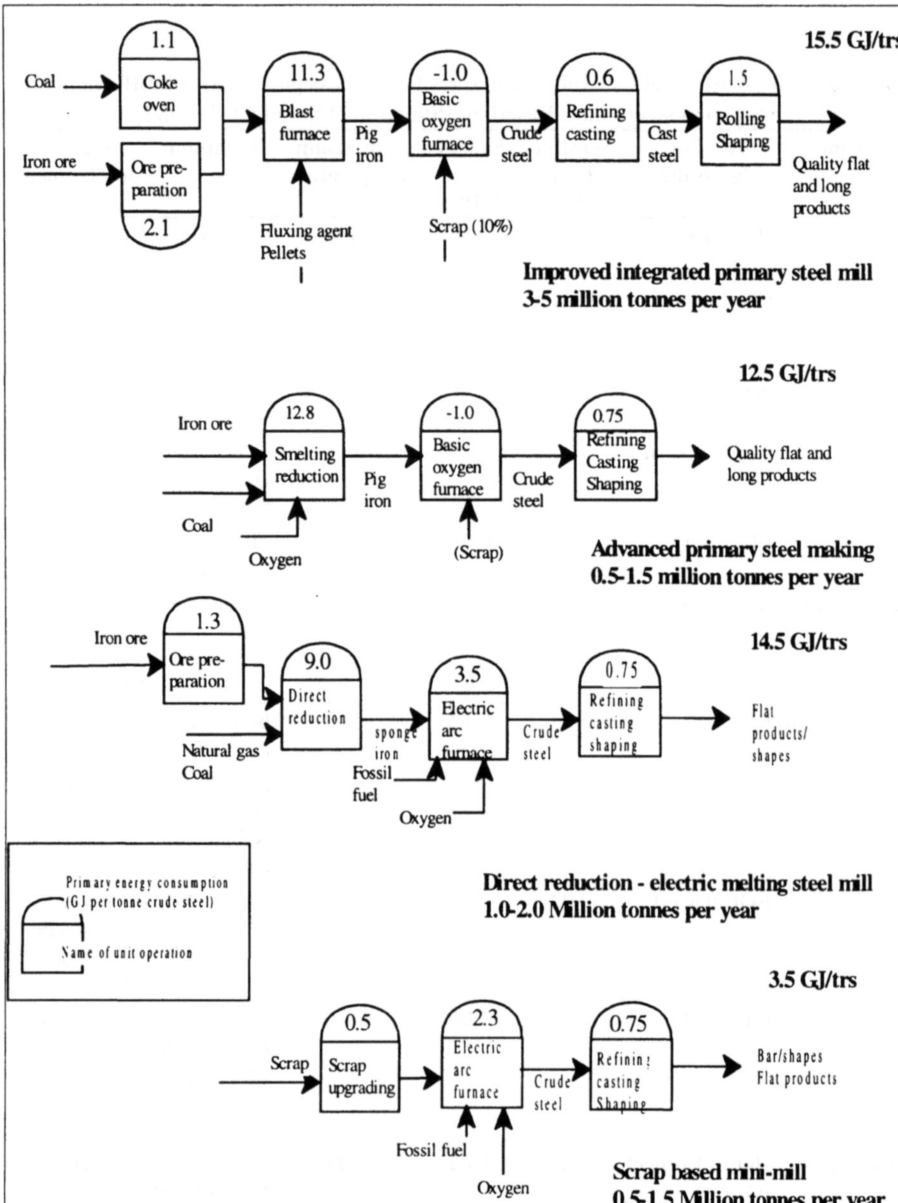

Figure 5.18: Flow sheets of future steel making processes. See Figure 5.5 for a comparison of current steel- making processes. The expected SEC is also given, expressed in GJ of primary energy per tonne of hot rolled steel. For an account of these data, see section 5 of this chapter.

Can the sensible heat of the cast steel can be recovered? As shown in Figure 5.6, about 25% of the thermal energy of liquid iron of 1550°C is due to the enthalpy of melting. Recovery of this energy is probably difficult, because the surface of the steel solidifies directly after the casting mold. Below 1500°C, the energy that can be recovered declines per tonne of steel approximately linearly with the temperature, namely by 0.7 MJ/°C. Cooling rates are higher at the higher temperatures.

The hot steel after the hot rolling mill is usually cooled by spraying water on the hot steel. The water, with a temperature of 70-80°C, can be recovered. However, considering that the temperature of the steel is about 800°C, this means a large loss of exergy. The heat can be upgraded by using a heat pump.

Two possible techniques for heat recovery of slabs between the caster and the hot rolling mill have been described in the literature. First, Kashima Steel Works in Japan makes use of radiant heat boilers placed above the slabs [Yoshida *et al.*, 1982]. A volatile medium, contained in heat pipes placed above the slabs, is vaporized and flows to a steam boiler, where it gives off its latent heat. Per tonne of slabs only 0.005-0.01 GJ of steam is generated. If we assume that heat recovery starts at the beginning of the horizontal part of the caster, at about 800°C, the maximum energy recovery potential is 0.6 GJ/tcs. The first radiant heat boiler was installed as early as 1980. The second possibility, which offers a larger recovery efficiency, is direct transfer of heat to steam in the slab cooler boiler [Maier and Angerer, 1986]. The slabs are conveyed through the boiler while giving off heat, which is used to generate steam of 40 bar and 450°C. With an entry temperature of 900°C and an exit temperature of 300°C, 0.32 GJ of steam per tonne of crude steel can be produced. Investment costs of about US$25 million for an installation that can process 2.2 million tonnes per year have been reported [Maier and Angerer, 1986]. The payout time is estimated at 6 to 8 years [Maier and Angerer, 1986]. A system like this has been installed at a steel mill in Eisenhüttenstadt (Germany) [Torlet, 1997].

<u>Future processes</u> In section 6.4 we concluded that melting will remain necessary to shape steel. Therefore, at least one heating and cooling step will be necessary. Thus, hot steel, slag, and gas will always be produced. Casting operations in future processes will differ from current continuous casting in several respects. First, strip casting eliminates hot rolling and reheating. Thin cast slabs require hot rolling, but the slabs are charged directly to a soaking furnace, to ensure uniform heat distribution, and then to the rolling mill. The temperature drop between casting and reheating is much smaller. Second, the production speed is much higher, because thinner steel is cast. Third, cooling time is much shorter. Heat recovery might increase the cooling time, because the temperature gradient between the hot steel and the surface of the heat exchanger is smaller than between hot steel and ambient air[1].

---

[1] The surface of the heat exchanger will radiate heat also. The emissive power of a surface depends on the temperature to the fourth power. When the temperature of the hot steel is 1600°C and the temperatureof the surface of the heat exchanger is 1400°C, 60% of the heat emitted by the steel is returned by the heat exchanger surface. When these temperatures are 1600°C and 800°C, respectively, only 10% of the heat is returned. The more heat is returned, the longer the cooling time.

What does this mean for the development needs of recovery techniques? When thin slab casting is applied, heat should be recovered at two places: after the caster and after the rolling mill. Strip casting requires heat recovery at one place. To achieve a high exergetic efficiency, heat should be recovered at high temperatures and in a large temperature range. It could well be that different heat recovery devices should be used for different temperature ranges. In the low-temperature range, an organic medium can be used. By making use of the heat of evaporation, a high heat capacity can be obtained. As described above, these devices are already commercial for conventional continuous casting. At higher temperatures, molten salts may be applied, although their corrosive nature puts high demands on the materials. As far as we know, hardly any R&D being undertaken concerning the high-temperature heat recovery of cast steel.

Conclusion concerning waste heat recovery at high temperatures Many heat recovery techniques are available, for both gaseous and solid streams. Implementation of these techniques has not been achieved, mainly because of the high investment costs involved. One of the topics of R&D should therefore be to make heat recovery more profitable by recovering a larger part of the heat at higher temperatures. For future steel-making processes the recovery of the heat of the cast steel over the whole temperature range from 1600°C to environmental temperature is a big challenge. Recovery of heat in the low temperature range can probably be developed first, since this technique is already available for continuous casting. Recovery at higher temperatures still requires much R&D.

5.6.6 CONCLUSIONS ON THE LONG-TERM ENERGY-EFFICIENCY IMPROVEMENT

This final section is an overview of the expected SECs of future steel-making processes. Furthermore, we discuss to what extent the exergy losses have been reduced and what needs to be done to achieve a further reduction. Finally, we estimate the future energy consumption from steel making.

Figure 5.18 gives an overview of four future steel making routes and the expected SEC, expressed as GJ primary energy per tonne of hot rolled steel.

The first process route is an improved version of the blast furnace route. The SECs are taken from Worrell *et al.* [Worrell *et al.*, 1993]. Most of the techniques that have to be applied to achieve this potential have already been demonstrated, and they can be added to the process without major adaptations. Although larger improvements are possible by using newer techniques, it is likely that many integrated steel mills will be adapted in this way, because it involves only evolutionary changes.

The second process route is an advanced primary steel-making route incorporating an efficient smelting reduction process and strip casting. The SEC is 34% lower than the SEC of the current best-practice integrated mill. All techniques have been proven on a pilot-plant scale and are expected to be commercially available within 15-20 years. The major driving forces are the lower environmental impact and the large reduction in production costs.

The third process route depicted in Figure 5.18 is based on an EAF combined with direct reduction of iron ore according to the AREX process and strip casting. The AREX process is the most efficient commercial DR process available. To account for energy- efficiency improvement of the AREX process, we assume that the SEC of the future AREX process is 0.5 GJ/tcs lower than that of the current process. This improvement is the same as can be achieved in the blast furnace.
The fourth process route is advanced scrap-based electric melting in combination with strip casting. The energy demand for melting scrap is lower than for melting DRI, because DRI contains slags that have to be heated and refined.

All techniques required for these processes can probably be made commercially available within 20 years. Implementation will take a considerably longer period. In the next century all process routes may be used side by side. The choice of the process will depend on (a) geographical factors, such as the availability of natural gas or cheap electric power; and (b) market factors, such as the availability of high-quality scrap and the demand for specific steel products; and (c) the development of the price of steel products. The development of the technologies described is taking place almost completely in the iron and steel sector and depends little on developments in other sectors. However, governmental support is not uncommon. The development of nearly all smelting reduction processes has been supported by the national government.

Is a further reduction in SEC to be expected in the longer term? We discuss the major energy losses of the future processes.
The *advanced primary steel making route* has a SEC that is about 6 GJ/trs higher than the theoretical value, as discussed in Section 5.3. Where does this energy go, and can this loss be avoided?
First, heating and melting are required also in advanced process. The energy of solid streams, namely blast furnace slag, BOF slag, and hot steel, equal to about 2-2.5 GJ/trs, is still lost. Waste heat recovery from the slags is technically possible. No practical means is available for recovering all the heat losses from hot steel. For a further improvement of the SEC, R&D needs to focus on heat recovery from hot steel over the whole temperature range from 1600°C to environmental temperature, including the heat of melting.
Second, the energy losses associated with oxygen production are not accounted for in determining the minimum energy demand. On the basis of primary energy carriers, about 1.2 GJ/trs is required. Oxygen production may become more efficient, for instance by applying membranes. The possibility of using oxygen-enriched air instead of pure oxygen should also be investigated. A negative effect of this option is that the volume of gas that has to be compressed and heated increases.
Third, not all energy of the export gas of the SR process can be recovered. In the model we used to credit for the export gas, it is assumed that in the best case 40% of the enthalpy is lost in the conversion to electricity. In some SR processes, steam is produced by heat exchange with the high temperature gas. Besides the loss that occurs during this heat exchange, it is assumed that 65% of the enthalpy of the steam is lost in the steam turbine.

Fourth, refining, casting and shaping require about 0.75 GJ/trs. Refining – removing the last impurities and adjusting the composition to that of the desired steel – requires 0.5 GJ/trs. Strip casting requires 0.15 GJ/trs, and the remainder is for shaping. These values are expressed in primary energy units, whereas the demand is for electricity. The end-use demand may be reduced by several control and optimization measures, e.g. adjustable speed drives.
Finally, the energy input to the SR process is higher than could be expected on the basis of evaluating the reduction of hematite alone. This can be explained by investigating the exergy losses that occur in the SR process itself. Many chemical reactions are responsible for the high temperature heat: coal gasification, postcombustion, reactions of metals with oxides. The production of heat inherently results in an exergy loss. Exergy is also used to melt other components in the ore, like silicon oxide, which makes up 30-40% of the ore. The resulting molten slag layer is important in the SR process. A better understanding of the function of the slag layer, is likely tolead to better control of slag formation, resulting in reduction of the exergy loss.

The SEC of the *future scrap melting process* is 3.5 GJ/trs, 2.4 GJ/trs higher than the minimum energy demand for melting, achieved in an EAF. Half of this additional energy demand is due to the fact that the input to the EAF differs from pure iron. Scrap upgrading and refining both demand 0.5 GJ/trs. Charging the furnace with a raw material with the exact composition of the desired product would diminish this energy demand. The other half of the higher energy demand is due mainly to the conversion of fossil fuels to electricity. Scrap melting in 100% fossil fuel-fired furnaces has been investigated, but so far these furnace has not been more efficient than EAFs. Efficient utilization of the large volume of waste gas in the furnace itself, may be an option to decrease the SEC. The SEC of future EAFs may be further reduced when a way is found to recover the heat of the hot steel. Theoretically, making steel from scrap requires little energy.

In the introduction to this chapter we referred to a study described in a report of the World Energy Council (WEC) that explored future energy demand for steel making under different scenarios [WEC, 1995]. Assuming a business-as-usual development scenario energy demand might grow from 18.6 EJ in 1990 to 25.4 EJ in 2020. According to WEC, when advanced technologies are implemented, the demand in 2020 might be 19.5 EJ. Assuming successful development, the techniques described in this chapter will probably become commercially available before 2020. The diffusion of these techniques will take place in the decades following market introduction. To make a projection of the potential future primary energy demand for steel making we make two more assumptions. First, after 2020 steel demand will grow by the same rate as assumed in the WEC study for the period 1990-2020, namely 1.7% a year [WEC, 1995]. In addition, growth in developing countries is assumed to be 4% a year. Second, the ratio of primary steel to secondary steel remains at the 1990 level of 70:30 [WEC, 1995]. On the basis of these assumptions, world primary energy demand for steel making in 2050 will be 20.8 EJ. Since it can be expected that the SEC will be further reduced by evolutionary changes, and that the use of scrap will increase, future energy demand can be projected to be below the 1990 level. If advanced heat recovery techniques are developed and adopted, an even greater reduction in the energy demand can be achieved.

5.7 Discussion

In this section we comment on the methodology, the choices we made and the uncertainty of the data. The discussion of the methodology will elaborate on the discussion presented in the chapter dedicated to the paper and board industry.

We opt for a definition of the energy service that allows us to include the recycling of steel scrap. Recycling and reprocessing of scrap has a much lower SEC than primary steel making, and is therefore an important energy-efficiency improvement option. However, it cannot be concluded that all steel should be made according to this process. First, the resources of scrap are not sufficient to meet demand if all steel were to be produced from scrap: world steel demand will grow, it is impossible to collect all steel at the end of its lifetime, and that quality of the scrap is not homogeneous. Second, the product mix of a scrap-based mill is different from that of an integrated mill. With the introduction of thin slab casting, this difference has been eliminated to some extent. To circumvent these problems we presented the potentials for energy-efficiency improvement in integrated mills and scrap-based mills separately.
An even broader definition can also be considered, for instance one that includes the production of other materials that can replace steel. We realize that this might result in large energy-efficiency improvements in the long term. Studies that compare the energy requirement for the production of different materials have been published [Boustead and Hancock, 1979; Worrell *et al.*, 1994]. An assessment of the energy-efficiency improvement potential as a result of material substitution requires additional information about and analysis of expected developments in energy efficiency and in the demand for different products, competition between products, and the emergence of new products

The selection of techniques was performed on the basis of the results of the exergy analysis. Three groups of techniques can be distinguished. The first group consists of techniques that avoid at least one heating and cooling step. The second group is made up of techniques that reduce the temperature level required in different processes. The third group contains techniques that recover and apply heat at high temperatures.
Noteworthy is the lack of technologies that involve a completely different way of steel making. There do not seem to be any technologies to make steel at lower temperatures. The reduction of iron ore to iron hardly proceeds at low temperatures; the opposite reaction is favored thermodynamically. It can be concluded that the reason no technologies have been found for reducing iron oxide at low temperatures is that no practical ways of achieving this have been discovered.

In theory, the reduction of hematite with carbon is very efficient. The Gibbs free energy of that reaction is 6.8 GJ/tonne Fe, very close to the Gibbs free energy of decomposition of hematite into the elements, which is 6.6 GJ/tonne Fe. From an energy point of view there is no reason to look for other reductants. Nevertheless, it has been proposed to use hydrogen [Gretz *et al.*, 1991]. One advantage of using hydrogen is that no carbon dioxide is formed. Of course, this is true only when hydrogen is made without the use of fossil fuels. The use of hydrogen does not entail energy saving in itself. To produce 1 tonne of pure iron, about 650 Nm^3 hydrogen is required theoretically. This equals about 6.5 GJ/tonne. This amount is of the same order of magnitude as the minimum energy demand when coke is used.

Now we comment on the accuracy of the input data and on the way in which we had to convert data to make a comparison possible.

We had to rely on figures presented by developers that were based on the results of pilot-plant experiments or were design values. Data on new techniques are rarely supported by other sources of information. Nevertheless, it is possible to make a rough check of the data for SR processes by calculating the expected demand for coal and oxygen.

The main energy input of SR processes is non coking coal. It can be expected that the coal demand is higher than the coke demand in blast furnaces, as coal still contains 20-30% weight (wt) [Hoffman, 1992] volatile matter that has to be heated and evaporated. On the other hand, the ash content is a few percent lower [IISI, 1982]. The coal demand of SR processes is up to 30% higher than the coal demand of the blast furnace, which is in line with what can be expected.

On the basis of stoichiometric ratios, the oxygen demand (in tonnes) can be determined to be between 90% (for complete conversion to carbon monoxide) and 180% (for complete conversion to carbon dioxide) of the coal demand. SR processes with a high degree of postcombustion have a higher oxygen demand than processes with a low degree of post combustion. Oxygen is not only provided externally, but it is also generated within the process by the reduction of iron ore. Depending on the composition of the ore, the oxygen released per tonne of Fe is about 300-500 kg. The oxygen released in the prereduction shaft is usually not available for coal combustion. It can be determined that the reported oxygen requirements for second-generation SR processes are well in line with the value that could be expected on the basis of coal requirement, degree of postcombustion and degree of prereduction. Differences in the reported and the calculated oxygen demand equal 0.05-0.1 GJ/tonne Fe (primary energy), or less than 1% of the SEC.

On the basis of the foregoing analysis, we can state that the input data on oxygen and coal consumption of all SR processes are consistent with what can be expected from the stochiometric oxygen requirement and the differences in composition of coking and noncoking coal. We expect that variations in these input data are so small, that they do not affect our conclusions with regard to energy-efficiency improvement potential of SR processes.

Data on the investment costs of SR processes are subject to many underlying factors that cannot be assessed easily. For instance, there may be differences resulting from local prices

of equipment and the year to which the valuta relate. There seems to be a relation between investment costs and complexity of the process. The CCF, HIsmelt, and the AISI processes are less expensive than the more complex DIOS and COREX processes. We evaluated the effect that varying the investment costs of the SR processes would have on the production costs of hot metal. It was shown that the production costs with all second generation SR processes are in all cases lower than in a blast furnace.
For comparison of the SECs, we converted the energy carriers to primary energy carriers using a low and high case and a simple model for in-house electricity production. We showed in Figure 5.13 that the SECs of SR processes depend strongly on the way in which the export gas is utilized. SR processes with high export gas production are not always more efficient than the blast furnace process. Careful consideration should be given to the matter of what to do with the export gas. We assumed that all export gas is converted to electricity in a combined cycle plant. Other applications of the export gas may be considered as well. For instance, it can be used as fuel for a fuel cell, as reducing gas in DRI processes, or recycled at high temperatures to the melter. These options should be investigated to find the optimum use of energy for the production of iron (and electricity as a by-product).

5.8 Conclusions and recommendations

In this chapter we have analyzed the potential for the improvement of energy efficiency in the iron and steel industry that can be realized in the long term. We used exergy analysis to show that the main exergy losses in an integrated steel mill are due to the use of high temperatures. On the basis of the results of this analysis we concluded that long-term energy-efficiency improvement should be directed toward reducing these losses by (a) avoiding intermediate heating and cooling steps; (b) reducing the temperature required in various process steps; and (c) recovering and applying heat at high temperatures.

The focus in this chapter was on smelting reduction processes, which avoid coke making and ore agglomeration, and on near-net-shape casting techniques, which avoid or reduce the need for reheating before rolling. By a combination of these techniques, the SEC might be brought down from the current best-practice figure of 19 GJ/trs to 12.5 GJ/tcs, or a reduction of about 35%. The production costs of steel strip from a future integrated mill that uses smelt reduction and strip casting are far below those from a current integrated mill. Both smelting reduction and strip casting are likely to be available within two decades.
Direct reduction has a lower energy requirement than reduction of ore in an SR process, mainly because melting is avoided. However, subsequent melting remains necessary to shape the steel. Because of the low carbon content, DRI has to be melted in an EAF. The SEC of production of steel in the DRI-EAF route is about 2 GJ/trs higher than that of the SR-BOF route.

Electric arc furnaces can make steel from a 100% scrap charge, thus avoiding the need for iron ore reduction. The SEC of steel making of current best-practice EAF mills is about 7 GJ/tcs expressed in primary energy carriers, using a 40% efficiency of electricity generation. This may come down to 3.5 GJ/tcs by the use of more efficient melting furnaces, more efficient casting and shaping techniques, and assuming a 60% efficiency of electricity generation. Steel mills with an EAF have changed considerably over the past decade; they are now competitive with integrated steel mills in the production of flat products, a market that has previously been the monopoly of integrated steel mills. The use of scrap only for the production of steel is not possible, because not enough scrap is available and the quality of scrap is not sufficient to make all steel products. In the future different routes to produce steel will continue to exist side by side.

For all process routes, a further reduction of up to 2.5 GJ/trs can be achieved when techniques will become available for recovering and applying the high-temperature heat of hot steel and slag. Several concepts of slag heat recovery have been developed. Because of the high investments, none of these concepts has been commercially applied. Heat recovery of the hot steel at temperatures below 800°C is a commercial technology. R&D should be directed at recovering heat at higher temperatures, including recovery of the heat of melting. No such technology is under development.

The selected energy-efficient techniques described in this chapter will probably become available before 2020. The diffusion of these techniques will take place in the decades following the market introduction. During this period the techniques will probably be improved, which may result in higher energy efficiency. It can be projected that when all the steel in the world is produced according to the most efficient processes, world energy demand for steel making will stabilize or even decline. In this projection it is assumed that the current ratio of primary to secondary steel making will still be applicable and that world steel production will grow by 1.7% a year on average. In addition, growth in developing countries is assumed to be 4% a year. Further reductions in energy demand can be achieved when advanced heat recovery techniques are developed and adopted and when the use of scrap is increased.

New techniques are being developed within the iron and steel industry itself. However, governmental support is not uncommon. Nearly all smelting reduction processes are being developed with a form of financial support from the government. The main driver for the development of new techniques is a reduction in production costs. Improvement in energy efficiency can contribute to this. The role of the government in improving energy efficiency in the iron and steel industry is still limited. Several areas may be the subject of governmental policy:

- Financial support for the development of energy-efficient technologies;
- Encouraging iron and steel companies to implement the most efficient techniques, e.g. through voluntary agreements;

- Providing an efficient and effective scrap recycling system and stimulating the maximum use of scrap by iron and steel companies;
- Encouraging research to further improve energy efficiency, e.g. by developing techniques to recover and apply high-temperature heat and processes to make steel directly from iron ore.

Acknowledgements We would like to thank K. Meijer (Hoogovens Staal), L. Hofer (VAI), E. Nieuwlaar and W.C. Turkenburg (Utrecht University) for providing information, suggestions, and comments on this study. The exergy analysis is based on earlier work by T. Ros. We thank The Netherlands Organization for Scientific Research for its financial support.

Abbreviations

AISI	American Iron and Steel Institute
BF	Blast Furnace
BOF	Basic Oxygen Furnace
CCF	Cyclone Converter Furnace
cwt	hundredweight
DIOS	Direct Iron Ore Smelting
DR-EAF	Direct Reduction-Electric Arc Furnace
DRI	Direct Reduced Iron
EAF	Electric Arc Furnace
ERS	Environmental Reference System
HIsmelt	High Intensity Smelting
IISI	International Iron and Steel Institute
JISF	Japan Iron and Steel Federation
LCR	Liquid Core Reduction
LD	Linz-Donawitz
LHV	Lower heating value
OHF	Open Hearth Furnace
PRS	Prereduction Shaft
SEC	Specific Energy Consumption
SR	Smelting Reduction
SRV	Smelting Reduction Vessel
tcs	tonne of crude steel
thm	tonne of hot metal
tpd	tonne per day
tph	tonne per hour
tpy	tonne per year
tpi	tonne pig of iron
trs	tonne of hot rolled steel
thm	tonne hot metal

TSC	Thin Slab Casting
VAI	Voest Alpine Industrieanlage

CHAPTER 6

FIXING ATMOSPHERIC NITROGEN WITH LESS ENERGY[1]

6.1 Introduction

Ammonia has become one of the largest chemical products since its first production in the beginning of this century [Appl, 1992a]. Figure 6.1 shows that the world production of ammonia increased more than a 20-fold in the period 1950-1990 to about 100 million tonne N. The fossil energy consumption for the production of nitrogen fertilizers is dominated by the production of ammonia and can be estimated to be between 3 and 4 EJ[2] in 1992, or 3-4% of the fossil energy globally consumed by industry in total [3].
Reduction of the energy demand for the production of fertilizers is important in the light of both environmental and economic considerations. Many studies have analysed the opportunities for the reduction of the energy demand on component level (e.g. [Laciak and Pez, 1988; Deshmukh, 1993] [Dybkjaer and Gram, 1984]), on plant level (e.g. [Livingstone and Pinto, 1983; LeBlanc, 1984; Grotz and Grisolia, 1992]), and on a more aggregated level (e.g. [Heath *et al.*, 1985; Appl, 1992a; Dybkjaer, 1995a; Worrell *et al.*, 1995]). All these studies have a relatively short time frame of up to one decade. Insight into the opportunities on a longer term is required to be able to formulate energy strategies for this sector and to be able to decide whether the development of specific technologies should be stimulated.

This chapter aims at identifying and characterizing options that can contribute to a reduction of the energy required for atmospheric nitrogen fixation, in order to estimate the long-term potential for energy-efficiency improvement and assess R&D-priorities. A systematic approach will be used which has already been applied to the paper and board industry [Beer *et al.*, 1998] and the iron and steel industry [Beer, 1998]. The method – described in chapter 2 of this book – consists of three steps. First, the potential for improvement of the energy efficiency is determined on the basis of an exergy analysis of the currently prevailing processes. Second, an inventory is made of options that can contribute to an improvement of the energy efficiency in the long term. Third, the options are characterized by estimating the future energy consumption and costs, and by evaluating

[1] Co-author: K. Blok, Department of Science, Technology and Society, Utrecht University

[2] Assuming that the average specific energy consumption is somewhere in between 30-40 GJ/tonne NH_3 (including feedstock energy consumption) and that the world ammonia production in 1992 was 94.2 million tonne [FAO, 1996].

[3] The global primary energy use for industry was 134 EJ in 1992 [WEC, 1995]

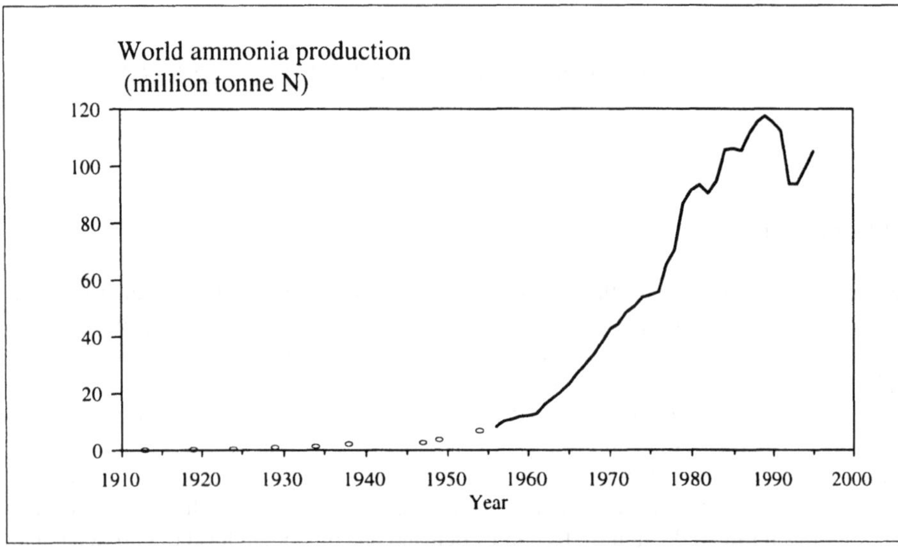

Figure 6.1: Development of the world ammonia production. Data from 1910 to 1970 are taken from [Honti, 1976]; Data from 1970 to 1995 are taken from [FAO, 1996].

the current stage of the development, and the technical change required to bring the technology to commercialization.

In Section 2 the history of the production of nitrogen fertilizers and the development of the specific energy consumption (SEC) over time are described. Descriptions of state-of-the-art production processes are given in Section 3. In Section 4 the energy service for further analysis is selected. Section 5 deals with the minimum SEC for the production of nitrogen fertilizers. The values are compared to the SEC of state-of-the-art processes in order to determine the maximum potential for energy-efficiency improvement. In Section 6 exergy analyses of state-of-the-art ammonia and nitric acid processes are presented. Based on the results of the exergy analyses, an inventory is made in Section 7 of options that can contribute to an improvement of the energy efficiency of nitrogen fixation. Section 8 gives a characterization of the selected technologies. The focus is on chemical processes. However, options for biological nitrogen fixation are also briefly discussed. Section 9 presents a discussion of the results and in Section 10 conclusions are drawn and recommendations for R&D decision and policy makers are given.

6.2 The production of nitrogen fertilizers over the past 100 years

The objective of this section is to give an overview of the processes that have been used to fix nitrogen in the past century. There are few chemical processes of which the history has been so extensively described as for the ammonia synthesis. This historical overview is mainly based on [Mittasch, 1951; Ihde, 1964; Haber, 1971; Topham, 1985; Tamaru, 1991].

At the end of the nineteenth century it became clear that the conventional resources of nitrogen, ammonium sulphate – obtained as by-product from the coke and town-gas production out of coal – and Chile saltpeter – obtained from naturally occurring deposits of sodium nitrate –, were not sufficient to meet the growing demand for fertilizers. At that time there were no technical means for the fixation of atmospheric nitrogen. Fundamental understanding of the basic principles of the reaction of nitrogen and hydrogen was still lacking. Over and above this, physical chemistry was still in its childhood. The existence of chemical equilibria, for instance, had just been discovered. Research on nitrogen fixation was concentrated in academic laboratories. Industry became involved during the first decade of this century.

Three processes for the fixation of nitrogen have been commercialized. The first to come to commercialization was the cyanamide process, in Italy in 1905. In this process calcium or barium carbide (CaC_2 or BaC_2) reacts at a temperature of 700-1000°C with atmospheric nitrogen to form the corresponding cyanamide ($CaCN_2$ or $BaCN_2$). The German chemical company BASF put considerable effort into the development. The maximum production was several hundreds ktonnes of nitrogen per year [Honti, 1976]. There were still a few cyanamide plants in operation after the second world war. The energy consumption was about 180 GJ/tonne N [Tamaru, 1991].

The second commercial nitrogen fixation process was the electric arc process. Atmospheric nitrogen is bound to oxygen at extremely high temperatures (above 3000°C), generated by an electric arc, to give nitrogen oxides. The first but unsuccessful attempt to demonstrate the feasibility of the process was made in the USA at Niagara Falls. In Europe, three types of furnaces were developed. The Birkeland-Eyde furnace (Norway), the Schönherr furnace (BASF, Germany), and the Pauling furnace (Austria). The power consumption of the Birkeland-Eyde furnace was 225 GJ_e/tonne N [Honti, 1976]. Because large-scale fossil fuel-fired power plants were not available at that time, electricity for the electric arc process was generated with hydro power. The maximum annual amount of nitrogen produced according to this process was about 350 ktonnes [Honti, 1976]. In the 1950s a plant was in operation in Wisconsin (USA) that provided the heat thermally, instead of electrically. The fuel consumption was 85 GJ/tonne N and the electricity demand 18 GJ_e/tonne. This was competitive with the pre-war synthetic ammonia plants. However, improvements in the last technology outdated the Wisconsin process, that was shut down in 1956 [Honti, 1976].

The third process for nitrogen fixation is the synthetic ammonia process, or the Haber-Bosch process. It is based on the catalytic conversion of hydrogen and nitrogen. Haber, a German scientist, started research into the equilibrium composition of a nitrogen-hydrogen-ammonia mixture in 1903. BASF took up the industrial development in 1908 and put Bosch in charge. In 1911 a pilot-plant was built, and in 1913 the first commercial Haber-

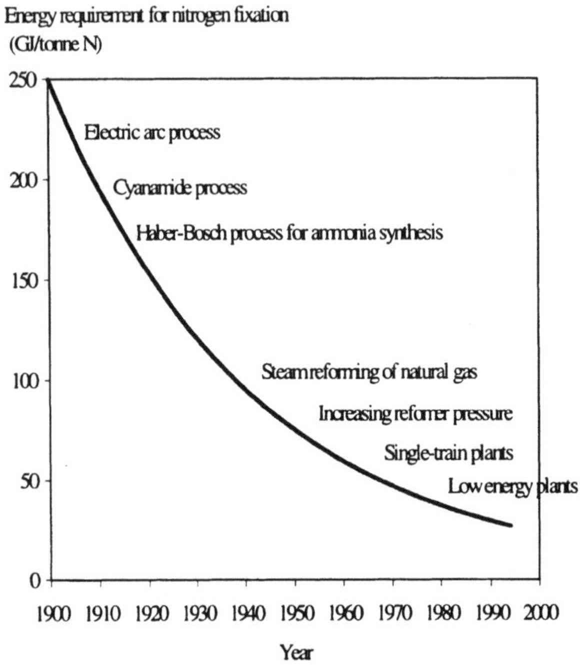

Figure 6.2: Development of the energy requirement for nitrogen fixation.

Bosch plant was put into operation in Oppau (Germany). Hydrogen was produced by reaction of coke with air and steam. The product gas, consisting of hydrogen and carbon monoxide, was called water gas. First, hydrogen was just cooled out. In 1914 BASF developed the catalytic reaction of carbon monoxide and steam to yield more hydrogen [Topham, 1985]. This reaction is now known as the water gas shift reaction. During World War I Germany was in need of large quantities of nitrogen-based explosives but cut off of fixed nitrogen supplies. The German government therefore financially supported the development of the Haber-Bosch process. After World War I, when the favourable economics of the Haber-Bosch process became apparent, similar processes were developed in other countries. In the USA the first successful plant based on a process similar to the Haber-Bosch was completed in 1921. In France, Grand Paroisse built a plant according to the Claude process (a process using a pressure of 1000 bar) in the 1920's. Other processes that were developed in that period were the Casale process, the Fauser process, the Mont Cenis process, and the NEC process. The processes differed mainly in operation conditions of the ammonia synthesis and in the way hydrogen was produced, although all processes were initially based on the production of water gas from coke. Later many new processes were developed. However, all were based on the original Haber-Bosch process for the ammonia synthesis. Nitrogen fixation according to the catalytic conversion of hydrogen and nitrogen soon dominated the market. In the period 1910 to 1950 the synthetic nitrogen fixation increased from about 0.5 million tonne N to about 4 million tonne N [Haber, 1971; Honti, 1976]. This increase was completely met by the Haber-Bosch process (and its derivates).

In the period 1950-1990 major improvements have been achieved in the energy-efficiency and the economics of the Haber-Bosch process. The capacity per plant increased from less than 500 tonnes per day in 1950 to 1,500 tonnes per day in 1990. In the 1950s, the first plants based on steam reforming of natural gas were introduced. A large improvement was obtained with the introduction of centrifugal compressors in the 1960s, which was important for the development of the energy-efficient single-train plants. Other major improvements were the increase in reforming pressure, the development of new CO_2-removal processes, low-temperature water gas shift reactors, the methanator, and advanced pressure vessel technology. In the 1980s several companies introduced energy efficient ammonia plants, with SECs of 27-29 GJ(LHV)/tonne NH_3 (33-35 GJ/tonne N). Examples are the AMV-process (ICI, first plant brought onstream in 1982) [Livingstone and Pinto, 1983], the Kellogg Low Energy Process (introduced by M.W. Kellogg in 1983) [Anonymous, 1997], the Braun Purifier Process (first unit built in 1966, improved since then) [Gosnell, 1997], and the Topsøe Low Energy Process [Topsoe, 1997]. The development in energy consumption of ammonia production is illustrated in Figure 6.2. In the period 1950-1995 the SEC decreased by 1.5% a year on average.

6.3 State-of-the-art production processes

In this section the state-of-the-art processes to produce the so-called straight N-fertilizers ammonia, urea, nitric acid and ammonium nitrate are briefly described. Straight N-fertilizers contain only the nutrient nitrogen. Nitrogen is often applied to the field in the form of a multi-nutrient fertilizers, that can also contain phosphor, potassium, or sodium. Because the energy demand for the production of multi-nutrient fertilizers from the straight fertilizers is small, we will not deal with that [Worrell and Blok, 1994]. At present, the production of all N-fertilizers starts with the production of ammonia.

6.3.1 AMMONIA SYNTHESIS

Figure 6.3 shows a block diagram of a typical contemporary ammonia plant using steam reforming of natural gas. In 1989 80% of the global ammonia capacity was based on steam reforming of natural gas [Worrell and Blok, 1994]. Such an ammonia plant can be divided into two parts: the synthesis gas plant and the ammonia synthesis plant.

In the synthesis gas plant natural gas is converted to a gas mixture consisting of H_2 and N_2. This works as follows. First, desulphurized natural gas is reformed with steam according to

$$CH_4 + H_2O \leftrightarrows CO + 3\,H_2 \ (\Delta G° = 151 \text{ kJ/mole}) \tag{1}$$

$$CO + H_2O \leftrightarrows CO_2 + H_2 \ (\Delta G° = -20 \text{ kJ/mole}) \tag{2}$$

The highly endothermic steam reforming reaction (1) takes place in tubes filled with a nickel-based catalyst, heated externally by burning additional natural gas. Half of the heat of combustion is transferred to the reaction mixture to raise the temperature to the level of the reformer exit and to provide in the heat of reaction. A large part of the other half is recovered in the convection zone of the furnace to generate steam, and to preheat combustion air, fuel and feed. The thermal efficiency on the basis of lower heating value of natural gas is 95% [Rostrup-Nielsen, 1994; Dybkjaer, 1995b]. The steam to carbon (mole) ratio is typically between 2.5 and 4.0. Excess steam above the stoichiometric ratio is required to prevent carbon deposition on the catalyst. Typical reaction conditions (T=800-900°C, p=25-35 bar) are a compromise between thermodynamic constraints (chemical equilibrium), kinetic constraints (reaction rate), and costs (size of equipment). Since the conversion of methane does not proceed completely at these conditions, a secondary reformer is necessary. In this reactor partial oxidation of the remaining methane takes place at 1000°C according to

$$CH_4 + 0.5\,O_2 \leftrightarrows CO + 2H_2 \ (\Delta G° = -86 \text{ kJ/mole}) \tag{3}$$

followed by the water gas shift reaction (2). Oxygen is supplied with compressed air. The amount of air is controlled to give a stoichiometric syngas (i.e. N:H=1:3). The secondary reformer is an autothermal reformer, i.e. no external heat supply is required. The gas mixture that leaves the secondary reformer is stripped from all oxygen compounds as these can later de-activate the ammonia synthesis catalyst. First, CO is converted with steam to CO_2 and additional H_2 (reaction 2) in a two step shift-converter. Next, CO_2 is removed by chemical or physical means. Finally, the remaining CO is converted back to methane by letting reaction (1) proceed in the opposite direction over a nickel-based catalysts at 250-350°C. The result is a gas containing hydrogen and nitrogen in the ration 3:1, and some argon and methane. This gas mixture is called synthesis gas, or, in short, syngas.
In the past decade several concepts have been developed in which part of the reforming duty is shifted from the primary to the secondary reformer. The aim was to reduce the stress on the catalyst tubes, hence extend their lifetime. Examples are the Braun Purifier Process (Brown & Root), the AMV- and LCA-process (ICI), and the Kellog Reformer Exchanger System (KRES) (M.W. Kellogg Company).

Once the syngas has been produced, the catalytic formation of ammonia according to

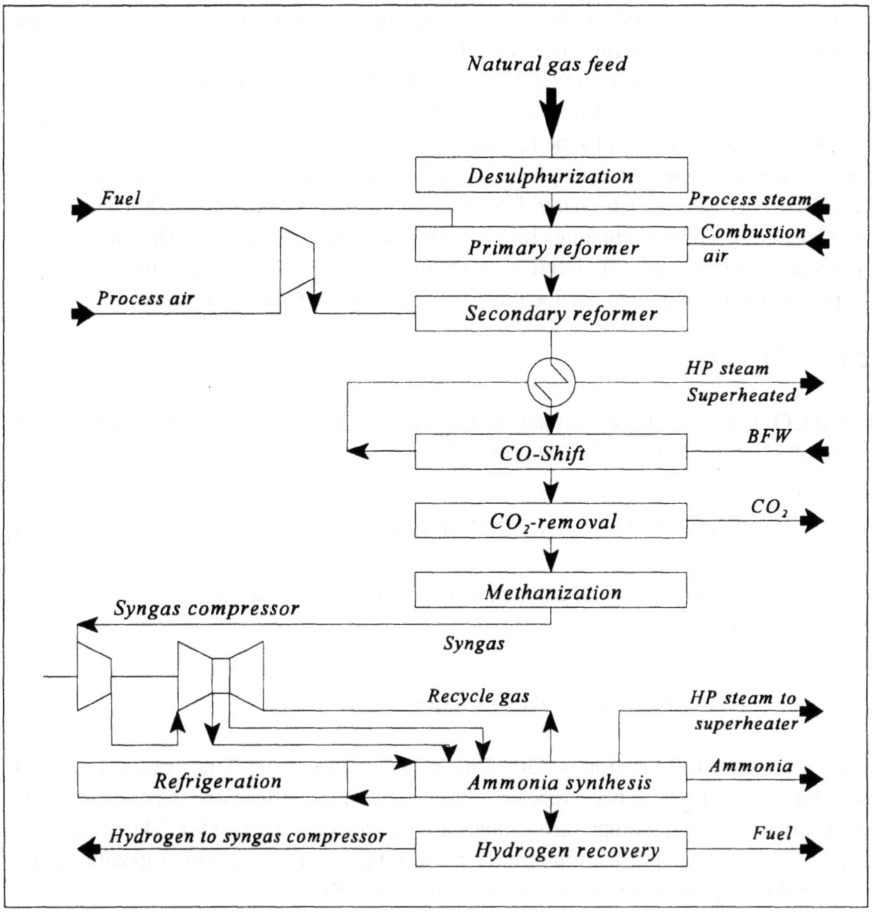

Figure 6.3: Block diagram of a typical state-of-the-art ammonia plant using steam reforming of ammonia.

$$N_2 + H_2 \leftrightarrows 2\ NH_3\ (\Delta G° = -33\ \text{kJ/mole};\ -1.2\ \text{GJ/tonne N}) \qquad (4)$$

can take place. The reaction is favoured by high pressures and low temperatures. The ammonia product is separated from the reaction mixture by condensation. The condensation temperature varies from ambient to -25°C depending on the gas pressure. At ambient temperatures water or air can be used, at lower temperatures refrigeration using ammonia as medium is usually applied. Practical synthesis pressure is not only dictated by the equilibrium composition, but also by the size of the equipment, operation costs, and the catalyst activity. The temperature of the exothermic reaction is controlled at about 450°C

by cooling between the catalyst beds, generating high-pressure steam. Modern ammonia synthesis plants operate at pressures of 80 to 200 bar. Ammonia concentration in the product gas of about 15% is achieved. Higher concentrations are possible, but are usually avoided since ammonia itself deactivates the catalyst. After recovery of ammonia the gas is recompressed and recycled to the reactor.
The market for complete ammonia plants is dominated by five companies: Haldor Topsøe (Denmark), M.W. Kellogg Company (USA), Brown & Root (USA), ICI (UK), and Krupp Uhde (Germany). These companies have developed their proprietary design of some (or all) parts of the plant, and hold license for the other parts. The lay-out of the plants may differ per company, and even per location, but the basic scheme is the same.

6.3.2 Urea production

NH_3 and CO_2 react to form carbamate in a continuously operating synthesis reactor at a pressure of 120 to 250 bar and 180 to 200°C according to

$$2NH_3 + CO_2 \leftrightharpoons NH_2COONH_4 \text{ (carbamate)} \quad (5)$$

followed by the decomposition of carbamate into urea in several decomposition stages with decreasing pressure:

$$NH_2COONH_4 \leftrightharpoons NH_2CONH_2 + H_2O \quad (6)$$

The ΔG° for the overall reaction is 230 kJ/mole (8.2 GJ/tonne N). Since excess ammonia is used, but also to prevent back-reaction, reactants and carbamate are removed from the product mixture from decomposition stages and recycled to the reactor. The production process is finished by solidification of urea in a prilling tower or vacuum evaporation plant and a granulation process [VDI, 1990] [Hoffmeister, 1993].

Typical capacity of urea plants is between 600 and 1500 tonnes per day [VDI, 1990]. Urea plants are usually located closely to ammonia plants, since the raw materials NH_3 and CO_2 are both products of an ammonia plant. The SEC of an urea plant is about 2.3 GJ/tonne urea (4.9 GJ/tonne N), mainly required in the form of high pressure steam [Anonymous, 1997]. Snamprogetti (Italy) is the main supplier of urea production technology.

6.3.3 Nitric acid production

Nitric acid is usually made by mixing gaseous ammonia with air, yielding nitric oxide according to

$$NH_3 + 1.25\ O_2 \leftrightharpoons NO + 1.5\ H_2O \ (\Delta G° = -355 \text{ kJ/mole; } -25 \text{ GJ/tonne N}) \quad (7)$$

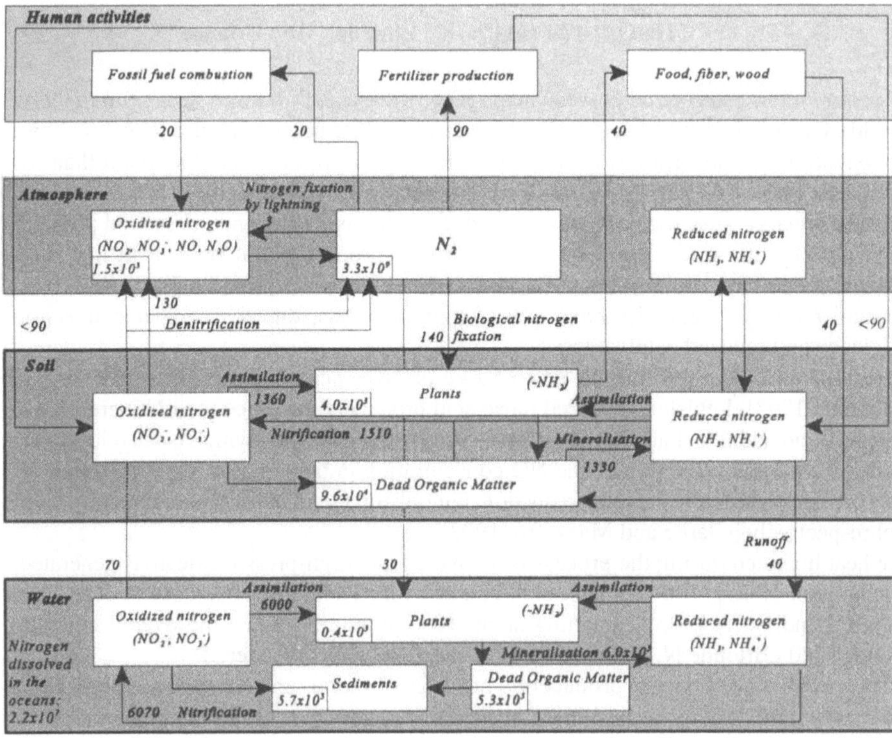

Figure 6.4: Simplified scheme of the nitrogen cycles and the effect of human activities. Three environmental compartments are given: atmosphere, soil and water, and the 'compartment' of human activities. In each compartment the nitrogen reservoirs are indicated by a light grey rectangle.. The number in the left corner gives the amount of nitrogen that is contained in the reservoir in million tonnes of N. Arrows represent flows of nitrogen. The unit of the numbers is million tonnes of N per year. The scheme is derived from [Smil, 1997]. The numbers are taken from [Ayres, 1994; Kinzig and Socolow, 1994].

over a platinum-rhodium catalyst at 800-950°C. To minimize decomposition of NO into nitrogen and oxygen long catalyst contact times are avoided and the gases are cooled rapidly at the converter exit. NO is further oxidized according to

$$2\,NO + O_2 \leftrightarrows 2\,NO_2 \;(\Delta G° = -71 \text{ kJ/mole}; -2.5 \text{ GJ/tonne N}) \qquad (8)$$

Subsequently, NO_2 is absorbed in water

$$3\ NO_2 + H_2O \leftrightarrows 2\ HNO_3 + NO\ (\Delta G° = -8.3\ kJ/mole;\ -0.3\ GJ/tonne\ N) \quad (9)$$

Processes can be categorized as weak acid (50-65 wt% acid) or direct strong (up to 99 wt % acid). Furthermore, nitric acid technology can be either mono-pressure (equal pressure for oxidation and absorption) or dual-pressure (higher pressure for absorption than for oxidation). Mono-pressure processes operate either at 0.3-0.6 MPa or 0.7-1.2 MPa. Dual-pressure processes use the first pressure range for the oxidation and the second pressure range for the absorption process. Mono-pressure processes that operate in the high-pressure range have an investment costs advantage over dual pressure processes. Dual-pressure processes generally have lower operation costs, due to a higher conversion efficiency and reduced catalyst loss.
Nitric acid process used to emit up to 50 kg of NOx per tonne of HNO_3 [Clarke and Mazzafro, 1993]. Legislation forced nitric acid producers to reduce their NO-emissions. In general, an additional absorption tower is used to recover NO, which is recycled to the process. A further reduction of the NO-concentration in the exhaust can be obtained by selective or nonselective catalytic reduction, but only on the expense of ammonia or fossil fuel respectively [Clarke and Mazzafro, 1993].
The heat liberated during the process is recovered and high-pressure steam is generated. Per tonne of HNO_3 (100 wt%) about 0.8 tonne of high-pressure steam can be recovered [Worrell and Beer, 1995], equalling about 10 GJ/tonne N. Since the overall enthalpy change is 20 GJ/tonne N, about 50% of the reaction heat is recovered.
All processes are net energy producers. The SECs of mono- and dual-pressure processes are comparable [Anonymous, 1988]. At present the most energy-efficient process is the Uhde mono-pressure process (0.55 Mpa). The SEC of this process is -2.2 GJ/tonne HNO_3 (100%) (-9.9 GJ/tonne N) [Anonymous, 1988]. Typical capacities vary from 100 to 500 tonnes per day, although plants with capacities of1,000 tonnes per day have been built [Weatherly, 1997]. The main suppliers of nitric acid technology are: Grand Paroisse (France), Stamicarbon (DSM, the Netherlands), Krupp Uhde (Germany), Weatherly (USA) and Espindesa (Spain).

6.3.4 AMMONIUM NITRATE PRODUCTION

Ammonium nitrate is produced in three steps. First, nitric acid is neutralized by ammonia:

$$HNO_3 + NH_3 \leftrightarrows NH_4NO_3\ (\Delta G° = -87\ kJ/mole;\ -3.1\ GJ/tonne\ N) \quad (10)$$

The heat released during this exothermic reaction is generally used to evaporate water in the reactor and concentrate the reaction mixture to 83-87 wt% ammonium nitrate [Weston, 1993]. Excess heat is removed and used for acid preheating, ammonia evaporation or steam generation. Part of the heat of reaction becomes available in the form of low-pressure steam.

The second step is concentration of NH_4NO_3 to more than 99% in vacuum evaporators. Finally, solid granular product is formed in a prilling tower or granulation process. Nitric acid technology is offered by many companies, e.g. Uhde (Germany), ICI (UK) and DSM (the Netherlands). Ammonia nitrate process are net energy producers. The SEC of a typical ammonium nitrate process is about -0.6 GJ/tonne NH_4NO_3 (-1.7 GJ/tonne N) in the form of low-pressure steam [Worrell and Blok, 1994].

6.4 Selection of the energy service

In this section energy service is selected that will be used for the assessment of long-term options to improve the energy efficiency. An energy service is defined as the product of human activity obtained by the use of energy meant to satisfy a human need. However, before we can select an energy service, we need more insight into the ways nitrogen is fixed and enters plants. Figure 6.4 gives a simplified representation of the reservoirs and flows of nitrogen in and between the atmosphere, the soil and water.

Plants can take up nitrogen in the form of ammonium (NH_4^+) or nitrate (NO_3^-) dissolved in the soil solution, through the process of assimilation. Nitrogen in the terrestrial system is predominantly present in dead organic matter, which gives off nitrogen very slowly as reduced nitrogen. This process is called mineralization. The amount of naturally available nitrogen in the soil is not sufficient for intensive agriculture. To avoid nitrogen depletion, crop rotation and manure fertilization used to be applied. Nowadays, synthetic nitrogen fertilizers are used to supplement the demand for nitrogen of crops. Part of the applied nitrogen in synthetic fertilizers will never reach the plant due to immobilization in organic matter, leaching, denitrification in anaerobic situations, and vaporization. The amount of nitrogen that reaches the plant varies from 0.1 to 0.9 tonne N per tonne of applied N, depending mainly on precipitation, temperature, type of soil, and method of application [Kropff and Spitters, 1990].

Unfortunately, the most abundant source of nitrogen, atmospheric nitrogen (about 3,000,000 million tonne of N) is inaccessible to most life forms [Kinzig and Socolow, 1994]. Some specialized microorganisms, e.g. blue-green algae and some bacteria, can fix atmospheric nitrogen. Annually about 170 million tonne N is fixed in this way; 140 million tonne N enters the soil and 30 million tonne enters waters. Of this amount, about 40 million tonne is due to human cultivation of food and wood. Human activities caused an increase of the annual nitrogen fixation from about 130 million tonnes in pre-industrial times to about 280 million tonnes N in 1990. Besides the cultivation of plants, food and wood, the use of synthetic fertilizers contributes 90 million tonnes N to this increase, and combustion of fossil fuel adds another 20 million tonnes [Ayres, 1994]. The disruption of the nitrogen cycles by human activities results in a build-up of nitrogen in the compartments, which may have serious effects on ecosystems (e.g. acidification, eutrophication, too high nitrate-concentration of ground waters) and on the N_2O-concentration in the atmosphere (affecting the stratospheric ozone concentration and contributing to the enhanced greenhouse effect) [Ayres, 1994; Voet *et al.*, 1996]. We will not deal with these problems here. For

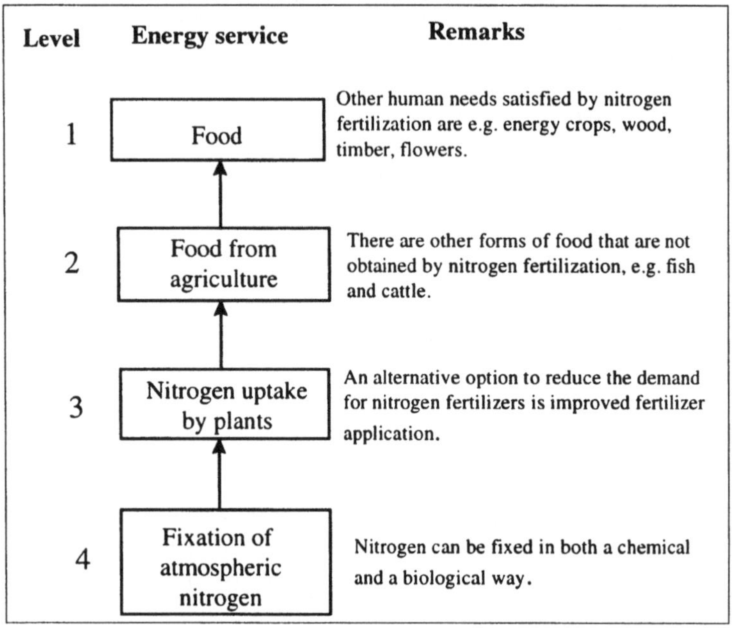

Figure 6.5: Hierarchy of energy services that can be used to describe nitrogen fixation.

discussions of causes, interactions and solutions the reader is referred to e.g. [Kinzig and Socolow, 1994; Voet *et al.*, 1996; Smil, 1997].

The energy service for our analysis can be selected at various hierarchic levels. One way to depict this is as follows, see Figure 6.5. At the highest level that we consider are the human needs that are eventually satisfied by nitrogen fixation. We focus on the human need "food", which can be obtained by several human activities. One level lower is food from agriculture, which is the practice of large-scale soil cultivation to grow more food than will grow in a natural way. The agricultural food production can be enhanced by increasing the amount of nitrogen that is taken up by the plant, which is an energy service at the third level. At the fourth level are synthetic nitrogen fertilizers and biologically fixed nitrogen. Intermediate and lower levels can be considered.

The objective of our analysis is to find and characterize technologies and options that can contribute to a reduction of the energy required for nitrogen fixation. This can be achieved by both more energy-efficient production of synthetic nitrogen fertilizers and enhanced use of biological nitrogen fixation. Therefore, we describe the energy service as *fixation of atmospheric nitrogen*.

The main focus of our analysis will be on options to improve the energy efficiency of the production of synthetic nitrogen fertilizers in both existing and new processes. Enhanced use of biological nitrogen fixation will also be discussed, albeit less-detailed than chemical nitrogen fixation. Options that prevent the loss of fixed nitrogen from the terrestrial system to water or atmosphere are left out of consideration. An estimate of the potential to reduce the demand for nitrogen fertilizers by improving the fertilizer application is given by Worrell [Worrell *et al.*, 1995]. Worrell describes several options, e.g. application of the recommended fertilizer level, increased use of manure, and improved fertilizer spreader practice, that have a technical potential to reduce the demand for nitrogen fertilizers in the Netherlands by 44% with reference to the 1988 fertilization rate. In other countries this saving is probably lower, since the fertilizer application rate is very high in the Netherlands [Worrell *et al.*, 1995]. Also on higher levels alternatives are available.

6.5 Theoretical specific energy consumption

This section starts with determining the theoretically minimum specific energy consumption (SEC), i.e. the amount of energy required to fix one tonne of nitrogen regardless the process. We present several thermodynamic values for selected nitrogen fertilizers and discuss the applicability of each quantity to be used as minimum SEC. Then, the minimum energy requirement for the main production processes of nitrogen fertilizers is determined. Finally, the amount of energy required for biological nitrogen fixation is discussed.

6.5.1 THEORETICALLY MINIMUM SEC

To break the triple bond in N_2, 33 GJ/tonne N is required. This amount is not the theoretically minimum SEC, because energy is released when new bonds are formed. An overview of thermodynamic properties of nitrogen fertilizers is given in Table 6.1. The enthalpy of formation (ΔH_f°) and the Gibbs free energy of formation (ΔG_f°) refer to the formation of the compound out of the elements at standard conditions (T=25°C, p=1 bar). Because the enthalpy of formation is negative for all compounds, heat is released in the formation of the compounds from the elements. The Gibbs free energy is equivalent to the maximum amount of work that can be obtained from the reaction when a device exists that can perform the reaction reversibly. In other words, a change in Gibbs free energy is that part of the change in enthalpy that can be obtained as work[1]. The Gibss free energy of formation is smaller than the enthalpy of formation for all compounds in Table 6.1, indicating that the amount of work that can be obtained from the formation reaction is smaller than the amount of heat. Work is preferred over heat, because it can in theory be converted completely into another form of energy. The Gibbs free energy of formation is therefore a better measure of the theoretically minimum energy demand than the Gibbs free energy of formation. However, not all elements required for the production of nitrogen

[1]The Gibbs free energy makes use of the second law of thermodynamics that states that it is not possible that the only result of a process is that an amount of heat is given off by a reservoir and converted to work.

fertilizers are freely available. Nitrogen and oxygen are available in the atmosphere, but the element hydrogen is hardly present in nature. To produce hydrogen, energy is required, which is not accounted for in ΔH_f° and ΔG_f°.

The use of exergy avoids this problem because the most stable compounds in the environment are used as reference for calculating thermodynamic properties of compounds[1]. Exergy is the best measure for the theoretically minimum SEC, because it gives the lowest amount of work required to change the naturally occurring compounds into the desired compounds. Table 6.1 shows that the exergy of all nitrogen fertilizers is positive, which indicates that to make nitrogen fertilizers from the most stable compounds in nature net energy is required, although for nitric acid this value is fairly small. For completeness, the lower heating value and the higher heating value are given[2]. The heating values are often used to indicate the energy consumption of commercial plants. Especially the LHV is used to indicate the energy content of fuels.

Table 6.1: Thermodynamic properties of nitrogen fertilizers and intermediate materials (in GJ/tonne N and, in parenthesis, GJ/tonne product). All properties are derived from [Weast, 1983] and [Nieuwlaar, 1996]. For the calculation of exergy, T_0=25°C and p_0=1,01 kPa are used as reference conditions.

compound	ΔH_f°		ΔG_f°		exergy		LHV		HHV	
Ammonia ($NH_{3\,(g)}$)	-3.3	(-2.7)	-1.2	(-1.0)	24.1	(19.8)	22.6	(18.6)	27.0	(22.2)
Nitric acid ($HNO_{3\,(l)}$)	-12.4	(-2.8)	-5.8	(-1.3)	3.2	(0.7)	a	a	a	a
Urea ($CO(NH_2)_{2\,(s)}$)	-11.9	(-5.6)	-7.1	(-3.3)	24.5	(11.4)	19.4	(9.1)	22.3	(10.4)
Ammonium nitrate ($NH_4NO_{3\,(s)}$)	-13.1	(-6.6)	-6.6	(-2.3)	10.5	(3.7)	a	a	a	a

[a] Nitric acid and ammonium nitrate decompose on heating into water, nitrogen and oxygen. The decomposition may be violent, like a detonation: the change in enthalpy is relative small, the entropy change is large. Since no reaction with oxygen occurs, no heating values are determined.

---

[1]Exergy expresses the quality of energy and can be defined as the shaft work or electrical energy necessary to produce a material in its specified state from materials common in the environment, in a reversible way, heat being exchanged only with the environment at temperature T_0 [Riekert, 1974]. The material common in the environment, the reference compound, can differ depending on the reference system that is used. We use the reference system described by Nieuwlaar [Nieuwlaar, 1996]. Nieuwlaar uses pure liquid water as reference compound for hydrogen. Other reference systems may use water vapour in air or sea water as reference compound for hydrogen

[2]The lower heating value (LHV) is the energy released when a compound reacts with oxygen to carbon dioxide, water vapour and nitrogen. When the water vapour is allowed to condensate, account should be made for the heat of vaporization. Including the latter gives the higher heating value (HHV).

6.5.2 Theoretically minimum SEC for the main production processes of nitrogen fertilizers

The theoretically minimum SEC for the major production processes of nitrogen fertilizers are determined as the difference between the exergy and enthalpy respectively of the reactants and the products of the overall reactions that take place in the processes (see Table 6.2). Section 4 gives more details on the processes. All values in Table 6.2 are negative. The negative reaction exergy gives the maximum potential for exergy recovery when the reaction is performed reversibly. The negative reaction enthalpy means that all reactions are exothermic, thus heat is produced. It is the maximum potential for heat recovery. No additional energy inputs are taken into account.

Table 6.2: Change in enthalpy and exergy for the production of ammonia fertilizers according to the stoichiometric reactions (GJ/tonne N and, in parenthesis, GJ/tonne product). All values are given at standard conditions (25°C, 1 bar) and with water in the vapour phase.

Production of:	Exergy change	ΔH°	Overall reaction
Ammonia	-1.6 (-1.3)	-2.7 (-2.2)	Ammonia production by steam reforming of natural gas [a] $0.4422\ CH_4 + 0.6155\ H_2O + 0.6407\ Air = NH_3 + 0.4424\ CO_2 + 0.0060\ Ar$ ($Air = 0.2099\ O_2 + 0.7804\ N_2 + 0.0003\ CO_2 + 0.0094\ Ar$)
Nitric acid	-33.0 (-7.3)	-37.7 (-8.4)	Catalytic combustion of ammonia to form nitric oxide production and absorption of nitric dioxide in water: $3\ NH_3 + 5.25\ O_2 = 2\ HNO_3 + 3.5\ H_2O + NO$
Urea	-0.3 (-0.1)	-3.2 (-1.5)	Reaction of ammonia with carbon dioxide: $2\ NH_3 + CO_2 = NH_2CONH_2 + H_2O$
Ammonium nitrate	-3.1 (-1.1)	-5.2 (-1.8)	Neutralization of nitric acid by ammonia: $NH_3 + HNO_{3\ (aq)} = NH_4NO_3$
Ammonia including energy content of methane	24.7 (20.0)	27.3 (22.5)	Account is made for the HHV and the exergy respectively of methane.
		22.6 (18.6)	Account is made for the LHV of methane

[a] It is assumed that natural gas consists completely of methane. In practice, the methane content can vary.

Since methane is a commercial fuel, it is illustrative to give the enthalpy and exergy changes of the production of ammonia from methane with the enthalpy and exergy of methane included. These values are given in the row under the dotted line in Table 6.2. Some attention needs to be paid to the theoretically minimum SEC of ammonia production according to the steam reforming process of methane. This is, in exergy terms, 24.7 GJ/tonne N. This consists of the exergy of methane of 26.3 GJ/tonne N and the reaction exergy of -1.6 GJ/tonne N[1]. The value of 20.8 GJ/tonne NH_3 (LHV) (25.3 GJ/tonne NH_3 (HHV)) is often used as minimum SEC for the production of ammonia (see e.g. [LeBlanc, 1984; Dybkjaer, 1995a; Radgen, 1996]). However, this value is not the minimum SEC of the reaction, but the amount of methane that is stochiometrically required to produce ammonia.

Table 6.3: Comparison of the exergy loss of the most energy-efficient plants and the theoretical exergy loss of the stochiometric reaction for the production of nitrogen fertilizers. In the last two columns the exergy to produce the raw materials and the exergy of the methane to produce ammonia are taken into account. (GJ/tonne N and in parenthesis GJ/tonne product).

Product	Exergy loss of in production route consisting of the most energy-efficient plant[a]	Exergy requirement of the most energy-efficient plant including production of raw materials and exergy of methane[b]	Theoretically minimum exergy requirement (= exergy of product)
Ammonia	9.4 (7.2)	34.0 (28.0)	24.1 (19.8)
Nitric acid	27.2 (6.0)	34.5 (7.7)	3.1 (0.7)
Urea	12.0 (5.6)	36.5 (17.0)	24.5 (11.4)
Ammonium nitrate	21.4 (7.5)	34.3 (12.0)	10.5 (3.7)

[a] Data for the most efficient plants are based on: Ammonia (Kellogg KRES and KAAP [LeBlanc, 1996]), nitric acid (Uhde mono-pressure process [Anonymous, 1990a]), ammonium nitrate (Hydro Agri [Worrell and Beer, 1995]), and urea (Snamprogetti [1997]).

[b] Note that the actual exergy requirements (column 2) minus the minimum exergy requirement (column 3) are not equal to the exergy loss of producing the product and ammonia (columns 1). This is because other products are formed, which are not accounted for as loss, e.g. CO_2 or steam. In ammonia production these products account for 0.6 GJ/tonne N, in nitric acid production for 3.6 GJ/tonne N, and in ammonium nitrate production for 0.5 GJ/tonne N. Ammonium nitrate gets only 50% of its nitrogen from nitric acid, thus a correction of 0.5*3.6=0.9 GJ/tonne N for by-product formation in nitric acid production is required. The CO_2 formed in ammonia production is used in urea production, thus no correction is needed.

[1]The difference with the exergy of ammonia is caused by the exergy of carbon dioxide that is also formed in the reaction.

6.5.3 Actual exergy requirements of processes to produce nitrogen fertilizers

Table 6.3 gives the difference in exergy between output and input of the most energy-efficient processes. Table 6.3 also shows the exergy required to make a product including the exergy required to make the raw materials and the exergy of methane. These figures can be compared with the exergy of the products, since these values give the theoretically minimum exergy required to make a product. The difference is the potential for improvement and varies from 9.4 GJ/tonne N for ammonia to 27.2 GJ/tonne N for nitric acid. Due to the use of exergy the figures in the table differ from the energy figures based on LHV, which are generally used in the literature. For example, the exergy requirement of the most efficient process to produce ammonia is 28.0 GJ/tonne NH_3, whereas this value is 27.2 GJ/tonne NH_3 based on the LHV of methane.

6.5.4 Biological nitrogen fixation

Besides chemical methods for nitrogen fixation, there is also a biological route to fixation of nitrogen. We make an estimate of the energy requirement of this route.
Since about 100 years it has been know that some microorganisms are able to bind atmospheric nitrogen [Leigh, 1991]. Because of its agricultural importance, the most well-known is *Rhizobium*, a bacterium living in symbiosis with leguminous plants, like bean and pea. *Rhizobium* induces the formation of nodules in the root of the plant. In turn for its nitrogen donation, the bacterium receives energy in the form of carbohydrates from the plant. The plant controls the number and the activity of the bacteria living in its root nodules, to ensure that the growth of the plant is not being limited [Beringer *et al.*, 1982; Wordragen, 1996].

The energy for biological nitrogen fixation is supplied in several steps. First, the carbohydrate, usually glucose, is oxidized. About 40% of the energy of oxidation of glucose (2880 kJ/mole) is used to make adenine triphosphate (ATP) from adenine diphosphate (ADP) [Atkins, 1983]. ATP acts as the energy storage in living beings. The energy released on hydrolysis of ATP to ADP is 30.5 kJ/mole [Stryer, 1981]. Oxidizing one mole of glucose can generate as much as 38 moles of ATP [Atkins, 1983]. In plants nitrogen is converted to ammonium according to

$$N_2 + 6\,e^- + 12\,ATP + 12\,H_2O \rightarrow 2\,NH_4^+ + 12\,ADP + 12\,P_i + 4\,H^+ \qquad (11)$$

where P_i is a phosphate-group [Stryer, 1981]. The energy requirement in the form of ATP is 13 GJ/tonne N or about 35 GJ/tonne N in the form of glucose. The reaction for nitrogen fixation can differ depending on cellular conditions. The amount of ATP required ranges from 12 to 24 mole/mole N_2 [Leigh, 1991]. Therefore, the range in energy requirement is 35 to 70 GJ/tonne N in the form of glucose. When we compare this with the values given in Table 6.3, we can conclude that the lower range is about equal to the energy requirement of chemical nitrogen fixation.

6.6 Exergy analysis

The goal of this section is to locate the operations with the largest exergy losses and to assess to what degree these losses can be avoided. We start by giving an overall picture of the exergy losses that occur during the production of ammonium nitrate and urea, starting with natural gas. Then, we focus on the processes with the largest exergy losses, namely the production of nitric acid and ammonia. We present exergy analyses of state-of-the-art energy-efficient processes to produce ammonia and nitric acid. We give a review of exergy analyses of ammonia production processes that have been presented in the literature over the past 25 years. Unfortunately, no analysis of a modern syngas plant has been published. As we indicated in section 3 of this article, the technology for the production of ammonia has improved considerably over the past decades. Since the starting point of our analysis should be a state-of-the-art process, we make an exergy analysis of such a syngas plant ourselves.

6.6.1 EXERGY LOSS IN THE PRODUCTION OF UREA AND AMMONIUM NITRATE

Table 6.4 and Table 6.5 present the exergy losses that occur during the production of ammonium nitrate and urea. The data are based on the currently most efficient processes for the production of ammonia (Kellogg KRES and KAAP [LeBlanc, 1996]), nitric acid (Uhde mono-pressure process [Anonymous, 1990a]), ammonium nitrate (Hydro Agri [Worrell and Beer, 1995]), and urea (Snamprogetti [1997]).

When we look at Table 6.4 of the exergy flows in the production chain of ammonium nitrate, we see that the exergy loss in the ammonia synthesis of 8.9 GJ/tonne N is of the same order of the loss in the nitric acid synthesis of 9.1 GJ/tonne N. The exergy loss in the ammonium nitrate synthesis is much smaller (1.2 GJ/tonne N). Table 6.5 reveals that the exergy loss that occurs in the urea synthesis are about one third of the loss in the ammonia synthesis. Of the 36.5 GJ/tonne N exergy input, 8.8 G/tonne N is lost in the ammonia synthesis and 3.2 GJ/tonne N in the urea synthesis. The remainder is passed to the product urea (and carbon dioxide).

Since the production of ammonia and nitric acid are the processes with the largest exergy losses, the remaining part of this section will focus on these processes. The emphasis will be on ammonia production, since ammonia is the raw material for all nitrogen fertilizers

Table 6.4: Exergy balance for the production of ammonium nitrate (GJ/tonne ammonium nitrate and GJ/tonne N).

		mass	Exergy in			mass	Exergy out	
		(t/t)	(GJ/t NH_4NO_3)	(GJ/t N)		(t/t)	(GJ/t NH_4NO_3)	(GJ/t N)
1. ammonia production	CH_4	0.3	11.9	34.0	NH_3	0.4	8.4	24.0
					CO_2	0.5	0.2	0.6
					Loss		3.3	9.4
	Sum		11.9	34.0	Sum		11.9	34.0
2. nitric acid produciton	NH_3	0.2	4.4	12.6	HNO_3	0.8	0.5	1.5
	Electricity		0.0	0.1	MP steam	0.7	0.8	1.8
	LP steam	0.1	0.1	0.2	Loss		3.3	9.5
	Sum		4.5	12.9	Sum		4.5	12.9
3. ammonium nitrate production	NH_3	0.2	4.2	11.9	NH_4NO_3	1.0	3.7	10.5
	HNO_3	0.8	0.5	1.5	LP steam	0.1	0.2	0.5
					Loss		0.9	2.5
	Sum		4.7	13.4	Sum		4.7	13.4
	Total input[a]		12.0	34.3	Total loss		7.5	21.4

[a] Total input is the sum of all energy inputs minus the energy outputs.

Table 6.5: Exergy balance for the production of urea (GJ/tonne urea and GJ/tonne N).

		mass	Exergy in			mass	Exergy out	
		(t/t)	(GJ/t urea)	(GJ/t N)		(t/t)	(GJ/t urea)	(GJ/t N)
1.ammonia production	CH_4	0.4	15.9	34.0	NH_3	0.6	11.2	24.0
					CO_2	0.7	0.3	0.6
					Loss		4.4	9.4
	Sum		15.9	34.0	Sum		15.9	34.0
2. urea production	NH_3	0.6	11.2	24.0	Urea	1.0	11.4	24.5
	CO_2	0.7	0.3	0.7	Loss		1.2	2.6
	HP steam	0.7	1.1	2.2				
	Electricity		0.1	0.2				
	Sum		12.7	27.1	Sum		12.7	27.1
	Total input[a]		17.0	36.5	Total loss		5.6	12.0

[a] Total input is the sum of all energy inputs minus the energy outputs.

6.6.2 Exergy analysis of ammonia production processes

In the past several exergy analyses of the production of ammonia have been performed. We review a number of these analyses to locate the operations with the major exergy losses and to assess whether there has been a reduction in exergy loss in specific unit operations. In Table 6.6 the results of five exergy analyses of ammonia plants have been summarized.

The exergy losses in the reformer sections are about 7-9 GJ/tonne N and have hardly changed over the years. Expressed as the share of the total losses, however, the exergy losses in the reformer have increased considerably. It is not surprising that the processes in the reforming section are responsible for the largest exergy losses. After all, here natural gas is burnt to provide in the heat demand of the endothermic steam reforming reactions and to generate steam. During combustion about 30% of the exergy content of a fuel is lost [Nieuwlaar, 1988].

The exergy losses in the syngas purification and the ammonia synthesis (including associated equipment) reduced over the years. The exergy losses in the syngas purification are dominated by the losses that occur in the CO_2-removal. The introduction of better solvents reduced both the power requirement for recycling and the heat requirement for regeneration [Dybkjaer, 1995a]. The losses in the ammonia synthesis are due to heat transfer, recycling, purge flow and cooling. The synthesis pressure has been reduced over the years. As a result, the compression power has decreased. On the other hand, a low pressure has a negative effect on the partial ammonia pressure at equilibrium. Less ammonia can be cooled out using cooling water, thus more refrigeration power is required [Radgen, 1997]. The recycle power increases also, because large gas volumes have to be handled. Since the exergy losses in the ammonia synthesis loop increase and in the compressors decrease, the overall effect of a lower synthesis pressure on the exergy use is limited.

Because the largest exergy loss occur in the reforming section, this loss is investigated more closely. The analyses presented in Table 6.6 are all based on syngas plants that use a fired primary reformer. In Section 3 we have indicated that several processes have been developed that avoid the use of a fired primary reformer. Although these processes differ in the way this is achieved, the general approach is that the heat of the secondary (autothermal) reformer is used to provide in the heat demand of the primary reformer. Since our analysis should be based on a state-of-the-art process, we will perform an exergy analysis of a very modern syngas production plant.

Table 6.6 : Overview of exergy analyses of ammonia production processes. All values are expressed in GJ/tonne N [Riekert, 1974; Cremer, 1980; Szargut *et al.*, 1988; Dybkjaer, 1995a; Radgen, 1997].

Source	Riekert, 1974		Cremer, 1980		Szargut, 1988		Dybkjaer, 1995		Radgen, 1997	
Representative for a plant of:	end 1960s		mid 1970s		early 1980s		mid 1980s		specific plant built in 1992	
Energy (LHV) input	43.5		44.1		42.5		35.5		39.1	
Exergy input	44.1		48.2		44.0		37.1		43.4	
Exergy loss	19.2		18.5		18.6		12.7		13.1	
Primary reformer	6.8	36%	5.3	29%	4.4	23%	5.9	47%	4.6	35%
Steam generation					1.6	8%	2.9	23%	1.6	12%
Secondary reformer	1.4	8%	1.4	8%	1.7	9%	incl. in primary reformer		2.3	17%
Syngas purification	1.1	6%	1.9	10%	3.4	18%	0.8	7%	1.1	8%
NH3 Synthesis			0.7	4%	0.6	3%	1.8	14%	1.3	10%
Heat exchangers }			3.0	16%	1.9	10%	}			
Compressors }	9.9	52%	2.3	12%	1.6	8%	0.6	5%	1.5	11%
Heat loss			1.0	5%	0.1	1%				
Purge/stack			1.8	12%	3.5	19%	0.4	3%	0.6	5%
Other losses			0.7	4%	0.6	3%	0.2	2%	0.1	1%
Exergetic efficiency[a]	58%		62%		56%		66%		69%	

[a] The exergetic efficiency is the ratio of the exergy of the useful products (ammonia, carbon dioxide as far as used for urea production, steam) and the exergy expenditure (natural gas, electricity, steam).

The plant we will analyse is based on the Kellogg Reformer Exchanger System (KRES) [LeBlanc, 1996] [Czuppon *et al.*, 1996]. The heart of a KRES syngas plants consists of two reformers: a gas-heated steam reformer and an autothermal reformer. The feed stream is split into a part that flows to the autothermal reformer and a part that flows into catalyst tubes in the gas-heated reformer. In the autothermal reformer the main reaction is partial oxidation of methane. Oxygen enriched air is used for this reaction. The hot gas that leaves the autothermal reformer is used to supply in the heat of the steam reforming reaction that takes place in the gas-heated reformer. The configuration of KRES eliminates the need for a primary reformer that is fired. However, the heat generated of the hot gas is not sufficient to meet in the heat demand of the gas-heated reformer.

Preheating of the feed in a separate furnace is necessary. This furnace is fired by natural gas.

Advantages of KRES compared to a conventional steam reforming process are lower capital and operation and maintenance costs, improved energy efficiency, and less space requirement. KRES has been successfully used in an ammonia plant in Canada since 1994 [Czuppon *et al.*, 1996]. This plant was retrofitted to include both KRES and the Kellogg Advanced Ammonia Process (KAAP) for efficient ammonia synthesis. The SEC of this plant is 27.2 GJ(LHV)/tonne NH_3 (33.0 GJ/tonne N) [Czuppon *et al.*, 1996].

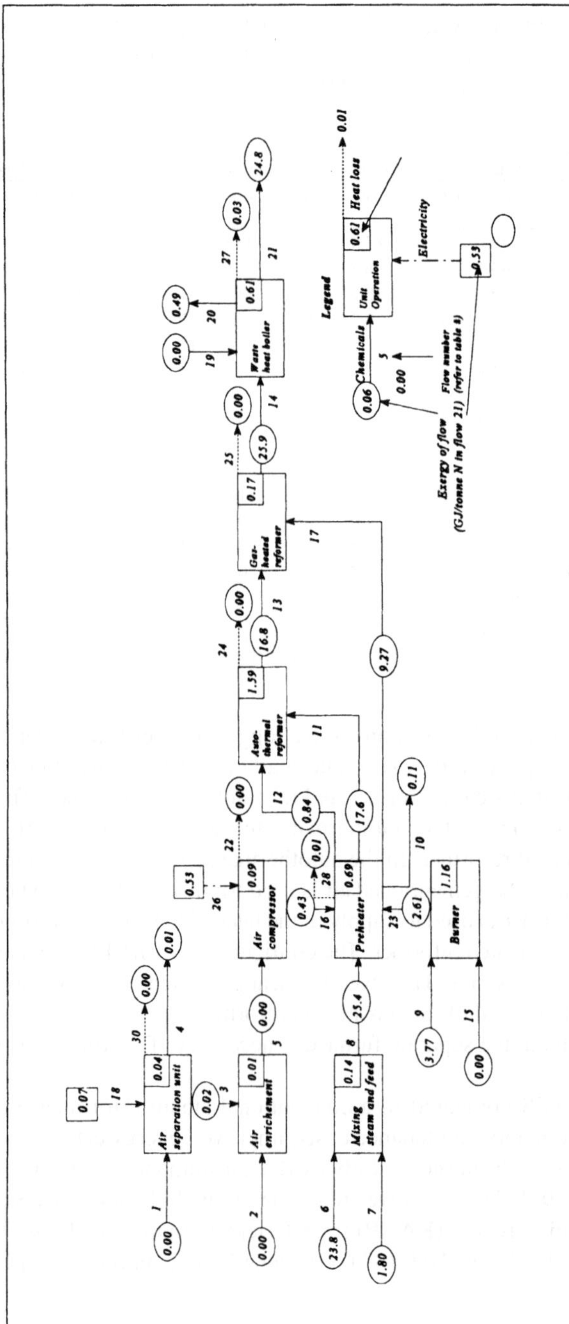

Figure 6.6: Flowsheet of the KRES syngas plant. Exergy of flows and exergy loss in unit operations are also indicated.

Table 6.7: Basic specifications of the KRES-plant for ammonia synthesis gas production.

Unit operation	Specifications
Oxygen plant	Cryogenic separation; 100% pure oxygen produced; Assumed power consumption 280 kWh/tonne [Hendriks, 1994].
Air enrichment	Oxygen content in air to the ATR varied from 28 to 32 mol% in test [Czuppon *et al.*, 1996]. To balance material flows an oxygen content of 27.4% mol% is required. We assumed that the reactor also performs well at this value, which lies slightly below the test range.
Air compressor	Multi-stage compression with intercooling to 30°C; pressure ratio of each stage 3.0; stage efficiency 90%; 10% of the energy input is converted to heat; energy required for cooling is not taken into account; power requirement and outlet temperature calculated with ENERPACK [parameters proposed by ENERPACK].
Mixing steam and feed	Steam to carbon ratio in tests varied from 3.0 to 4.0. Balancing material flows resulted in a steam to carbon ratio of 3.5. The split of the feed between the gas-heated reformer and the autothermal reformer varied in the test from 25/75 to 30/70. We used 35/65 to balance the material flows. The effect of using a ratio lower than the test range is that less heat is generated in the autothermal reformer and more heat is required in the gas-heated reformer. This is compensated by assuming preheat temperatures that are higher than that used in the test. Air and autothermal reformer-feed are preheated to 660°C (reported range 600-640°C). The gas-heated reformer-feed is preheated to 555°C (reported range 500-550°C).
Fired pre-heater	Enriched compressed air and feed/steam are preheated in the radiation and convection zone of a furnace fired by natural gas and with 10% excess air. Heat transfer efficiency is 97%. Stack temperature is 80°C. [Own assumptions].
Autothermal reformer	Dry gas composition is specified [Czuppon *et al.*, 1996]. Both catalytic partial oxidation and steam reforming take place in the reactor. Reported exit temperature is about 1000°C [LeBlanc, 1996]. We used 1060°C to match the heat balance of the autothermal reformer.
Gas-heated reformer	Shell and tube design. Open tubes filled with catalysts [LeBlanc, 1996]. Pressure drop over the exchanger 217 kPa [Czuppon *et al.*, 1996]. Reported exit temperature 660-680°C [Czuppon,]. We used 670°C.
Waste heat boiler	Exit temperature of syngas set at 450°C, which is the temperature of the first step of the shift converter. Firetube boiler.25% of the heat is lost to the environment. pressure drop 5% of ingoing gas pressure [Kreith and West, 1997]. Steam pressure 50 bar [Kreith and West, 1997].

The exergy analysis has been carried out by using the software package ENERPACK [Nieuwlaar, 1996]. For our analysis the environmental reference system (ERS) described in ENERPACK has been used. Specifications of temperatures, pressure and dry synthesis gas composition were based on data from M.W. Kellogg [Czuppon *et al.*, 1996; LeBlanc, 1996]. Ranges for the steam to carbon ratio, split of feed between gas-heated reformer and autothermal reformer, and oxygen content of the combustion air were also given by these sources. The exact values of the last three variables were chosen so that the material balances of all unit operations were complete. All flows are normalized to 1 tonne of nitrogen in the synthesis gas leaving the gas-heated reformer. Table 6.7 gives the basic specifications of the plant and an account for data and assumptions. Table 6.8 gives specifications of all material flows.

Figure 6.6 specifies the exergy of all flows and the exergy loss of all unit operations. The total exergy input is 30.0 GJ/tonne N and the exergy of the useful products (steam and syngas) is 25.3 GJ/tonne N. The difference is exergy loss and amounts to 4.7 GJ/tonne N, of which 4.5 GJ is internal loss and 0.2 GJ is external loss. This exergy loss means a reduction of about 4 GJ/tonne N compared to the exergy loss of the reformer section of the modern ammonia plant that Dybkjaer [Dybkjaer, 1995a] has analysed (see Table 6.6).

We will discuss the main internal losses per unit operation. The largest internal exergy losses (1.6 GJ/tonne N) occur in the autothermal reformer. In this reactor the chemical exergy of natural gas is converted to chemical and physical exergy of the reaction products by partial oxidation. When we take into account all reactions that take place in the autothermal reformer (reactions 1-3) the difference in chemical exergy between reactants and products at standard conditions is about 3.6 GJ/tonne N. This amount is partially converted to physical exergy, which is manifested in a temperature rise from 933 to 1334 K. The increase in physical exergy can be calculated to be 2.0 GJ/tonne N. The remaining part of the chemical exergy change is lost. It can be expected that this is largely due to the irreversible conversion of chemical exergy to physical exergy. This exergy loss can be estimated by multiplying the Gibbs energy of reaction, determined at reaction conditions, by the Carnot factor [Hinderink *et al.*, 1996]. The Gibbs energy of the overall reaction in the autothermal reformer can be determined to be 6.2 GJ/tonne N[1]. When we assume an average reactor temperature of 1200 K, the Carnot factor is 0.25, resulting in an exergy loss of 1.6 GJ/tonne N. Since this equals the remaining part of the exergy loss, it can be concluded that all internal exergy loss in the autothermal reformer is due to the exergy loss

[1] The Gibbs energy of reaction can be approximated by: $\Delta G_{T_r} = \Delta H° - T_r\,(\Delta S° - R \ln (P_r/P_0)^{\sum\upsilon})$, where: ΔG_{T_r} = Gibbs free energy of reaction (kJ/mole), $\Delta H°$ = standard enthalpy change of reaction (kJ/mole), T_r = temperature of reaction (K), $\Delta S°$= standard entropy of reaction (kJ/mole/K), R= gas constant (kJ/mol/K), P_r = reaction pressure (kPa), P_0 = reference pressure, $\sum\upsilon$= sum of stoichiometric coefficients [Hinderink *et al.*, 1996]. Note that when the reaction is performed at reference pressure and temperature, the Gibbs energy of reaction equals the standard chemical exergy change.

that occurs in the conversion of chemical to physical exergy at the prevailing temperature in the reactor.

Table 6.8: Temperature, pressure, volume and composition of all material flows in the KRES syngas-plant. When no value is given for a compound, this compound is not present in that flow. The flow numbers refer to the flows in Figure 6.6.

#	name	T	p	Exer-gy	volume	Composition (weight)							
		K	kPa	GJ/t N	t/t N	CH_4	H_2	CO_2	N_2	CO	Ar	O_2	H_2O
1	Air	298	101.3	0.0	0.58			0.005	0.737		0.013	0.226	0.02
2	Air	298	101.3	0.0	1.18		0.000	0.005	0.737		0.013	0.226	0.02
3	Oxygen	298	4053.0	0.0	0.13						0.000	1.000	
4	Nitrogen	298	4053.0	0.0	0.45			0.006	0.951		0.017		0.03
5	Enriched air	298	4053.0	0.0	1.31			0.004	0.664		0.011	0.303	0.02
6	Natural gas	298	4053.0	23.8	0.58	0.778			0.222				
7	Steam	524	4053.0	1.8	1.80								1.00
8	NG + steam	402	4053.0	25.4	2.38	0.191			0.054				0.76
9	NG fuel	298	810.6	3.8	0.08	1.000							
10	Stack	353	101.3	0.1	1.49			0.158	0.687		0.012		0.14
11	NG + steam to ATR	933	4053.0	17.6	1.55	0.191			0.054				0.76
12	Heated compressed air	933	4053.0	0.8	1.31			0.004	0.664		0.011	0.303	0.02
13	ATR syngas	1334	4053.0	16.8	2.86	0.001	0.028	0.156	0.334	0.081	0.005		0.40
14	GHR syngas	943	3850.4	25.9	3.69		0.040	0.171	0.271	0.106	0.005		0.41
15	Combustion air	298	101.3	0.0	1.41			0.005	0.737		0.013	0.226	0.02
16	Compressed air	400	4053.0	0.4	1.31			0.004	0.664		0.011	0.303	0.02
17	NG + steam to GHR	828	4053.0	9.3	0.83	0.191			0.054				0.76
19	Boiler feed water	298	101.3	0.0	0.48								1.00
20	High pressure steam	538	5066.3	0.5	0.48								1.00
21	Cooled syngas	723	3657.8	24.8	3.69		0.404	0.171	0.271	0.106	0.005		0.41
23	Gas to pre-heater	2109	101.3	2.6	1.49			0.158	0.687		0.012		0.14

The second largest exergy loss (1.2 GJ/tonne N) occurs in the burner, where natural gas is completely combusted. The chemical exergy is converted to physical exergy in the form of hot gases, during which about 30% of the chemical exergy is lost. The chemical exergy of the combustion gases is very small. The physical exergy is transferred to compressed air and feed to increase the temperature of these flows in the preheater. The exergy loss due to irreversibilities in heat transfer is 0.7 GJ/tonne N[1].

In the waste heat boiler 1.1 GJ/tonne N of physical exergy of the hot syngas is partly transferred to water to raise steam. The exergy of the steam is only 0.5 GJ/tonne N, the remaining 0.6 GJ/tonne N is lost due to inefficient heat transfer as a result of large temperature differences.

Table 6.9 summarizes the main exergy losses in the KRES-syngas plant. This configuration reduces the exergy loss by about 4 GJ/tonne N to 4.7 GJ/tonne N compared with conventional syngas production. Nevertheless, the largest exergy loss is because of the conversion of chemical to physical exergy, required to generate heat to drive the steam reforming reaction.

Table 6.9: Overview of exergy loss in KRES syngas plant.

		GJ/tonne N
External exergy loss		0.2
Internal exergy loss, of which:		4.5
	conversion of chemical exergy to physical exergy	2.9
	irreversibilities in heat transfer	1.2
	others	0.6
Total exergy loss		4.7

6.6.3 Exergy analysis of a nitric acid plant

Szargut *et al.* describe an exergy analysis of a 1980-nitric acid plant [Szargut *et al.*, 1988]. Table 6.10 summarizes the results of this analysis. The plant is a dual-pressure process producing a weak acid (67%). The converter operates at 0.115 MPa and the absorber at 0.5 MPa.

[1] We assumed that the combustion gases reach the adiabatic flame temperature. The temperature difference between the hot combustion gases and the preheated gases is therefore on the order of 700-800°C. In practice, this temperature difference is not attained, because combustion and heat transfer occur simultaneously, which results in a lower temperature of the combustion gases. The effect of a lower temperature is that the exergy loss due to heat transfer is smaller, and due to conversion of physical to chemical exergy is larger. The exergy loss of the burner and preheater together remains the same.

Table 6.10: Exergy analysis for a 100 tpd nitric acid (67%) plant [Szargut *et al.*, 1988]. All values in GJ/tonne N in nitric acid.

Input	Exergy	Output	Exergy
Ammonia	13.83	Useful outputs:	
Electricity	1.97	HNO_3 (67%)	1.21
Water	0.03	Steam (4.2 Mpa, 400°C)	3.63
		Exergy losses in:	
		Converter	4.94
		Absorber	2.21
		Steam production	1.97
		Other heat exchangers	0.69
		Turbine and compressor	0.93
		Other losses	0.24
Total	15.83	Total	15.83

Although the analysis is based on more than a 15 year old data, the plant is very efficient. Modern nitric acid plants have an HP steam recovery of 4 GJ /tonne N (exergy content) [Anonymous, 1988]. The electricity requirement of about 0.14 GJ_e/tonne N of a modern plant is much lower than that of the plant analysed by Szargut. The difference between the old plant used in the analysis and modern nitric acid plants is that in-plant energy use is more optimized in the last case.

The exergy analysis shows that the converter is responsible for the largest exergy losses. In the converter the chemical exergy of ammonia (13.8 GJ/tonne N) is converted to chemical exergy (3.4 GJ/tonne N) and physical exergy (5.8 GJ/tonne N) of the combustion products with a temperature of 1200 K. The remainder is lost due to the irreversible combustion of ammonia (4.3 GJ/tonne N) and heat lost to the environment (0.3 GJ/tonne N). The exergy losses in the absorber are due to solution of gas in water. Generation of steam in a series of heat exchangers where the converter exit gas is cooled causes exergy losses due to heat exchange over a temperature gradient.

It can be concluded that the main option to reduce the exergy losses of nitric acid production is controlling the conversion of ammonia to NO, so that the reaction proceeds with a higher degree of reversibility.

6.7 Options to improve the energy efficiency of nitrogen fixation

6.7.1 Selection of energy efficient options

In this section we will select options that can contribute to an improvement of the energy efficiency of nitrogen fixation, based on the results of the exergy analyses. We separate the discussion into a part concerning the syngas production, a part concerning the rest of the plant, and a part concerning nitric acid production.

We determined the exergy losses of the production of ammonia syngas in a KRES-plant to be 4.7 GJ/tonne N. Almost all exergy losses are due to the conversion of chemical exergy to physical exergy needed to drive the steam reforming reaction. The exergy losses due to partial combustion of natural gas, in the pre-heater and the autothermal reformer, are 2.9 GJ/tonne N. Heat transfer, in the preheater and the waste heat boiler, accounts for 1.3 GJ/tonne N. The exergy losses can be reduced by several approaches.

The first approach involves *optimizing the KRES-plant*. At several places losses occur that might be avoided by another lay-out of the plant or by changing the process conditions. For instance, the generation of steam by heat exchange with the syngas of 670°C is not very efficient. This heat can also be used to replace part of the pre-heating of the reactants in the fired pre-heater. A pinch analysis is a useful tool to optimize the energy consumption of the plant. Heat exchangers for the temperatures and pressures used in the plant are available [Shah, 1997]. Plant optimization can bring the exergy loss down by up to 1.3 GJ/tonne N, the exergy loss due to heat transfer. The actural reduction of the exergy loss depends on the temperature difference over the heat exchanger. Since the technologies for plant optimization are already available we will not deal further with this option.

The second approach to reduce the exergy loss involves options that make *better use of the chemical exergy of a fuel*. In the ammonia syngas production, a large exergy loss is the result of using a high-quality fuel for a process that requires heat of a much lower quality. This exergy loss can be reduced by matching the process that donates exergy, in this case combustion of natural gas, to the process that accepts exergy, in this case steam reforming of natural gas. This is referred to as exergy matching.

One option to improve the match is to adapt the exergy donating process, so that the exergy losses reduce. This can be achieved by e.g. burning natural gas in a gas turbine to generate electricity and use the exhaust gases to supply in the heat for the steam reforming reaction. The chemical exergy of the fuel is best used when the exhaust gases leave the turbine at a temperature as low as possible. The exit temperature of modern industrial gas turbines is about 600°C. Although this temperature is too low to provide completely in the heat demand of steam reforming, gases of this temperature can be used to preheat feed and fuel. An even better match can be achieved by adapting the steam reforming process so that the heat demand can be satisfied completely by the heat of the exhaust gases of a gas turbine. The currently prevailing process conditions for steam reforming of natural gas, of about 800-900°C and 25-35 bar are an optimum considering the volumes of gas to be handled and the production of hydrogen. At these conditions about 85-90% of the methane is converted to hydrogen, depending on the amount of steam. Hydrogen production is favoured by high temperatures, low pressures and high steam content. To obtain a better match with the exergy of the exhaust gases of a gas turbine, the temperature at which steam reforming is performed should be lowered. Lowering the temperature would result in a lower conversion of methane to hydrogen. This effect can be compensated by reducing the pressure, but this would mean that the size of the installations has to be increased, which is economically unfavourable. The methane conversion can also be increased by adding more steam to the reaction mixture. This is already common practice, also to prevent

carbon formation. However, additional steam consumption increases the energy requirement.
Quite another approach that allows lowering the reaction temperature without reducing hydrogen production is avoiding equilibrium to be achieved by adding reactants or removing products. Especially the option of removing hydrogen is interesting, because this can also mean a reduction in the energy demand for syngas purification. This can be achieved by using membrane reactors and by using a physical or a chemical sorbent.
A better exergy match can also be attained by *providing the heat for steam reforming in another way than through using the combustion gases of a fuel.* An option is coupling the High-Temperature Nuclear Reactor (HTR) to the syngas plant [Kugeler and Schulten, 1989; Fujimoto *et al.*, 1992; Hada *et al.*, 1992]. The HTR is still under development, but is expected to be available around 2010. The HTR produces helium at about 900°C, which can be used to supply in the heat demand of the steam reforming reaction. However, it is expected that the economics of the process are unfavourable compared to conventional syngas production [Smit and Beer, 1995]. The use of nuclear energy might introduce several problems that are beyond the scope of this chapter, for instance risk for proliferation of missile materials and storage of nuclear waste. Over and above this, the use of nuclear energy does not result in an improvement of the energy efficiency, although it can result in a reduction of the emissions of carbon dioxide. Therefore we will not deal with this option further.

Quite a different approach to reduce the exergy loss of syngas production is to use *another method for hydrogen production.* Currently, steam reforming is the most energy-efficient method for large-scale hydrogen production [Czuppon *et al.*, 1993]. Hydrogen-rich industrial gases, for instance blast furnace gas, have been used for ammonia production. However, these flows are often contaminated with dust an extensive cleaning is required. Nowadays, these gases are usually used as fuel in the plant where they are produced, because this is more cost-effective. Water electrolysis has also been used for hydrogen production, but the energy demand is considerably higher than for steam reforming of natural gas [Anonymous, 1990b; Czuppon *et al.*, 1993]. This option might be of interest when electricity is produced from non fossil-fuel sources, for instance by hydro power. Production of hydrogen from biomass has also the advantage of avoiding the use of fossil fuels. Evaluation of four biomass gasifiers showed that the requirement for biomass, expressed in GJ/GJ hydrogen (HHV), is 15 to 35% higher than that for steam reforming of natural gas [Williams *et al.*, 1995]. In the near future producing hydrogen from biomass will remain more expensive than hydrogen from steam reforming of natural gas. [Williams *et al.*, 1995]. However, Williams *et al.* [Williams *et al.*, 1995] expect that a rise in natural gas prices will make biomass-derived hydrogen cost-competitive with hydrogen produced by steam reforming of natural gas by the year 2010. In conclusion, neither water electrolysis nor biomass can reduce the energy demand for hydrogen production. If one wants to avoid the use of fossil fuels, especially biomass is an interesting alternative. In this chapter we will not deal further with these options, because we are in the first place interested in improving the energy efficiency.

The exergy losses in the remainder of the ammonia plant are due to the syngas purification, mainly CO_2-removal, and the ammonia synthesis loop itself. We have not analysed these exergy losses in detail. The discussion on opportunities to improve this part of the plant will therefore be more general than the preceding discussion on the syngas plant.

The exergy losses in the *syngas purification* can be reduced by improving the separate operations, better integration of the operations, and by reducing the need for purification. Many improvements have been achieved in the separate operations [Dybkjaer, 1995a], and there is no reason to assume that no further improvements will be made. Better integration can be achieved by minimizing differences in temperature and pressure between operations. Since the associated reduction of exergy losses is small, we will not deal with these options separately. Reducing the need for purification can be achieved by producing a syngas without oxygen-compounds that can poison the ammonia catalysts, or by developing catalysts that are less sensitive to poisoning. Since the first commercial introduction of the Haber-Bosch process iron-based catalysts have been used for the ammonia synthesis. The search for new catalysts has been very extensive, see e.g. Aika and Tamaru [Aika and Tamaru, 1995] for a review, and has resulted in several adaptations to the conventional iron-based catalyst and at least one new catalyst, based on ruthenium. Although the new catalysts have improved characteristics on other aspects, e.g. activity, their sensitivity to poisoning by oxygen-compounds has hardly been improved [Appl, 1992a]. At present, there is no indication that a commercial ammonia catalyst will be developed that is less inhibited by oxygen-compounds than the traditional catalyst [Anonymous, 1994].

The exergy losses in the *ammonia synthesis loop* occur mainly in the syngas compressor, the recycle compressor and the ammonia recovery unit. The losses in the syngas compressors can be reduced by synthesizing ammonia at a lower pressure. The position of the chemical equilibrium of the ammonia synthesis (reaction 4, page 173) is shifted to ammonia at higher pressures and lower temperatures. In industrial practice temperatures of 350-450°C and pressures of 80-200 bar are used. The ammonia concentration at these conditions is about 15%. With commercial catalysts it is technically possible to synthesize ammonia at 350°C and 40-50 bar, the pressure of the syngas plant [Appl, 1992a]. However, at this low pressure the reaction rate would decrease significantly. If the effect of the low reaction rate is not compensated for, the ammonia concentration in the reaction mixture will be very low. The energy demand for recycling the gas volumes, which are larger at lower pressure, and for ammonia recovery from such large flows would nearly offset the savings from the lower synthesis pressure [Appl, 1992a]. The ammonia concentration can be increased by using a more active catalyst or by extending the reaction time by enlarging the catalyst volume. The last option is very costly but technically feasible, and is therefore not considered further.

Concerning the first option, ruthenium-based catalysts with a 20 times higher activity than the conventional catalysts are commercially available [Appl, 1992a; Czuppon *et al.*, 1996; Nielsen, 1997]. M.W. Kellogg uses a ruthenium-based catalyst in its advanced ammonia process (KAAP) in Canada since 1992. This has not resulted in an energy saving, but in an increase in ammonia production capacity [Czuppon *et al.*, 1996]. Haldor Topsøe was

also developing a ruthenium-based catalyst [Nielsen, 1997]. However, the expected energy saving is small, on the order of 0.1 GJ/tonne NH_3 [Nielsen, 1997]. The power demand for the syngas compressor reduces, but the power demand for the refrigeration compressor increases almost equally. Considering this, and the high price, the unproven lifetime and the narrow operating temperature range of the ruthenium-based catalyst, Haldor Topsøe decided to base its future designs on traditional catalysts [Nielsen, 1997].
A low-pressure catalyst may well result in an energy saving when the need for recycling can be eliminated. This can be achieved when equilibrium is avoided to establish by removing selectively the product ammonia. We discuss the opportunities to do so by using membrane reactors and by using chemical or physical sorbents.

The exergy loss in the production of nitric acid is mainly due to the irreversible combustion of ammonia. It does not seem to be possible to achieve a considerable reduction of these losses using the conventional nitric acid plant. Parameters like catalyst type, temperature, pressure, gas velocity and converter design have been chosen carefully to ensure an optimum conversion efficiency [Clarke and Mazzafro, 1993]. Side reactions, like the direct combustion of ammonia to nitrogen and water or the decomposition of nitric oxide into nitrogen and oxide, can reduce the conversion efficiency when these parameters are changed. Enhanced recovery of the heat of reaction might result in a higher steam production, or another use for the excess heat may be found. However, this requires the development of improved materials for the construction of heat exchangers, because condensation of corrosive liquid nitric acid can damage the heat exchanger material.
It can be concluded that to reduce the exergy loss of nitric acid formation from ammonia a method must be found to control the conversion of ammonia into nitric oxide. This method should enable the recovery of (part of) the chemical reaction exergy in the form of a high-quality energy carrier, like electricity. We investigate two options in more detail: conversion of ammonia in a fuel cell and in a gas turbine.
Another option is to produce nitric oxides not from ammonia, but by the direct oxidation of atmospheric nitrogen. It is known that at high temperatures nitric oxides are formed. The electric arc process, for instance, applied this method to make nitric oxide. During combustion processes nitric oxides are also formed. We assess whether these or other processes can be used for commercial nitric acid production.

In the previous discussion we have almost completely limited ourselves to improving the current production process of nitrogen fixation. In section 2 we indicated that the focus will be on improving current processes. However, alternatives to the current practice may result in a considerable reduction of the energy requirement for producing nitrogen fertilizers. Therefore, we also consider these alternatives, albeit in less detail than the improvements to the current processes.
The first alternative involves *other routes of chemical fixation of atmospheric nitrogen*. In the past, several processes have been commercialized to the direct oxidation of nitrogen, like the electric arc furnaces and the Wisconsin process. At that time, the energy demand was higher than that of the Haber-Bosch process. However, it may be that current

technology can reduce the energy consumption. Since these process still require high temperatures, hence lead almost inevitable to exergy losses, we also discuss the developments in the chemistry aimed at splitting nitrogen at low temperatures.
The second alternative to current production methods is *avoiding the use of nitrogen fertilizers at all.* It is known that some micro-organisms can fix atmospheric nitrogen. It might be possible to bio-engineer crops in such a way that they can fix nitrogen themselves. Another possibility is to extend the symbioses that now exists between nitrogen-fixating microorganisms and legumes to other crops. We discuss these options very briefly to investigate their opportunities to reduce the energy demand of nitrogen fertilization.

We have summarized all options for improving the energy efficiency of nitrogen fixation in Table 6.11. In the next Section we characterize specific options. In the table it is indicated whether the option is dealt with and, if so, in which section.

Table 6.11: Options to improve the energy efficiency of atmospheric nitrogen fixation.

Option to improve the energy efficiency	Dealt with in section:
AMMONIA SYNGAS PLANT	
• Optimizing syngas plant	not dealt with
• Exergy matching: Lowering heat demand for steam reforming	section 6 8.1
• Exergy matching: gas turbine	section 6 8.1
• Other sources of hydrogen	not dealt with
REMAINDER OF AMMONIA PLANT	
• Improving syngas purification	not dealt with
• Low pressure ammonia synthesis, combined with reduced power demand for recycling and product recovery	section 6.8.1 and 6.8.1
NITRIC ACID PRODUCTION	
• Enhanced heat recovery	not dealt with
• High-quality energy recovery - gas turbine	section 6.8.2
• High-quality energy recovery - fuel cell	section 6.8.2
• High-temperature direct oxidation of nitrogen	section 6.8.2
ALTERNATIVE NITROGEN FIXATION METHODS	
• Biological nitrogen fixation	section 6.8.3
• Alternative chemical nitrogen fixation	section 6.8.3

6.7.2 Collection of Information

We give a brief account of the way the information was collected. The identification and characterization of new techniques started with a search for relevant literature, performed in two ways. First, the following literature databases were searched: Environline/Energyline, Chemical Abstracts, Applied Science and Technology Index, and Carl Uncover. Second, volumes of journals specific to nitrogen fertilization and chemical engineering were scanned to identify emerging techniques. Of the following journals the volumes from 1990 to 1996 were scanned: Nitrogen, Fertilizer Focus, Fertilizer International, Catalysis Today, Hydrocarbon Processing, Industrial & Engineering Chemical Research. The selection of journals was based on the results of the database search. Much new relevant literature, both papers and books, was found by checking the references of the literature found by both methods.

Simultaneously with the literature search, the following developers of processes for nitrogen fertilizer production were contacted: Haldor Topsøe, The M.W. Kellogg Company, Brown & Root, ICI, Krupp Uhde, and Weaterly. The aim was to obtain the most recent data on their processes.
We checked all data for accuracy and reliability by consulting experts, and by making our own calculations and judgments, and by obtaining evidence from other sources.

6.8 Characterization of options to improve the energy efficiency

The aim of this section is to characterize options that can improve the energy efficiency of nitrogen fixation in the future on the following features: specific energy consumption, costs, stage of development, and degree of technical change required to implement the technology or option compared to the current processes for nitrogen fixation. Concerning the last characteristic, we distinguish three categories of required technical change. First, techniques that require an *evolutionary change* imply a continuation of the trend in technological development. No changes in the way the energy service is performed are expected, and the effects on the following aspects are small or negligible: performance, process parameters, quality and nature of the products, the purchasing and supplying industry, and the plant layout. Second, a *major change* is required when at least part of the energy service is performed according to a new principle, the performance of the process increases more than one can expect by trend extrapolation, and there are considerable effects on the other aspects. Finally, a *radical change* is required when a new energy service arises or all aspects change to a large extent. For a more extensive description of this categorization, see chapter 2 of this book.
We deal first with the production of ammonia. Then, we deal with the production of nitric acid. Finally, alternative routes to the fixation of atmospheric nitrogen are considered.

6.8.1 AMMONIA SYNGAS PRODUCTION

In Section 6 we concluded that the most important way to reduce the exergy losses in the ammonia syngas production is making better use of the chemical exergy of the fuel. In this Section we elaborate on options to improve the exergy match between the exergy donating process, combustion of natural gas, and the exergy accepting process, steam reforming of natural gas. The match can be improved by combustion of natural gas in a gas turbine to generate electricity, and using the exhaust gases to provide in the heat demand for steam reforming. A further improvement of the match can be attained by lowering the temperature of the heat required for steam reforming, which can be achieve with membrane reactors. First, we discuss the use of membrane reactors for syngas production. Then, we deal with the application of gas turbines.

Membrane reactors

Before we deal with application of membrane reactors for syngas production, we introduce some definitions and parameters and give general characteristics of the most important membranes. A membrane can be defined as a selective barrier between two phases through which transport of a material can take place due to a driving force [Mulder, 1993]. In gas separations this driving force is usually a pressure difference. A feed stream is passed over the membrane, upon which part of the gas crosses the membrane and can be taken away by a sweep gas. The resulting flow is called the permeate. The part of the feed gas that does not cross the membrane leaves the unit and is called the retentate. The partial pressure of the desired component in the feed gas should be higher than in the permeate. A membrane reactor can have the membrane and the reactor coupled in one unit, or have them separated. When a catalyst is used to promote the reaction in a coupled membrane reactor, one can speak of a catalytic membrane reactor [Zaman and Chakma, 1994]. The membrane material may be mixed or coated with a catalyst, or may be catalytically active itself.

Two parameters to characterize the performance of membranes are selectivity and permeability [Mulder, 1996]. Permeability indicates how well a gas is transported through a membrane at given pressure and temperature. Selectivity is a measure for the degree of separation of two gases and is equal to the ratio of the permeabilities of two gases. For gas mixtures the separation factor is used to indicate the selectivity. When only one component can pass the membrane it is called permselective.

Table 6.12 presents several types gas of separation membranes. Polymeric as well as inorganic materials can be used to make membranes for gas separation.

Polymer membranes are commercial, but can only be applied at low temperatures and low pressures. Since many chemical reactions take place at elevated temperatures to ensure a sufficient reaction rate, polymeric membranes cannot be used for these reactions. The other membranes in the table are made of inorganic materials. The mechanical strength and the thermal stability is higher than of polymeric membranes, making their application in processes that operate at elevated temperature and pressure possible [Mulder, 1993].

Table 6.12: Selected parameters and characteristics for gas separation membranes.

Type	Materials	Application and merits	Main limitations to application	Stage of development
Polymer membranes	Different types of polymers, e.g. polyacrylamide, polysulphone or polyurethane (1)	Low-temperature separation processes (1,3)	High temperatures and high pressures(2,5)	Commercial for several gas separations, e.g. hydrogen from ammonia purge gas (3)
Porous membranes	Alumina, zirconia, titania Carbon, Glass (4)	Separation of small molecules from larger molecules (6)	Low selectivity (2) Brittleness (5)	Commercial with pore sizes from 40 Å to 10μm (4); Lab-scale membranes with pores smaller than 10Å are available (5,7)
	Zeolites (e.g. alumina)	Selective removal of compounds depending on the pore size (7)	Brittle, low selectivity and permeability, difficult to produce defect-free zeolites (7)	Lab-scale tests of membranes (8)
Dense membranes	Palladium or Pd-alloys (5-7)	Permselective hydrogen removal and dosing (6)	Low permeability, brittleness, expensive, CO and sulphur affect permeability and catalytic properties (6)	Lab-scale tests of membranes (8)
	Vanadium, Zirconium, Tantalum, Niobium (5)	Permselective hydrogen removal and dosing (5) Less brittle and higher permeability than Pd (5)	High surface resistance to hydrogen transport (5)	Lab-scale tests (5); Prototype of base metal (V, Ti, Nb) support with Pd-coating (9,10).
Composite membranes	Thin layer of dense membrane on porous support (2)	Selective removal of hydrogen or oxygen (6) Stronger than dense membranes (6)	Dense material tends to break loose from the support at repetitive thermal cycles (5)	Commercial for Pd and Pd-alloy composite membranes of several μm, used for small-scale high-purity hydrogen production (7)

Notes: The literature on membranes is extensive. The following references can be used to find the given information: (1) [Drioli and Giorno, 1996], (2) [Mulder, 1996], (3) [Koros and Flemming, 1993], (4) [Soria, 1995], (5) [Armor, 1995], (6) [Zaman and Chakma, 1994], (7) [Saracco *et al.*, 1994], (8) [Veen, 1997], (9) [Edlund and Peterson, 1995], (10) [Bryan, 1996].

Porous inorganic membranes have a high permeability, but the selectivity is usually low [Zaman and Chakma, 1994]. Porous membranes can be made symmetric and asymmetric. In the last case the density of the membrane decrease from the surface to the inside. In 1995 about twenty companies worldwide offered inorganic porous membranes, with pore sizes in the range of 40 Å to 10 μm [Soria, 1995]. Zeolite membranes have pores of few Å. Unfortunately, pure zeolite membranes are very difficult to produce, selectivity and permeability are low, and they are fragile [Saracco *et al.*, 1994]. Incorporating the zeolite in a polymer matrix has been proposed, but this makes high-temperature application impossible.

Dense membranes have a high selectivity for only one specific gas, but the permeability is low. The high selectivity of palladium (Pd) for hydrogen enables its use in several commercially interesting reactions, like dehydrogenations and syngas production. However, as shown in Table 6.12, several disadvantages hamper possible application. Pd-alloys, e.g. with silver or yttrium, are less temperature sensitive and less susceptible to poisoning than pure palladium, but the permeability may be lower [Saracco *et al.*, 1994] [Armor, 1995]. Other metals, such as tantalum (T), vanadium (V), zirconium (Zr) and niobium (Nb) may be an alternative. It has been proposed to make a membrane of a support of one of the last materials coated with a thin layer of palladium [Bryan, 1996].

Composite membranes offer a solution to the low permeability of dense membranes, and may also reduce the costs. Several methods are under development to make a very thin (≪10μm), defect-free, layer of Pd-alloy on a (porous) support, that adheres to the surface on repeating process and temperature cycles [Armor, 1995].

Membrane reactors for syngas production

In this section we discuss the applicability of a membrane reactor for syngas production and the problems that may be expected. Furthermore, we present possible configurations of syngas plants incorporating a membrane and assess the potential for energy-efficiency improvement.

Membrane *steam reforming* has been extensively described, see for instance [Adris *et al.*, 1991; Alqahtany *et al.*, 1993; Lægsgaard Jørgensen *et al.*, 1995; Barbieri and Maio, 1997; Barbieri *et al.*, 1997; Sogge and Strøm, 1997; Strøm *et al.*, 1997].

The hydrogen production by steam reforming of natural gas is limited by chemical equilibrium. In a membrane steam reformer the same or even a higher methane conversion as in a conventional reformer can be attained at a lower temperature. The change in heat requirement for the steam reforming reaction with temperature is very small, assuming the same conversion. As a consequence, any reduction in fuel demand by lowering the temperature of steam reforming stems from a reduction in energy demand for preheating fuel and feed. The overall effect on the SEC depends on the efficiency of heat recovery. An advantage of a lower temperature is that the thermal stress on the reformer tubes is reduced, increasing their lifetime. Furthermore, when the permeate is pure hydrogen, gas purification is no longer required. However, both lowering the temperature and selective removal of hydrogen enhance the risk of coke formation, which may de-activate the catalyst. This can partially be avoided by increasing the steam to carbon ratio.

Both porous and dense membranes can be used in a membrane steam reformer. Porous membranes have a high permeability, but show a low selectivity. Gas purification is still required. Dense membranes, on the other hand, have a higher selectivity for hydrogen, but a lower permeability. Porous membranes are further developed than dense membranes. Composite membranes may be a good alternative, provided that the permeability is high enough and the membrane layers do not separate upon repeated heating.
Several configurations of membrane steam reforming can be considered. Figure 6.7 gives three examples of membrane reactors where the reactor and the membrane are coupled. Figure 6.7a shows a membrane steam reformer that is natural-gas fired, like in the conventional primary reformer. Figure 6.7b and c show configurations that use a gas-heated reformer, resembling the KRES syngas plant. An important design aspect is the type of membrane. A dense membrane would produce a hydrogen permeate of high purity. The retentate can be sent to a shift reactor for additional hydrogen production (Figure 6.7b) or be used as a fuel (Figure 6.7c). The plant of Figure 6.7.b has a high methane to syngas conversion, but the retentate requires purification, which increases the investment and the O&M-costs of the plant. In plant of Figure 6.7c the retentate can first be expanded to drive the permeate compressor, but the methane to syngas conversion will be low. Strøm *et al.*[Strøm *et al.*, 1997] analysed the specific energy consumption of the membrane steam reformers illustrated in Figure 6.7a and 6.7b. The temperature of steam reforming is about 600°C and the pressure 40 bar. The SEC is determined to be 28.7 GJ/tonne NH_3 for the configuration of fig. 6.7a. The SEC of the plant in fig.6.7b varies from 26.4 to 26.9 GJ/tonne NH_3, depending on oxygen content of the air to the secondary reformer and the nitrogen surplus in the syngas. The heat demand of the steam reforming reaction can only be met when oxygen-enriched air is used. The power demand for the oxygen plant is included in the calculation of the SEC. The difference in SECs between the two configuration is mainly due to losses caused by underfiring. No data are available for the SEC of the plant presented in fig. 6.7c.
Strøm *et al.* did not study a conventional steam reformer, but we can compare the data with the SEC of the most efficient ammonia plant of 27.2 GJ/tonne NH_3. The SEC of the fired membrane reactor is higher than this value. The SEC of the gas-heated membrane reactor is about 0.5-1 GJ/tonne NH_3 lower than that of the current most efficient ammonia plant. It can be determined that heating the reformer feed to 600°C instead of 900°C results in a reduction of the energy requirement of about 1.5 GJ/tonne NH_3 (1.8 GJ/tonne N). Because part of this heat can be recovered, the outcome of the study of Strøm *et al.* seems reasonable. It should be noted that the reduction of the SEC diminishes when more heat is recovered in the conventional syngas production.

The investment costs of a membrane steam reformer are determined by the costs of the membrane itself and the costs of the installations. One expert estimated the costs for a Pd-membrane on ceramic support at 5,000 US$/m^2, and the surface area required for a membrane steam reformer for a 1,000 tpd ammonia plant at up to 1000 m^2 [Veen, 1997]. Thus, total membrane costs for such a plant would be 5 million US$. For comparison, the investments for a conventional 1,000 tpd ammonia plant are about 100 million US$(1986)

[Chauvel and Lefebre, 1989]. The preparation of syngas accounts for 61% of these costs, hence 60 million US$. The costs of the membranes are less than 10% of these costs. However, since it is probably necessary to replace the membranes several times during the lifetime of the reactor, the costs for the membranes over the years will be larger. Whether the costs of the installations are lower than of a conventional steam reforming plant depends largely on the selectivity of the membrane. When the syngas purification can be eliminated, the application of membranes will probably result in lower investment costs. When syngas purification is still required to some extent, it cannot be said beforehand how the investment costs of a new process will relate to the conventional process.

Membrane reactors for syngas production are still in the stage of applied research. Many barriers regarding material science, catalysis and chemical engineering have to be overcome before they will be commercial [Saracco *et al.*, 1994].

Although membranes are already used in the industry, and the syngas reactions do not change by introducing a membrane reactor, the idea of avoiding the limits posed by chemical equilibrium has not been applied in the industry so far. In principle, the membrane reactor can be incorporated in an existing plant, although considerable adaptations are required. Process parameters like temperature and pressure change. The effects on the quality and the nature of the product and on the purchasing and supplying industries are small. To summarize, we can say that changing from the state-of-the-art technology for syngas production (e.g. KRES) to membrane steam reforming requires an evolutionary to major technical change.

Supplying the heat for steam reforming by high-temperature CHP

The exergy analysis of the KRES-plant revealed that 1.2 GJ/tonne N of exergy is lost in the burner due to the combustion of natural gas to generate heat of about 1000°C. The chemical exergy of natural gas will be used more efficiently when it is first burnt in a gas turbine that generates electricity, and when the hot combustion gases are used to provide the heat for the steam reforming reaction. A gas turbine that is well-suited for this application is Asea Brown Boveri's GT26, that combines a high electric efficiency (38%) with a high exhaust temperature of 608°C [Anonymous, 1993]. This temperature is sufficient to replace the preheater. To raise the temperature of the gas to 1000°C required for steam reforming, partial oxidation of natural gas in the autothermal reformer is still required though. When a gas-heated membrane steam reformer is developed that can be operated at 600°C, the exhaust gases of a gas turbine can be directly used to supply in the heat demand. In that case, an autothermal reformer is no longer necessary to generate a hot, partially reformed syngas. It should be noted that the capacity of the gas-heated steam reformer should be increased to balance the avoided reforming capacity of the autothermal reformer. Furthermore, nitrogen should be added to the system in another way than with the combustion air for partial oxidation, as is done in conventional steam reforming.

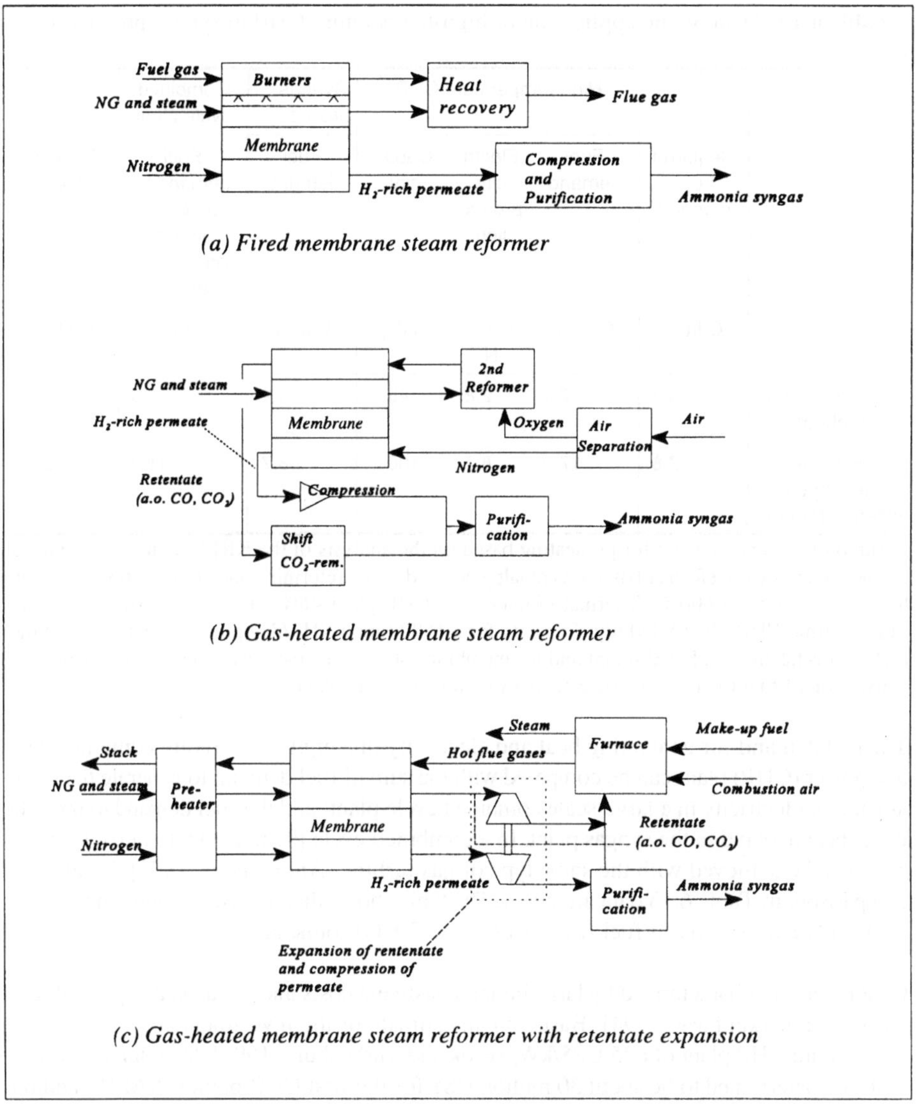

Figure 6.7: Three examples of simplified flow sheets for ammonia syngas plants incorporating a membrane steam reformer (a) fired membrane steam reformer; retentate purged after heat recovery [Strøm *et al.*, 1997]; (b) gas-heated membrane steam reformer; retentate shifted and CO_2 removed, permeate compressed; combined stream purified [Strøm *et al.*, 1997]; (c) gas-heated membrane steam reformer retentate expanded to drive permeate compressor; retentate used as fuel [Sogge and Strøm, 1997].

Table 6.13: Data on the application of high-temperature CHP in syngas production.

		High-temperature CHP			Syngas process	Combined cycle plant	
	Required heat demand	Fuel demand	Electricity production	Capacity	Fuel demand	Fuel demand to produce same amount of electricity	Overall fuel savings
	GJ/t N	GJ/t N	GJ_e/t N	MW_e	GJ/t N	GJ/t N	GJ/t N
Replacement of preheater	3.4	7.4	2.8	45	4.2	4.9	1.7
Gas-heated membrane steam reformer	7.8	17.0	8.7	100	8.7	11.2	2.8

Assumptions: Heat demand for preheating based on the analysis of the KRES-plant; Heat demand for membrane steam reformer based on enthalpy demand steam reforming and shift reaction including heating reactants to 600°C; Thermal efficiency of CHP-plant 46% and electric efficiency 38%; running time CHP-plant 8000 h/y; Capacity for a 1500 tonne NH_3 /day plant; Fuel demand of the preheater is taken from KRES-plant and for membrane steam reformer based on 90% efficiency for conversion of fuel to heat; electric efficiency combined cycle plant 58%.

The fuel demand for generating heat and electricity in a high-temperature combined heat and power (CHP) plant can be compared with the sum of fuel demand to generate the same amount of electricity in a large-scale combined cycle plant[1] and the fuel demand to provide in the heat demand in the syngas plant. In a combined cycle plant an electric efficiency of 58% can be achieved with the same type of gas turbine [Anonymous, 1993]. To allow a comparison, in Table 6.13 data are given. The table shows that the use of high-temperature CHP can result in overall fuel savings of 1.7 to 2.8 GJ/tonne N.

CHP plants are characterized by large initial investment costs and a reasonable profitability [Blok and Turkenburg, 1994]. Based on an estimate of the investment costs for a high-temperature CHP plant of 625 US$/$kW_e$ [Blok and Turkenburg, 1994], the total investment costs are determined to be about 30 million US$ for the first CHP-plant (45 MW_e) and 60 million US$ for the second plant (100 MW_e). For comparison, the investment costs for a plant for ammonia syngas production (1500 tonne NH_3/day) by gas-heated membrane steam reformer are estimated at 100 million US$ [Strøm *et al.*, 1997]. Blok determined the pay-back period of industrial CHP installations of 20 MW_e in the Netherlands at less than 5 years [Blok, 1993].

[1]A combined cycle plant consists of a gas turbine, a waste heat boiler to raise steam and a steam turbine.

CHP in general is a technology that is widely applied to provide in the steam demand of industries. Application of CHP to provide in the process heat demand is rather new. One example is the application of high-temperature CHP in a refinery in Denmark to supply in part of the heat demand [Bodewes and Lugten, 1992]. Direct coupling to a process requires a high reliability of the turbine. Back-up equipment may be required. Implementation of high-temperature CHP requires replacement of the furnace and can therefore only be done when a new plant is built. Retrofitting existing plants is not a big issue, since there are only a few gas-heated reformers at present and no membrane steam reformers at all. High-temperature CHP can be considered when new plants are built. However, since the ammonia plant is changed into a co-producer of ammonia and electricity, we consider that implementation of high-temperature CHP requires a major technical change compared to the gas heated reformer.

6.8.2 AMMONIA SYNTHESIS

The energy demand for the ammonia synthesis can be reduced by selective removal of the product ammonia in combination with synthesis at a low pressure. We discuss two options: membrane reactors and reactors using chemical or physical sorbents.

Membrane reactors in ammonia synthesis

Several polymeric materials that can separate NH_3 from N_2 and H_2 have been proposed and applied [Pez and Laciak, 1988]. Since all materials suffer from a low selectivity and a low permeability for ammonia and have a poor thermal stability, their applicability in the ammonia synthesis loop is limited. An alternative is to separate ammonia from nitrogen and hydrogen by means of capillary condensation in the pores of zeolites [Veen, 1997]. It has been know for years that alumina zeolites and carbon can be used as ammonia sorbents [Laciak and Pez, 1988]. Capillary action is used for separation of liquid mixtures by nanofiltration and microfiltration [Mulder, 1996]. Research on capillary condensation is being performed at the University of Bath, and is directed at finding the optimum pore size to achieve a good selectivity and permeation [Veen, 1997]. This research is conducted as part of a large European project to study the use of ceramic membranes for gas separation processes [Tennison, 1997]. The research is still in a very early stage of development and no applicable results are available.

The saving on energy demand by selective ammonia removal depends strongly on the selectivity and permeability of the membrane. A permselective membrane with a high permeability would allow that all nitrogen and hydrogen react and all ammonia is taken away immediately after it has been formed. The power demand for recycling and refrigeration for cooling out ammonia can be avoided. The power demand for syngas compression can be reduced by about 50%, when an advanced high-activity catalyst is used that allows a reduction of the reaction pressure from 100 to 50 bar [Appl, 1992b]. The power demand for the syngas compressor is 3 GJ/tonne N in a modern low-efficiency

ammonia plant [Appl, 1992b]. The reduction of the energy consumption is estimated at 3.5 to 4.5 GJ/tonne N. This energy becomes available as heat.
However, a membrane that is not permselective for ammonia and has a permeability that is too low to have all ammonia removed as soon as it has been formed results in a permeate that is not pure ammonia. The retentate still contains ammonia and hydrogen. As a consequence, ammonia has to be cooled out and hydrogen has to be recycled. These operations demand additional energy compared to the ideal case, thus the energy savings reduce.

The development of membrane reactors for application in the ammonia synthesis is in the stage of applied research. The use of capillary condensation to separate gas mixtures has not been demonstrated so far, although the idea of capillary action is used to separate liquids from mixtures. Although the configuration of the ammonia synthesis plant would change drastically, the layout of the other operations can remain the same. The steam system requires considerable adaptation. Implementation of a membrane reactor for selective removal of ammonia would require an evolutionary to major technical change compared to the current technology of nitrogen fixation.

Ammonia synthesis using sorbents for ammonia removal
Reactors based on chemical or physical removal of ammonia using sorbents have been proposed. The use of polymeric materials as ammonia sorbents has been described by Laciak *et al.* [Laciak and Pez, 1988]. Westerterp proposed two types of reactors that make use of solid or liquid sorbents to remove methanol from the methanol synthesis reactor [Westerterp and Kuczynski, 1987; Westerterp *et al.*, 1988; Westerterp *et al.*, 1989; Westerterp, 1993]. Westerterp's concept with solid sorbents encounters major technical problems with the handling and pumping of the solids. Handling and pumping of liquid are well-known operations, thus technical problems are expected to be smaller with a selective, liquid absorbent. The proposed configuration consists of a series of synthesis beds with interstage product removal by liquid absorbers. About 95% of the syngas is converted to product. The remaining small gas stream can be recycled or used for other purposes. The liquid absorbent stream has to be pumped around, but the power demand is much smaller than for recycling a gas stream.
The reduction in SEC that can be achieved by this reactor is dependent on the energy use for recycling of the reference process. For modern low-pressure processes this energy demand is 2-3 GJ/tonne N in the form of HP-steam. Westerterp assumes that the novel reactor operates at the same pressure, and that the recycle power can be reduced by a factor 2.5. Ammonia refrigeration is not necessary, but some energy is required for desorbing ammonia [Westerterp *et al.*, 1988]. Based on these assumptions, the reduction of SEC can be estimated to be 1-1.5 GJ/tonne N. This energy becomes available as high-pressure steam (exergy value 0.4-0.6 GJ/tonne N).
Research on ammonia sorbent materials takes place in many laboratories worldwide [Laciak and Pez, 1988; Pez and Laciak, 1988]. An example is the University of Twente, where the concept was tested successfully at laboratory scale [Westerterp, 1993]. However,

all R&D at the University of Twente on this type of reactors stopped in 1993, and has not been restarted so far [Westerterp, 1993]. These novel reactors can be seen as an evolutionary to major change in technical development compared to the current technology. The same arguments hold as for the membrane reactor.

6.8.3 Nitric acid production

The exergy analysis of the nitric acid plant pointed out that the irreversible combustion of ammonia to nitric oxide has a very low exergetic efficiency. The chemical exergy of ammonia is converted for a large part to physical exergy in the form of heat. This heat is not required in the process and exported as steam. We will evaluate two option to reduce this exergy loss. First, we assess the possibility of combusting ammonia in a combined cycle to produce electricity and nitric acid. Second, we assess the conversion of ammonia in a solid oxide fuel cell as a method to convert the chemical exergy of ammonia into nitric oxides and electricity.
Nitric oxides can also be produced by the direct oxidation of atmospheric oxygen at high temperatures. The high-temperatures can be achieved by combustion processes and by plasmas. We will investigate both options further.

Combustion of ammonia in a combined cycle plant
The idea is to combust ammonia at high pressure and expand the hot combustion gases in a gas turbine. Steam is raised in a waste heat boiler and expanded in a condensing steam turbine. Such a system, known as a combined cycle, is a well-accepted technology for the combustion of natural gas to provide electricity. In this case the products are electricity, but also nitric acid, produced by absorption of the cooled combustion gas in water. Since we have found no references for this option, we assess its feasibility on the basis of our own expertise.

Combustion of ammonia to produce nitric oxide makes different demands on the design of the gas turbine than combustion of natural gas does. We discuss three differences in some detail.
First, the combustion of natural gas to carbon dioxide and water in a gas turbine is the thermodynamically favoured reaction, whereas the combustion of ammonia should result in nitric oxide and water, instead of the thermodynamically favoured nitrogen and water. The combustion reaction should take place over a platinum catalyst to avoid nitrogen formation. The combustion chambers used for natural gas combustion are not directly suited for this catalytic reaction. However, this problem might be solved by developing a specially designed combustion chamber.
Second, typical operation conditions for a high-pressure ammonia conversion are 1.0 MPa and 940°C, [Clarke and Mazzafro, 1993], whereas that for natural gas combustion in a modern gas turbine are 1200°C and 1-3 MPa. Applying a higher pressure and temperature for the ammonia conversion would lead to a lower nitric oxide production. Modern gas turbines are therefore not well-suited for this purpose. However, about a decade ago gas

turbines that were generally available operated with a turbine inlet temperature of 900-1000°C and a pressure ratio of about 1 MPa. Such gas turbines may be suitable for the conversion of ammonia to nitric oxide [Krachtwerktuigen, 1986]. The electrical efficiency of these gas turbines is on the order of 25-30% (LHV), compared to 35-40% of a modern gas turbine. When such an older gas turbine is used in a combined cycle the electric efficiency can increase to about 42%.

The third difference is that to avoid decomposition of nitric oxide into nitrogen and oxygen, the combustion gases of ammonia have to be cooled very rapidly to about 150°C, whereas this is not required for the combustion gases of natural gas. Nevertheless, the expansion in a gas turbine is performed very rapidly to achieve a high efficiency. Gases leave the turbine with a velocity of 50 m^3/s and more [Krachtwerktuigen, 1986]. Assuming a typical size of a turbine (diameter 1 meter, length 1 meter), the time required for a gas molecule to pass through the turbine can be estimated to be 0.02 s. This time is short enough to ensure that only a small percentage of the nitric oxide decomposes [Gilbert and Daniels, 1948].

Apart from these differences, there are several other problems associated with combustion of ammonia in a gas turbine that will not be dealt with, for instance the risk of explosion of ammonia in air, and the formation of the aggressive nitric acid.

Assuming that all problems can be solved and an older gas turbine (electrical efficiency in combined cycle operation 42%) can be made suitable for catalytic ammonia combustion, the electricity production can be determined to be about 10 GJ_e/tonne N. In the most efficient contemporary nitric acid process about 10 GJ/tonne N of medium pressure steam is generated. Since the exergy of this steam equals 3.7 GJ/tonne N[1], the production of electricity is preferred from an exergy-point-of-view. The electric capacity of a combined cycle plant for a nitric acid plant with a typical capacity of 100 tpd can be determined to be about 2.5 MW_e.

The conversion of ammonia to nitric oxide and electricity in a gas turbine has not reached the stage of applied research yet. To solve the technical problems much R&D is required. The development of a new gas turbine is very expensive. However, using an older type of gas turbine might reduce the development costs considerably. Adaptations to the combustion chamber and further research to assess its practical feasibility are required.

This option is comparable with applying CHP in syngas production (section 8.1.5). Based on the same arguments as used there, especially the fact that a new plant is a co-producer of a chemical and of electricity, this option requires a major technical change compared to conventional nitric acid formation.

Conversion of ammonia in a solid oxide fuel cell

Solid oxide fuel cells (SOFC) consist basically of two electrodes and a solid electrolyte that can conduct an electric current and transport oxygen anions. The fuel reacts at the anode with oxygen anions; when ammonia is used as fuel the reaction products are NO_x and H_2O. At the cathode, oxygen is converted to oxygen anions. Temperatures in the range

[1] The quality of an energy carrier is the ratio between exergy and enthalpy. The quality of saturated steam of 40 bar is 0.37. Electricity has a quality of 1 by definition.

of 800-1000°C are required to achieve a sufficient oxygen anion transport. Solid oxide fuel cells offer the possibility to recover the reaction energy as electrical energy and to control the reaction at the anode by dosing the amount of oxygen at the reaction surface [Appleby and Foulkes, 1989].

Vayenas and Farr [Vayenas and Farr, 1980] reported that NO is the primary product of a solid oxide fuel cell with platinum electrodes and ammonia as a fuel. High NO selectivity could only be obtained at long residence times and low power output. However, at divergent process conditions up to 95% of the ammonia is converted to nitrogen. Since this research, performed at the end of the 1970's, the technique to make solid oxide membranes has been improved considerably. Notwithstanding the fact that coproduction of the commercially important nitric acid and electricity seems promising, we have found no more recent report of experiments on a solid oxide fuel cell operated on ammonia.

The $\Delta G°$ of the oxidation of NH_3 to NO is -355 kJ/mole NH_3 (see reaction 7). The maximum electricity production is therefore 25 GJ_e /tonne N. Due to electrochemical and material losses this maximum production cannot be achieved in practice. On top of this, energy is required to maintain the high temperature. Efficiencies of SOFCs operating with natural gas as fuel are 60% [Appleby, 1996]. When we assume that this efficiency can also be achieved with ammonia as fuel, the maximum practical electricity production is 15 GJ/tonne N. This is 50% morer than the enthalpy of the steam produced in current nitric acid processes (10 GJ/tonne N). When we compare the energy production on exergy basis, the fuel cell plant could produce 15 GJ/tonne N, compared with 3.7 GJ/tonne N of a current plant. The total electric capacity of a fuel cell for a nitric acid plant with a typical capacity of 100 tpd can be determined to be about 4 MW_e.

Large-scale solid oxide fuel cells are not commercially available at the moment, although many stacks in the kW class have been operated [Appleby, 1996]. A large diversity of SOFC technology is being developed worldwide [Appleby, 1996]. Two major developers that pursue commercialization of SOFC are Westinghouse and Ztek (both USA) [Moore, 1997]. Both companies expect to have the first SOFC commercial at the beginning of the next century [Moore, 1997]. The research activity on conversion of ammonia in a solid oxide fuel cell appears to be very low. It is not known if any research is ongoing at the moment. The R&D has still not reached the stage of applied research.

Long-term projections for the investment costs for SOFC power plants vary from 600 to 1200 US$/kW [Appleby, 1996]. When we use this cost range the investment costs for the SOFC in a 100 tpd nitric acid plant can be determined to be 2.5 to 5 million US$. However, to achieve these costs, the SOFC modules should be produced at considerably lower costs than at present. Improved and cheaper materials should be developed for the electrolyte and electrodes. Furthermore, the chemical engineering system of a SOFC should be made simpler [Appleby, 1996].

The production of nitric acid in a fuel cell deviates only from the conventional production route in that the combustion chamber is replaced by a fuel cell. However, since the integration of a fuel cell in an industrial process has never been demonstrated, we can see this development as a major change compared to the current technology.

Direct oxidation of atmospheric nitrogen at high temperatures
The direct oxidation of nitrogen to form nitric oxides has attracted the interest of many researchers during this century. In spite of all research activities, a process that could compete with the Haber-Bosch process has never been developed so far. In this section we will first review the processes that were once commercial. Then, we will evaluate whether these processes can improve the energy efficiency of nitrogen fixation when contemporary technology is used.

Nitric oxide is formed from nitrogen and oxygen either at high temperatures or in plasmas according to

$$N_2 + O_2 \rightleftarrows 2\,NO \quad (\Delta G° = 86.6 \text{ kJ/mole NO or } 6.2 \text{ GJ/tonne N}) \tag{12}$$

Calculation of the chemical equilibrium of air at 1 atm reveals that at a temperature of 2500°C the concentration of NO in equilibrium is only 2%, while at 1000°C this concentration is almost zero. Both the rate of reaction of the formation and deformation of NO increase with temperature. To avoid decomposition of NO at high temperatures the equilibrium concentration should be fixed by very rapid cooling of the gas mixture. The NO will then react to NO_2 according to

$$2\,NO + O_2 \rightleftarrows 2\,NO_2 \quad (\Delta G° = -34.8 \text{ kJ/mole NO or } -2.5 \text{ GJ/tonne N}) \tag{13}$$

The deliberate production of nitric oxides at high temperatures was a commercial process in the 1950's. A 40 tonne/day (100% HNO3) gas-fired process for the production of nitric acid was in operation in the USA for several years [Gilbert and Daniels, 1948; Foster and Daniels, 1951; Ermenc, 1956b; Ermenc, 1956a]. The Wisconsin process, named after the University where it was developed, used two beds filled with magnesium oxide pebbles. Air is preheated by passing through the first pebble bed, mixed with fuel gas, and then passed through the second pebble bed, which removes most of the heat. The maximum air temperature is about 2100°C. The flow of air is then reversed and the second bed becomes the preheating bed and the first bed is used to conserve the heat from the exit gases. Cooling in the pebble bed is very rapid. Nitric oxide concentrations higher than 1% have been obtained [Gilbert and Daniels, 1948]. Recovery of nitrogen oxides is done by adsorption on silica gel [Foster and Daniels, 1951]. The energy demand for the process was 4 GJ_e electricity and 19 GJ fuel per tonne of 100% HNO_3 (18 GJ_e and 85 GJ respectively per tonne N) [Ermenc, 1956b]. This high energy demand can be explained by the large volumes of air that have to be heated, heat losses through walls and from pebbles, and the energy demanded for adsorption and recovery of the produced NO.
Although the process performed technically according to expectation, the economics of the process were unfavourable in competition with the Haber-Bosch process and ammonia oxidation. Especially the NO_x recovery unit was expensive, making up 60% of the investment costs [Ermenc, 1956b]. Another main problem with this process was the life

time of the furnace materials. The material problem, the recovery of nitric oxides and the development of the Haber-Bosch process hampered further development.
Can the problems of this process be solved by current technologies? Since the 1950's refractory materials have been improved. Contemporary materials have a longer life time at high temperatures. Although higher temperatures may be applied, the concentration of NO will never exceed 2%, which is too low for economic recovery in an absorption tower. Other recovery methods, e.g. molecular sieves, might be used, although their technical and economical feasibility for this process has not been proven. Better heat recovery techniques can reduce the energy demand. However, no methods that can recover heat at 2100°C and simultaneously achieve a very rapid cooling are available. We can conclude that current technology cannot directly solve the problems of the Wisconsin-process. R&D is required to many aspects.

A second option to produce NO directly from the elements is by using *plasmas*. A plasma is a state of matter in which a significant number of the atoms or molecules are electrically charged or ionized. The behaviour of plasmas differs from that of liquids, solids or gases. Due to the high reactivity plasma chemistry offers new possibilities for chemical reactions. Plasmas can be produced by e.g. the release of chemical energy in flames, an electrical discharge, and by lasers [Smith, 1996]. Nitric oxide that is formed by lightning is based on plasma chemistry. Electric arc furnaces that were in operation at the beginning of this century used in fact plasmas to fix atmospheric nitrogen. The gas temperature in these electric arc furnaces was about 3000°C. The NO concentration in the product gas mixture was about 2%. The power consumption was in the order of 225 GJ_e/tonne N [Honti, 1976]. The reaction of nitrogen and oxygen in plasma can produce significant amounts of nitric oxide in the gas phase. Experiments performed in the 1960's and 1970's in electric discharge plasma reactors showed that NO-concentrations of 7-15% could be obtained [Timmins and Ammann, 1967; Venugopolan and Veprek, 1983]. There are many factors that determine the NO-concentration, like the electric current density and the pressure [Venugopolan and Veprek, 1983]. The reactor design may also influence the concentration, for instance by interaction of molecules with the reactor wall [Timmins and Ammann, 1967].
The SEC of NO formed in a plasma depends on many factors, like the geometry of the plasma reactor, the pressure, the electric current density, and the gas flow rate [Timmins and Ammann, 1967]. On the basis of several accounts in the literature on the energy use of nitric oxide formation in plasmas, we can estimate the SEC to be 180-300 GJ-electricity/tonne N. [Timmins and Ammann, 1967; Venugopolan and Veprek, 1983; Dudnikov *et al.*, 1988]. The SEC may be reduced when high-temperature heat can be recovered. Nevertheless, considering the large difference, it is highly unlikely that the SEC of HNO_3-production with plasma chemistry will ever come down to that of the contemporary production methods.

In conclusion we can state that the direct oxidation of N_2 and O_2 to NO at high temperatures or by plasma chemistry is technically possible, but can only be achieved at the costs of a high energy use. It seems justifiable to state that direct oxidation of NO neither at high temperatures nor by plasma chemistry offer an opportunity to improve the energy efficiency of nitrogen fixation.

Unintentional production of nitric oxides at high temperatures occurs during combustion processes. In section 4 we have shown that annually 20 million tonnes of nitrogen is fixed in this way, compared to 90 million tonne by the Haber-Bosch process. Considerable efforts are undertaken to reduce this amount, because nitric oxides contribute to acidification. The question may be asked whether it would be worthwhile to recover the nitrogen oxides and use them for nitric acid or ammonium nitrate production. A problem is that the sources are very diffuse and most sources emit small amounts of NO_x. By far the largest source of NO_x-emission is traffic, which is too diffuse to allow economic recovery of NOx. [RIVM, 1996]. Nowadays, NO_x produced in car engines is catalytically converted to N_2 in most industrialized countries.
The emission of NO_x of large-scale power plants is large enough to consider selective catalytic reduction of NO_x to N_2. The unabated NO_x-concentrations in combustion gases from power plants are too low (0.1% and lower) to allow economic recovery by the conventional absorption process that is being applied in the production of nitric acid. Another method should be developed. It can be estimated that the NOx production in a 600 MW_e coal-fired power plant is only about 10 ktonne N per year (assuming an unabated NO_x emission of 450 g/GJ_e). We can make a rough estimate that about 30% of the nitrogen that is fixed globally in combustion processes is fixed in large-scale electricity plants [OECD, 1997]. Based on this estimate, we can conclude that up to 2% of the global nitrogen fertilizer production could be avoided in principle by this option.

6.8.4 ALTERNATIVE ROUTES FOR ATMOSPHERIC NITROGEN FIXATION

Nature demonstrates that it is possible to fix atmospheric nitrogen at environmental conditions. Some plants are known to be able to fix atmospheric nitrogen at environmental temperature and atmospheric pressure. This so-called biological nitrogen fixation has been topic of extensive research for many decades. Nevertheless, the mechanisms underlying biological nitrogen fixation have still not been elucidated completely. In section 8.4.1 we will discuss briefly how biological nitrogen fixation can contribute to a reduction of the energy demand of nitrogen fertilisation. In section 8.4.2 we will discuss the developments in the chemistry of low-temperature nitrogen fixation.

Biological nitrogen fixation
Annually about 140 million tonne of atmospheric nitrogen is fixed by naturally occurring microorganisms, see Figure 6.4. Biological nitrogen fixation used to be the main way for nitrogen supply to plants before the rise of intensive agriculture, and still is for plants living on unfertilized lands. There are at least three ways biological nitrogen fixation can be used to decrease the demand for fossil fuels for the production of fertilizers:

1. Applying symbiotic systems in nature. Besides the symbiosis between *Rhizobium* and leguminous plants, other natural symbiotic associations between microorganisms and crops are worthwhile investigating [Beringer *et al.*, 1982]. For instance, the symbioses involving the blue-green algae and the symbioses between cyanobacteria and the water fern *Azolla*, which might be utilized to supply nitrogen to rice pads [Beringer *et al.*, 1982]. Furthermore, bacteria can be introduced into the soil additional to the natural occurrence. *Rhizobium* is already commercially available [Dickman, 1987; Leigh, 1991]. These bacteria are still natural, thus not genetically engineered.
2. Genetic engineering of cash-crops to enable symbioses. Manipulation of plants in such a way that they can undergo a symbiotic association with natural nitrogen-fixing bacteria seems very promising [Wordragen, 1996]. There is evidence that the processes that initialize the symbioses between *Rhizobium* and legumes are general to all plants [Wordragen, 1996]. The plant should be adapted so that it can recognize the *Rhizobium* bacteria and starts to divide its root cells instead of destroy the bacteria. Research is ongoing at the International Rice Research Institute in Manilla, in cooperation with Wageningen Agriculture University (NED) to investigate this option for rice [Wordragen, 1996]. This option is in the stage of applied research.
3. Genetic engineering of cash-crops to enable nitrogen-fixation. The idea is to manipulate cash-crops, like corn and wheat, in such a way that they can fix nitrogen themselves. This is a complicated route. It means the transfer of all genes involved in the process of nitrogen fixating from, for instance, *Rhizobium*, to a completely different organism. It is still a question whether this is possible. In addition, the environment of the plant should be such that nitrogenase, the enzyme that converts nitrogen into ammonia, can survive. Nitrogenase is extremely sensitive to oxygen [Postgate, 1987]. Naturally occurring nitrogen-fixing species have evolved a range of mechanisms to protect the enzyme against damage caused by oxygen. Considering this, and the fact that the mechanism of biological nitrogen fixation is not completely clear, it can be concluded that much R&D is required will this option ever become commercialized.

In Section 6.5 we determined the energy requirement for biological nitrogen fixation at 35-70 GJ/tonne N in the form of glucose. This glucose, that is provided by the plant, cannot be used anymore to supply in the energy requirement for the growth of the plant. When chemical nitrogen fertilization is applied, the plant does not need to provide this amount of energy for nitrogen fixation. However, this energy is not necessarily available for growth. Plants take up nitrogen from soil moisture predominantly in the form of nitrate. Converting nitrate to ammonium requires an amount of energy that is comparable with that of nitrogen fixation [Leigh, 1991]. Whether the highest yield can be obtained by nitrogen fertilizers or by nitrogen fixation can probably only be determined by field tests.

Although there is some experience with genetic engineering of food products, e.g. soy beans, it will probably take several decades before enhanced biological nitrogen fixation can be applied to major crops [Wordragen, 1996]. The influence on the yield of biological nitrogen fixation requires serious investigation. If this technology can be successfully developed, it can induce a reduction of the fossil energy use for nitrogen fertilization. However, much R&D will be required to attain this. Although biological nitrogen fixation is not new, applying this for nitrogen fertilization at a large scale has not been used before. Since this option does not need chemical nitrogen fixation, the effect on the fertilizers producing industry is considerable. Enhanced biological nitrogen fixation would mean a radical change compared to the current chemical nitrogen fixation.

Chemical nitrogen fixation at low temperatures

Since the 1960s it has been known that nitrogenase, the enzyme that catalyses the fixation of nitrogen in some living organisms, is a complex that consists of an organic structure with one or more metal atoms. Chemists have searched for a synthetic organic-metal complex that can split nitrogen at a low temperature ever since. The major problem was how one or two metal ions could supply the six electrons required to convert N_2 to $2N^{3-}$. Many complexes were proposed. However, in the first complexes that were synthesized dinitrogen was bound to a metal as a ligand, thus was not split. The bond length between the two nitrogen atoms was only slightly larger than that in free dinitrogen[1]. The bond length increased gradually with the progress of scientific development. It was only in 1995 that a organo-metal complex was found that could actually split dinitrogen at low temperatures (-35 to +30°C) and atmospheric pressure [Laplaza and Cummins, 1995]. This achievement is of no practical use as long as the nitrogen ions, that are bounded or bridged with an metal ion, cannot be liberated without producing dinitrogen. Protonation of dinitrogen bounded in an organo-metal complex to produce ammonia and hydrazine has been demonstrated by Shan *et al.* [Shan *et al.*, 1997]. Furthermore, the process should be cyclic, in other words the starting material, the organo-metal complex, has to be recycled. The latter can probably be achieved in an electrochemical way. The process described by Shan *et al.* was not cyclic. The organo-metal complexes were converted to other complexes.

The SEC of a process that fixes nitrogen at low temperature is hard to estimate. We know that at least six electrons are required to split the triple bond in dinitrogen. However, as long as the exact chemistry of a cyclic system is not known, we cannot determine the electrochemical potential. This is required to calculate the theoretical energy consumption for the reactions. Over and above this, some energy is required to drive the operations, e.g. pumps. We can use the energy demand for biological nitrogen fixation as a reference, which we estimated at 13 to 26 GJ/tonne N in the form of ATP (see section 6.5.4). We use ATP as reference and not glucose, because ATP is the electron donor. However, when an electrochemical approach is needed to make the system cyclic account should be made for the conversion of primary energy to electricity. Using a future electricity generation efficiency of 60%, the energy demand can be estimated at 22 to 43 GJ/tonne N. The lower bound of this range is lower than the SEC of contemporary ammonia production. However, since this type of processes has not been demonstrated on laboratory scale, no data are available to make a well-founded comparison.

[1] In this section we use dinitrogen for the element nitrogen (N_2) in order to distinguish it from the atom nitrogen (N).

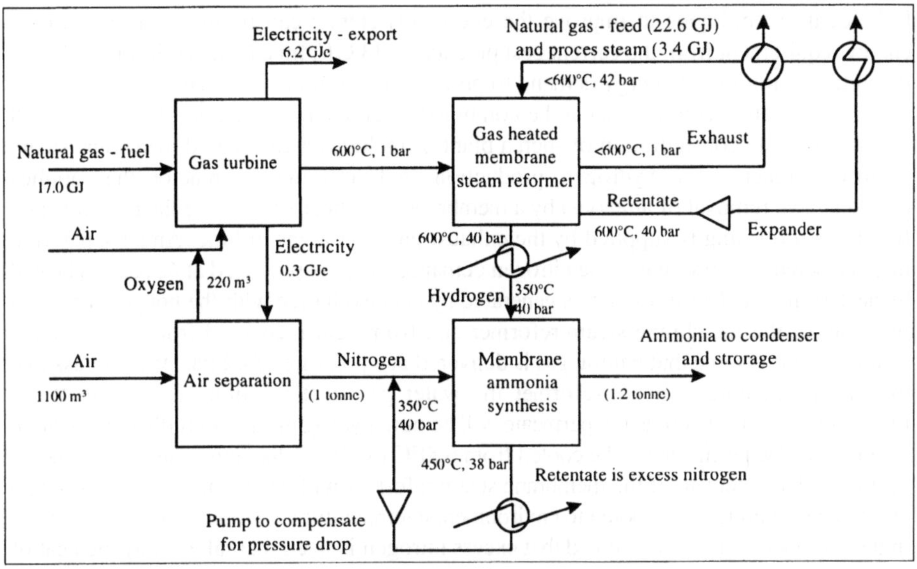

Figure 6.8: Flow sheet of a concept future ammonia plant.

Research on this topic takes place in many laboratories worldwide (see e.g. [Laplaza and Cummins, 1995; Fryzuk *et al.*, 1997; Leigh, 1997; Shan *et al.*, 1997]). Although serious progress has been made over the past 30 years, the research is still in the stage of fundamental research. Besides the need for more scientific understanding, there are several major hurdles that have to be overcome. For instance, the complex should not disaggregate upon chemical attack by hydrogen. It should also show a low air- and water sensitivity. Furthermore, the reaction of the bounded N to a useful product needs further research. Finally, the process should be made cyclic.

Most research to the low-temperature splitting of nitrogen is done as fundamental research, that is out of scientific curiosity without any direct application in mind. When one succeeds in developing a process for fixing atmospheric nitrogen at low temperatures, it would mean a completely different route of nitrogen fixation. Such a process would therefore be a radical change compared to the current technology.

6.8.5 Comparison of the options to improve the energy-efficiency of nitrogen fertilisation

In Table 6.14 we summarize the findings of this Section with regard to the energy consumption of the options. For comparison, the SECs of the currently most energy-efficient plants are given.

Individual technologies to improve the energy-efficiency can bring down the primary energy requirement from the current best practice of 33 GJ/tonne N to 30 GJ/tonne N. The theoretically minimum energy demand to produce ammonia is 24.1 GJ/tonne N.
Several separate technologies can be combined to an energy-efficient plant. Figure 6.8 gives a possible configuration of such a plant. It includes a gas-heated membrane steam reforming reactor for hydrogen production and a ammonia reactor that applies permselective removal of ammonia by a membrane. The heat for heating the reactants and for steam reforming is supplied by the exhaust gases of a gas turbine. An air separation plant is included to provide in the nitrogen demand, which is produced at 42 bar and heated by heat exchange. Process steam is generated by heat exchange with the hot permeate and retentate of the membrane steam reformer and from heat recovered from the ammonia synthesis. We assume that natural gas is delivered at a pressure of 42 bar. We also assume that in the membrane steam reformer the water gas shift reaction takes place. If the membrane is permselective, the permeate will be pure hydrogen, and no further purification is required. The permeate can be cooled from 600°C to 350°C, the temperature of ammonia synthesis. The retentate of the membrane steam reformer will be carbon dioxide and steam and can be expanded and cooled to environmental temperature. Carbon dioxide can be used in the urea synthesis. It is assumed that excess nitrogen is used that takes away the heat of reaction; after heat exchange the nitrogen is recycled. A small recycle pump is required to overcome the pressure drop. The pressure of both the steam reformer and the ammonia synthesis will be 42 bar. The heat generated in the plant is sufficient for heating the reactants and raising process steam. The heat of reaction of the ammonia synthesis is 3.3 GJ/tonne N, about equal to the process steam demand.

This concept of a future ammonia plant would have a natural gas demand of 39.6 GJ/tonne N and an electricity production of 6.2 GJ/tonne N. To produce the same amount of electricity in a combined cycle plant that uses the same type of gas turbine (see Section 8.1.3) with an electric efficiency of 58% would require 10.7 GJ natural gas. The amount of natural gas accounted to ammonia production becomes then 28.9 GJ/tonne N (23.8 GJ/tonne NH_3).
This concept shows the possibilities of achieving synergy when different technologies are applied. Using the heat of the exhaust of a gas turbine is only possible because the membrane reactor makes it possible to perform the steam reforming at a lower temperature.

The primary energy requirement to make nitric acid, including the natural gas requirement to produce ammonia, can be brought down from the current 26.8 GJ/tonne N to 6.4 GJ/tonne N when nitric acid is produced in a solid oxide fuel cell. A combination of the concept future ammonia plant and nitric acid production by ammonia conversion in a solid oxide fuel cell would result in a primary energy demand of 2.0 GJ/tonne N.

Table 6.14: Energy consumption of reference processes and of alternatives routes for nitrogen fixation.

	Natural gas	Electri-city	Heat	Total final energy	Total primary energy [a]
			(GJ/tonne N)		
Ammonia production - Reference plant	33.0			33.0	33.0
Membrane reactor for steam reforming	32-32.5			32-32.5	32-32.5
Membrane reactor for selective removal of ammonia	33.0		-3.5 to -4.5	28.5-29.5	29.8-30.5
Selective removal of ammonia using absorbents	33.0		-1 to -1.5	31.5-32	31.9-32.3
High temperature cogeneration - preheating	36.2	-2.8		33.4	31.3
Concept future ammonia plant					
High temperature cogeneration - membrane reactor for syngas and ammonia synthesis [b]	39.6	-6.2		33.4	28.6
Nitric acid production - reference plant	33.0	0.1	-9.0	24.1	26.8
Nitric oxide production in gas turbine	33.0	-10.0		23.0	15.3
Nitric oxide production in SOFC	33.0	-15.0		18.0	6.4
Fired high-temperature process	86	18		104	118
Plasma process		180-300		180-300	320-530
Alternative routes for nitrogen fixation					
Biological nitrogen fixation [c]					
Chemical low-temperature nitrogen fixation		13-26		13-26	23-46

[a] The conversion of electricity and heat to primary energy is based on an industrial CHP-plant running with an electric efficiency of 38% and a heat efficiency of 46%. These are the efficiencies of the gas turbine used for the high-temperature CHP for the steam reforming reaction (See Section 8.1.3). The exergy content is used to allocate the fuel input. Electricity has an exergy factor of 1 by definition, and the steam that is produced has an exergy factor of 0.4. To one unit of electricity, 1.77 units of fuel are allocated, and to one unit of heat, 0.71 units of fuel. For comparison, a modern

combined cycle power plant has an electric efficiency of 55-60%, which means that 1.82-1.67 units of fuel are required to produce one unit of electricity.

[b] The layout of the future plant is discussed in the text.

[c] The energy demand for biological nitrogen fixation is 35-70 GJ/tonne N in the form of glucose. Since this cannot be expressed in terms of final or primary energy carriers, no values are given in the table.

Figure 6.9 shows all options we have discussed ranked according to the degree of technical change and the stage of development. The high-temperature processes for nitrogen fixation, both fuel-fired as plasma-based, are not placed in the matrix, because they can probably not bring about more energy-efficient nitrogen fixation. The figure clearly shows that most options are still in the stage of applied research. Low-temperature chemical nitrogen fixation and genetical engineering of cash-crops require a radical change in the way the energy service is performed. This indicates that the time period to commercialization is long, and that there is no guarantee that the technologies will ever be commercialised. Co-production of nitric oxide and electricity in a combined cycle plant and a SOFC are not placed at the same place in the diagram, because the combined cycle plant can use off-the-shelf technology and the SOFC-technology has still to be demonstrated. Of all options, only the application of symbiotic systems in nature has been demonstrated. However, this option can only be used for a few leguminous plants.

It should be noted that there are many techniques to improve the energy-efficiency of current production routes that have not been included in this study. We have only dealt with long-term options. On the other hand, because realizing the long-term options results in considerable changes in processes, most short-term options we know at present cannot be applied when these long-term options have been adapted.

Most processes are in such an early stage of development that no reliable cost figures are available. Estimates of the investment costs for a membrane steam reformer for ammonia syngas production and for nitric oxide production in a SOFC showed that at present it cannot be said whether these new processes will have lower investment costs than the conventional processes.

6.9 Discussion

In this section we will comment on the choices we made, their consequences for the selection of technologies, and on the accuracy of the data.

First of all, it should be clear that due to our choice of the energy service – the fixation of nitrogen from the atmosphere – not all options to reduce the primary energy demand are included, for instance more efficient nitrogen fertilizer application.

The syngas plant that was used as the basis of the analysis is very energy efficient. In this plant the exergy losses are already halved compared to a conventional syngas plant. Nevertheless, of the exergy input of 30.0 GJ/tonne N still 4.7 GJ/tonne N is lost. A small reduction of the SEC can be obtained by optimization of the energy flows of the plant, e.g.

by avoiding heat exchange with a large temperature gradient. Large reductions can only be achieved by drastic changes. It should be noted, however, that the average SEC of ammonia production is above 30 GJ/tonne NH_3, thus in most existing plants there is still considerable room for improvement with short-term measures.
A consequence of the choice to analyse only the syngas plant is that it is not possible to study optimization of the complete ammonia plant. This choice is defendable by the fact that we had already a good insight into the exergy loss in the ammonia synthesis from previous exergy analysis. The main exergy losses occur due to compression of syngas. Nitric acid production has a very low exergy efficiency. Based on data for a 1980 plant, we concluded that improvement of this efficiency should mainly be directed at controlling the combustion of ammonia. Although the set of data is old, the conclusion also holds for contemporary plants, because the difference in SEC is small.The efficiency of electricity generation in power plants influences the potential for energy-efficiency improvement that can be obtained by applying high-temperature cogeneration. When this efficiency increases to 70% – a value that may be obtained in gas-fired combined cycle power plants – the fuel demand that should be allocated to ammonia production increases to 30.7 GJ/tonne N. At an efficiency of 58%, this was 28.9 GJ/tonne N. The potential for energy-efficiency improvement is about halved from 4.1 to 2.3 GJ/tonne N.

A SEC of 28.9 GJ/tonne N of the future plant is still 4.2 GJ/tonne N above the theoretically minimum energy demand for the production of ammonia according to the steam reforming process of methane of 24.7 GJ/tonne N. This difference is mainly due to the exergy loss that occurs during the partial combustion of methane. This loss can be avoided or reduced when the heat for the steam reforming reaction can be provided in another way or hydrogen can be supplied by another method. In Section 7 we briefly discussed the problems associated with the application of a High-Temperature Nuclear Reactor for providing the heat and with biomass for hydrogen production. Both options have there specific problems. However, in the long term these options can be considered to further decrease the fossil fuel demand.
The number of technologies that can reduce the energy demand for the largest energy-consuming step - the ammonia syngas production - is small. Over and above this, nearly all selected technologies are in an very early stage of development. This is remarkable because in two previous sector studies, to the long-term potential for energy efficiency improvement in the paper and board industry [Beer *et al.*, 1998] and the iron and steel industry [Beer, 1998], only measures were found that will be commercially available within about two decades. In these studies the main focus was also on the operations with the largest exergy losses. The difference in time period until commercialization of the selected technologies might be the result of the fact that in the ammonia syngas production a new technology has just been commercialized, whereas paper drying and iron production is still being done according to well-proven technologies. New technologies are being developed but have not been demonstrated so far. It might be concluded that in the ammonia synthesis an improvement of the energy efficiency by the introduction of new technologies has just been made, while in the other processes this can be expected in a relatively short time

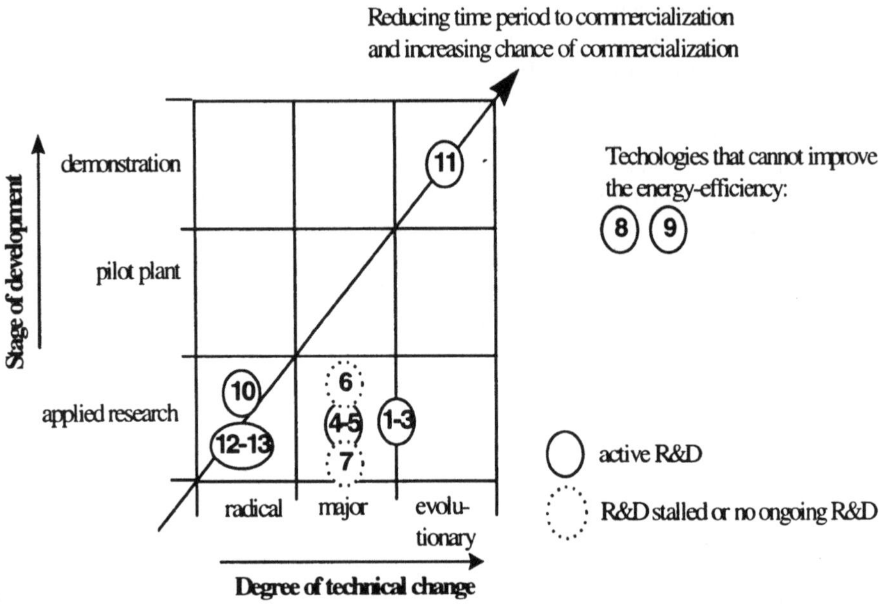

Energy-efficiency improvements to the ammonia process

1 Membrane reactor for steam reforming
2 Membrane reactor for selective removal of ammonia
3 Selective removal of ammonia using sorbents
4 High-temperature cogeneration - preheating feed and fuel
5 High-temperature cogeneration - membrane steam reformer

Energy-efficiency improvements to the nitric acid process

6 Nitric oxide and electricity production in a combined cycle plant
7 Nitric oxide and electricity production in a SOFC
8 Fired high-temperature direct oxidation processes
9 Plasma processes for direct oxidation of nitrogen

Alternative for nitrogen fixation

10 Low-temperature chemical nitrogen fixation
11 Using symbiotic systems in nature
12 Genetic manipulation of cash crops to enable symbioses
13 Genetic manipulation of cash crops to enable N-fixation

Figure 6.9: Comparison of selected options to improve the energy efficiency of nitrogen fixation on the degree of technical change required to replace the current technology and the stage of development.

period.

A result of the early stage of development is that pertinent data are hard to find. The data on SEC are in most cases based on own calculations and judgements. A modern ammonia plant is highly integrated with regard to energy flows. As a result, a reduction in the energy demand of a separate measure, can lead to a different reduction of the energy demand of the total plant. Lowering the temperature for steam reforming, for instance, might result in a shortage of steam to drive the syngas compressor. To judge these effects correctly, a complete plant analysis and optimization should be performed. This is beyond the scope of this study. An estimate of the energy demand of a plant that incorporated several measures gave an idea of possible synergetic effects. This study results in directions for development of new technologies that can bring about energy-efficiency improvement rather than in priority setting of R&D-needs for concrete technologies.

6.10 Conclusions and recommendations

At present, chemical nitrogen fertilizer production has a specific energy consumption that ranges from 26 to 39 GJ/tonne N fixed, depending on the type of fertilizer. The energy is mainly provided in the form of natural gas. Biological nitrogen fixation requires 35 to 70 GJ/tonne N, in the form of glucose.

The most efficient ammonia plant uses 33.0 GJ/tonne N, whereas the minimum energy demand is 24.1 GJ/tonne N. In the future, an ammonia plant that consists of a gas turbine to supply in the heat for the steam reforming reaction that takes place in a membrane reactor can have a SEC of 28.9 GJ/tonne N.

The energy recovery at the production of nitric acid from ammonia is now 9.9 GJ/tonne in the form of steam, while 34.5 GJ/tonne of N is liberated during the reactions. The energy recovery can be improved when ammonia is converted in a solid oxide fuel cell, generating 17 GJ of electricity. The direct oxidation of nitrogen at high temperatures does not have the potential to reduce the SEC for nitrogen fixation, although processes were commercial up to about 1955.

Recently, there has been considerable progress in splitting molecular nitrogen at low temperatures. However, coupling this process to a process where, for instance, the nitrogen atoms reacts with oxygen to NO instead of recombination of two nitrogen atoms to N_2, requires serious R&D. Furthermore, it is not clear at the moment whether such a process would be more energy-efficient than current chemical nitrogen fixation.

Enhancing biological nitrogen fixation can reduce the energy demand for nitrogen fixation. Except for making more use of symbiotic systems present in nature, all options require genetic engineering. These options are in a very early stage of development. The societal resistance against the use of genetic engineering might slow down the development and implementation. Furthermore, the effect on crop productivity is not exactly known yet.

The involvement of the nitrogen fertilizer industry in the development of the described technologies is small. Nearly all research takes place in non-commercial laboratories, although the development of membrane reactors has been taken up by some commercial firms. Governmental organizations can play a role in the stimulation of the development of technologies that can improve the energy-efficiency of nitrogen fertilization. Governmental interference can be justified given the fact that both environmental concern and ensuring sufficient food supply are responsibilities for the government. There are a number of areas that may be subject of governmental policy:

- Stimulating the demonstration of the use of hot exhaust of a gas turbine to provide in process heat demand;
- Stimulating the development and demonstration of the membrane reactor.
- Stimulating research to and the development of co-production of nitric oxide and electricity in a fuel cell. There may also be other industrial chemicals that can be co-produced with electricity in a fuel cell.
- Stimulating fundamental research to the direct oxidation of nitrogen and oxygen to nitrogen oxides.
- Evaluating the societal consequences of biogenetic engineering, and when the outcome of this discussion allows this, stimulating the research to enhance biological nitrogen fixation.

Acknowledgements - The author would like to thank S. Nonhebel (University of Groningen), H. van Veen (ECN), D.B. Klima (Weaterly Inc.), S.E. Nielsen (Haldor Topsøe), J. Gosnell (Brown & Root), J.R. Le Blanc (The M.W. Kellogg Technology Company), E. Nieuwlaar, E Worrell and W.C. Turkenburg (Utrecht University) for providing information, suggestions and comments on this study. The Netherlands Organization for Scientific Research is acknowledged for its financial support.

CHAPTER 7

CONCLUSIONS

In this book we examined the potential for improving the energy-efficiency of industrial processes in the long term. In the introduction a reference was made to the objective of the Factor 10 Club: improving the resource productivity by a factor of 10. This objective is used as guidance to evaluate the long-term potentials for energy-efficiency improvement.

The long-term potential for energy-efficiency improvement was studied for three industrial energy services: making a flat sheet of paper with certain specific properties from an intermediate material (pulp or waste paper); making steel with certain well-described properties; fixing atmospheric nitrogen as fertilizer. The analysis of paper production focuses on new processes for drying paper. The analysis of iron and steel production also takes recycling of steel scrap into account. Finally, the analysis of nitrogen fixation involves not only more energy-efficient chemical nitrogen fixation, but also completely different routes for nitrogen fixation, e.g. biological nitrogen fixation. This selection of energy services is chosen deliberately so that the scope of the new energy-efficient technologies broadens from paper production via steel making to nitrogen fixation.

In Table 7.1 the results of the case studies are summarized. The potential for energy-efficiency improvement is defined as follows:

$$\frac{\text{SEC of the currently most efficient plant} - \text{SEC of the future plant}}{\text{SEC of the currently most efficient plant}} \times 100\%$$

The SEC of the currently most efficient plant is the SEC of the plant that is currently in operation with the lowest SEC. The SEC of the future plant is the SEC of a plant in which all technologies that are described in this thesis have been implemented.

The potential for energy-efficiency improvement is limited by the theoretical minimum; therefore this value is also given. It can be concluded that in iron and steel production the maximum potential for energy-efficiency improvement is 65%, in ammonia production 27%, and in nitric acid production 88%.

A comparison of the long-term estimates of the potential for energy-efficiency improvement with the short-term estimates based on ICARUS leads to the conclusion that in paper and board production and in iron and steel production the long-term potential for energy-efficiency improvement is considerably larger. In the production of ammonia the short-term potential seems larger. However, this potential is based on a less efficient reference situation, namely a SEC of 40.7 GJ/tonne N for ammonia production. With regard to this reference SEC, the long-term potential for energy-efficiency improvement would have been 41%. Thus, it can be concluded that in all investigated processes the long-term potential for energy-efficiency improvement is more than twice as large as the short-term potential.

Table 7.1: Overview of the potential for energy-efficiency improvement in paper and board production, iron and steel making and chemical nitrogen fixation.

	Currently most efficient plant	Future pant	Minimum SEC	Long-term potential for energy-efficiency improvement	Short-term potential for energy-efficiency improvement[a]
Paper and board production	GJ/tonne paper				
specific heat demand	2.3-8.6	0.6-4.3	0.0	50-75%	28%
Iron and steel making	GJ/tonne hot rolled steel				
primary steel	19.0	12.5	6.6	34%	11%
secondary steel	7.0	3.5	0.0	50%	4%
Nitrogen fixation	GJ/tonne nitrogen				
ammonia	33.0	28.6	24.1	13%	16%
nitric acid	26.8	15.3	3.2	43%	-[b]

[a] The short-term potential for energy-efficiency improvement is the technical potential for the period 1990-2000 in the Netherlands, and is determined with ICARUS.
[b] No estimate is available for the short-term potential for the energy-efficiency improvement of nitric acid production.

This study showed that for the selected energy services technically opportunities to increase the energy productivity by a factor 2 can be identified: the same service can be provided by half of the current energy use. However, this is not sufficient to obtain the objective of the Factor 10 Club, which is to increase the resource productivity by a factor 10, hence to decrease the energy consumption for an energy service by 90%. The

conclusion is that to attain such factors much more is needed. Options to achieve this require re-thinking of the way societal functions are fulfilled or restructuring of infrastructure.
Restructuring infrastructure relates to the integration of energy flows between industrial plants and between industrial plants and the built. Energy and material cascading holds a large potential for further improvement of the energy productivity. As an example, it is conceivable that the heat recovered in a steel plant (see Chapter 5) can be used in a paper mill (see Chapter 4).
Rethinking the way societal functions involves, in this case, question like: do we cars to be made from defect-free primary steel or can we do with secondary steel also? or do we want to read the news from paper of from our computer screen? Whether such options are incorporated in the analysis depends on the way the energy service is defined. In this book we showed that a broad definition of the energy service results in more options to improve the energy efficiency. For instance, increased use of biological nitrogen fixation. However, it also resulted in more uncertainties in estimates on potential and costs and raises other question, for instance whether it can be justified to use genetically engineering.

Finally, we would like to make a few comments on the method used to assess the long-term potential for energy-efficiency improvement.

In our method *exergy analysis* is used to assess the potential for energy-efficiency improvement of state-of-the-art processes. Exergy analysis proved to be very useful in the analysis of the integrated steel mill and ammonia syngas production. Exergy analysis has few advantages over energy analysis in the case of paper making, because in the latter the temperature differences are small and almost no chemical modifications take place. An exergy analysis can be very time-consuming. It is therefore advised to make a preliminary assessment of the usefulness of an exergy analysis. It is concluded that particularly in the studies of the long-term potential for energy-efficiency improvement in iron and steel production and nitrogen fertilizer production, exergy analysis proved to be useful for giving give direction to the search for energy-efficient technologies.

The *identification and selection of technologies* formed another main part of the method described in this thesis. In some cases the results of the exergy analysis revealed a potential for improvement of the energy efficiency in an area in which we could not identify any technology. An example is heat recovery from hot steel. In these cases we had to think of new technologies, or point to the opportunities in these areas. It might well be that technologies are conceivable by experts involved in these specific processes. The principle of impulse drying, for instance, was thought of by a paper-making expert in the 1970s. It was only at the beginning of the 1980s that the first publications appeared in the international literature. If we had performed our analysis in the 1970s, impulse drying might not have been on our list. An expert who happened to know about the technology should have indicated its existence. In general, we can conclude that our energy-oriented approach does not point in the direction of new technologies that have been developed

specifically to improve the quality of the product or certain process parameters. Since such technologies might also lead to an improvement of the energy-efficiency, this could lead to omissions in our list of technologies. However, these omissions can be prevented to a large extent by consultation with experts. Nevertheless, it may well be that there are energy-efficient technologies conceivable which we have not identified.
Only a few technologies have been identified which are in an early stage of development. It is generally known that in connection with each new technology or product that is brought on the market, a large number of ideas are generated and investigated but abandoned before commercialization. To protect the exclusivity of these ideas, information on these technologies is generally not widely disseminated. Furthermore, even if information is available concerning technologies that are in an early stage of development, the potential for energy-efficiency improvement is often hard to assess. It can be concluded that the applicability of the method for identifying energy-efficient technologies that are in early stage of development is limited. Nevertheless, as shown in this thesis, most identified technologies can be commercialized in a period of up to three decades. The method does not allow to point out specific technologies that will be commercialized in more than 30 years. However, it is possible to assess the development of the energy-efficiency on the basis of an assessment of opportunities to reduce exergy losses.

In all three processes the potential for energy-efficiency improvement can be realized mainly by technologies that are specific to the sector, e.g. impulse drying, smelting reduction, and membrane steam reforming. For all sectors it can be concluded that high energy-efficiency can be attained when techniques are developed that can recover and apply heat supplied earlier to the process. In paper production, this can be achieved by recovering the latent heat from the evaporated moisture. In steel production, the energy-efficiency can be improved by recovering and applying the heat from red hot steel. In ammonia production, heat can be recovered from the hot syngas. Therefore, in this thesis besides making recommendations regarding the process-specific technologies, we recommend R&D to be focussed on heat recovery techniques.

There are at least four aspects that we did not consider but that affect our estimate of the potential for energy-efficiency improvement:

- The influence of new technologies on the energy consumption of other unit operations can affect the overall SEC of the process. For instance, reduction in the steam demand for paper making can result in a decrease in efficiency of the co-generation unit since it has to operate below its optimal heat/power ratio. In retrofit situations this effect might reduce the potential for energy-efficiency improvement. Optimization of the plant, which we did not consider, can counteract this effect. In newly built plants, the energy system can be optimized in the design stage, resulting in the optimum integration of all unit operations.
- The emphasis was on the unit operations with the largest exergy losses. As a consequence, we ignored developments in other unit operations. Since the analysed unit operations consumed on average 80% of the energy consumption

of the whole process, the impact of developments in other unit operations on the potential for energy-efficiency improvement is expected to be small compared to the operations we have analysed.

- The specific energy consumption of an industrial energy service might be reduced further by the continuing development of the technologies assessed. The energy efficiency of a new technology is not fixed at the status of the first commercial application. Later applications will probably have a higher efficiency, for instance because of feedback between practical experience and R&D.
- Opportunities for energy and material integration between industries are not investigated (see the discussion on the factor 10).

The overall effect of these four aspects on the potential for energy-efficiency improvement cannot be assessed accurately. We estimate that the effect of the last three aspects is small compared to the potential for energy-efficiency improvement identified in this thesis. Integration of energy flows between processes might result in a considerable improvement and should therefore be studied. However, to avoid a new barrier to the implementation of energy-efficient technologies, we recommend that integration between plants should only be considered after the efficiencies of the individual plants have been improved as far as possible.

REFERENCES

Chapter 1

Ayres, R.U. (1988), "Energy Inefficiency in the U.S. Economy: A New Case for Conservation", Dept. of Engineering and Public Policy, Carnegie Mellon University, Pittsburgh.

Beer, J.G. de, M.T. van Wees, E. Worrell and K. Blok (1994), "ICARUS-3 - The Potential of Energy Efficiency Improvement in the Netherlands up to 2000 and 2015", Department of Science, Technology and Society, Utrecht University, Utrecht.

Blok, K., W.C. Turkenburg, W. Eichhammer, U. Farinelli and T.B. Johansson (1996), "Overview of Energy RD&D Options for a Sustainable Future", European Commission, DG XII, Luxembourg.

Blok, K., E. Worrell, R. Cuelenaere and W. Turkenburg (1993), "The Cost-Effectiveness of CO_2 Emission Reduction Achieved by Energy Conservation", Energy Policy **21**, p656-667.

BP (1996), "BP Statistical Review of World Energy", BP, London.

CADDET (1997), "Internationally Data on Successfully Demonstrated Energy Efficiency Projects", Centre for the Analysis and Dissemination of Demonstrated Energy Technologies, Novem, Sittard, The Netherlands.

CEC (1992), "1992 Energy Technology Status Report", California Energy Commission, Sacramento.

ETSU (1984), "Energy Use and Energy Efficiency in UK Manufacturing Industry up to the Year 2000, volume 2", ETSU, Harwell, UK.

ETSU (1994a), "An Appraisal of UK Energy Research, Development, Demonstration & Dissemination", Strategic Studies Department, ETSU, Harwell, UK.

ETSU (1994b), *An Appraisal of UK Energy Research, Development, Demonstration & Dissemination,* HMSO, London.

Factor 10 Club (1994), Declaration of the Factor 10 Club.

FAO (1994), "Yearbook Forest Products 1993", Food and Agricultural Organization, Rome.

Farugui, A., M. Mauldin, S. Schick, K. Seiden and G. Wikler (1990), "Efficient Electricity Use: Estimates for Maximum Energy Savings", prepared by Barakat & Chamberlin, Inc. for the Electric Power Research Institute, Palo Alto.

Ficket, A.P., C.W. Gellings and A.B. Lovins (1990), "Efficient Use of Electricity", Scientific American **263**(3), p64-74.

Goldemberg, J., T.N. Johansson, A.K.N. Reddy and R.H. Williams (1985), "An End-Use Oriented Global Energy Strategy", Annual Review of Energy **10**, p613-688.

Goldemberg, J., T.N. Johansson, A.K.N. Reddy and R.H. Williams (1988), *Energy for a Sustainable World,* John Wiley, New York.

Grubb, M., J. Edmonds, P. ten Brink and M. Morrison (1993), "The Costs of Limiting Fossil-Fuel CO_2-emissions: A Survey and Analysis", Annual Review of Energy and Environment **18**, p397-478.

Houghton, J.T., L.G. Meiro Filho, B.A. Callander, N. Harris, A . Kattenberg and K. Maskell *(Eds.)* (1996), *Climate Change 1995 - The Science of Climate Change. Contribution of Working Group I to the Second Assessment Report of the International Panel on Climate Change*, Cambridge University Press, Cambridge.

IEA (1994), *IEA/OECD Scoping Study: Energy and Environmental Technologies to Respond to Global Climate Change Concerns,* OECD/IEA, Paris.

IEA (1996), *Comparing Energy Technologies,* OECD/IEA, Paris.

IEA/OECD (1997), "IEA Energy Technology R&D Statistics 1974-1995", Paris.

Jackson, T. (1991), "Least-Cost Greenhouse Planning - Supply Curves for Global Warming Abatement", Energy Policy **19**(1-2), p35-46.

Jochem, E. (1991), "Long-Term Potentials of Rational Energy Use - The Unknown Possibilities for Reducing Greenhouse Gas Emissions", Energy & Environment **2**(1), p31-44.

Koomey, J.G., C. Atkinson, A. Meier, J.E. McMahon and S. Boghosian (1991), "The Potential for Electricity Efficiency Improvements in the US Residential Sector", Lawrence Berkeley Laboratory, Berkeley.

Lovins, A.B. and L.H. Lovins (1991), "Least-Cost Climatic Stabilization", Ann. Rev. Energy Environ. **16**, p433-531.

Meadows, D., D. Meadows, J. Randers and W. Behrens (1972), "The Limits To Growth - A Report for the Club of Rome Project on the Predicament of Mankind", New York, Universe Books.

Melman, A.G., H. Boot and G. Gerritse (1990), "Energiebesparingspotentiëlen 2015 (Energy Conservation Potentials 2015)", Netherlands Organization for Applied Scientific Research TNO, Apeldoorn.

Mills, E., D. Wilson and T.B. Johansson (1991), "Getting Started: No Regrets Strategies for Reducing Greenhouse Gas Emissions", Energy Policy **19**(8-9), p526-542.

Nakicenovic, N. and A. Grübler (1993), "Energy Conversion, Conservation, and Efficiency", Energy - The International Journal **18**(5), p421-435.

Nakicenovic, N. (Study Director) (1995), "Global Energy Perspectives to 2050 and Beyond", WEC-IIASA, London.

Nilsson, L.J. and T.B. Johansson (1994), *Environmental Challenges to the Energy Industries*, in: *Expanding Environmental Perspectives*, L.J. Lundgren, L.J. Nilsson, *et al.* (ed.), Lund University Press, Lund.

PCAST (1997), "Federal Energy Research and Development for the Challenges of the Twenty-First Century", The President's Committee of Advisors on Science and Technology - Energy Research and Development Panel, Washington.

Reddy, A.K.N., R.H. Williams and T.B. Johansson (1997), "Energy After Rio - Prospects and Challenges", United Nations Development Programme, New York.

UNCED, 1992, “Agenda 21 : programme of action for sustainable development.Rio declaration on environment and development. Statement of forest principles : the final text of agreements negotiated by governments at the United Nations conference on environment and development (UNCED)" ,3-14 June 1992, Rio de Janeiro, Brazil.

UNFCCC (1992), *United Nations Framework Convention on Climate Change, Report of the Intergovernmental Negotiating Committee for a Framework Convention on Climate Change*, New York.

UNFCCC (1997), "Kyoto Protocol to the United Framework Convention on Climate Change", United Nations Framework Convention on Climate Change, Kyoto.

Watson, R.T., M.C. Zinyowera, R.H. Moss and D.J. Dokken *(Eds.)* (1996), *Climate Change 1995 - Impacts Assessment and Mitigation of Climate Change: Scientific-Technical Analysis. Contribution of Working Group II to the Second Assessment Report of the International Panel on Climate Change*, Cambridge University Press, Cambridge.

WCED, (World Commission on Environment and Development) (1987), *Our Common Future,* Oxford University Press, Oxford, UK.

WEC (1995), "Energy Efficiency Improvement Utilising High Technology", World Energy Council, London.

Worrell, E. (1994), "Potentials for Improved Use of Industrial Energy and Materials",PhD Thesis, , Utrecht.

Worrell, E., M. Levine, L. Price, N. Martin, R. van den Broek and K. Blok (1997a), "Potentials and Policy Implications of Energy and Material Efficiency Improvement", United Nations, New York.

Worrell, E., L. Price, N. Martin, J. Farla and R. Schaeffer (1997b), "Energy Intensity in the Iron and Steel Industry: A Comparison of Physical and Economic Indicators", Energy Policy **25**(7-9), p727-744.

Chapter 2

Ayres, R.U. (1988), "Energy Inefficiency in the U.S. Economy: A New Case for Conservation", Dept. of Engineering and Public Policy, Carnegie Mellon University, Pittsburgh.

Beer, J. de and K. Blok (1998), "Fixing Atmosperic Nitrogen With Less Energy", (To be published).

Beer, J. de, E. Worrell and K. Blok (1998a), "Future Technologies for Energy-Efficient Iron and Steel Making", Annual Review of Energy and Environment 3(1). p123-205.

Beer, J. de, E. Worrell and K. Blok (1998b), "Long-Term Energy-Efficiency Improvements in the Paper and Board Industry", Energy - The International Journal 3(1), p21-42.

Beer, J.G. de, M.T. van Wees, E. Worrell and K. Blok (1994), "ICARUS-3 - The Potential of Energy Efficiency Improvement in the Netherlands up to 2000 and 2015", Department of Science, Technology and Society, Utrecht University, Utrecht.

Blok, K. (1984), *Onbeperkt houdbaar - een toekomst met duurzame energie (Permanently Sustainable - A future with renewable energy),* Stichting Natuur en Milieu, Utrecht (In Dutch).

Blok, K., K. Burges, J.M. Bais and A.W.N van Dril (1994), "De NEEDIS Structuur (The NEEDIS Structure)", NEEDIS, Petten (In Dutch).

Blok, K., E. Worrell, R. Cuelenaere and W. Turkenburg (1993), "The Cost-Effectiveness of CO_2 Emission Reduction Achieved by Energy Conservation", Energy Policy 21, p656-667.

CADDET (1997), "Internationally Data on Successfully Demonstrated Energy Efficiency Projects", Centre for the Analysis and Dissemination of Demonstrated Energy Technologies, Novem, Sittard, The Netherlands.

CEC (1992), "1992 Energy Technology Status Report", California Energy Commission, Sacramento.

Chauvel, A. and G. Lefebvre (1989), *Petrochemical Processes,* Institut Français du Pétrole Publications, Paris.

Colombo, U. and O. Bernardini (1979), "A Low Energy Growth 2030 Scenario and the Perspectives for Western Europe", Report to European Economic Community, Brussels.

Dosi, G. (1982), "Technological Paradigms and Technological Trajectories", Research Policy 11, p147-162.

Dril, A.W.N. van, F.M.J.A. Diepstraten and M. Beeldman (1995), "SAVE-module produktiebedrijven : de modellering van energieverbruiksontwikkelingen (SAVE-module production firms: the modelling of the developments of energy use)", Netherlands Energy Research Foundation ECN, Petten.

Edmonds, J. A. and J. Reilly (1983), "Global Energy and CO_2 to 2050", Energy Journal 4(3), p21-47.

ETSU (1984), "Energy Use and Energy Efficiency in UK Manufacturing Industry up to the Year 2000, volume 2", ETSU, Harwell, UK.

ETSU (1994), *An Appraisal of UK Energy Research, Development, Demonstration & Dissemination,* HMSO, London.

ETSU (1997), "Energy Technology - The Next Step", European Commission - Directorate General for Energy DG XVII, Brussels.

European Commission (1996), "European Energy to 2020 - A Scenario Approach", Office for Official Publications of the European Communities, Luxembourg.

Farla, J., K. Blok and L. Schipper (1997), "Energy Efficiency Developments in the Pulp and Paper Industry. A Cross-Country Comparison using Physical Production Data", Energy Policy 25(7-9), p745-758.

Farla, J., R. Cuelenaere and K. Blok (1998), "Energy Intenisty and Structural Change in the Netherlands, 1980-1990", Energy Economics 20(1), p1-28.

Farugui, A., M. Mauldin, S. Schick, K. Seiden and G. Wikler (1990), "Efficient Electricity Use: Estimates for Maximum Energy Savings", prepared by Barakat & Chamberlin, Inc. for the Electric Power Research Institute, Palo Alto.

Ficket, A.P., C.W. Gellings and A.B. Lovins (1990), "Efficient Use of Electricity", Scientific American 263(3), p64-74.

Fulkerson, W., D.B. Reister and J.T. Miller (1989), "Energy Technology R&D: What Could Make A Difference?", Oak Ridge National Laboratory, Oak Ridge (Tennessee).

Goldemberg, J., T.N. Johansson, A.K.N. Reddy and R.H. Williams (1985), "An End-USe Oriented Global Energy Strategy", Annual Review of Energy 10, p613-688.

Goldemberg, J., T.N. Johansson, A.K.N. Reddy and R.H. Williams (1988), *Energy for a Sustainable World,* John Wiley, New York.

Gool, W. van (1995), "Exergy Analysis of Industrial Processes", Energy - The International Journal 17(8), p791-803.

Greentie (1997), "Greentie", IEA/Greentie, Sittard, the Netherlands.

Grubb, M., J. Edmonds, P. ten Brink and M. Morrison (1993), "The Costs of Limiting Fossil-Fuel CO_2-emissions: A Survey and Analysis", Annual Review of Energy and Envrionment 18, p397-478.

Hein, L.G. and K. Blok (1995), "Transaction Costs of Energy Efficiency Improvement", Presented at Proc. of the 1995 ECEE Summer Study: Sustainability and the Reinvention of Government- A Challenge for Energy Efficiency, The European Council for an Energy-Efficient Economy, Stockholm. Sweden.

Hinderink, A.P., F.P.J.M. Kerkhof, A.B.K. Lie, J. De Swaan Arons and H.J. van der Kooii (1996), "Exergy Analysis With a Flowsheeting Simulator - II. Application; Synthesis Gas Production From Natural Gas", Chemical Engineering Science 51(20), p4701-4715.

Howarth, R.B., L. Schipper, P.A. Duerr and S. Strom (1991), "Manufacturing Energy Use in Eight OECD Countries; Decomposing the Impacts of Changes in Output, Industry Structure and Energy Intensity", Energy Economics 13, p135-142.

IEA (1987), *Energy Conservation in IEA Studies,* OECD/IEA, Paris.

IEA (1994a), *IEA/OECD Scoping Study: Energy and Environmental Technologies to Respond to Global Climate Change Concerns,* OECD/IEA, Paris.

IEA (1994b), "SLT/CERT Annual Questionnaire for Country Submissions", IEA, Paris.

IETC, UNEP (1997), "maESTro", United Nations Environmental Programme - International Environmental Technology Centre, Osaka, Japan.

Jackson, T. (1991), "Least-Cost Greenhouse Planning - Supply Curves for Global Warming Abatement", Energy Policy 19(1-2), p35-46.

Kaiser, V. (1993), *Industrial Energy Management,* Editions Technip, Paris.

Kemp, R., F. van der Loo, C. Freeman, L. Soete, I. Miles, M. Boden, P. Street, K. Smith, O. Wicken and H. Anderson (1994), "Technology and the Transition to Environmental Stability", Maastricht Economic Research Institute on Innovation and Technology, University of Limburg, Maastricht.

Koomey, J.G., C. Atkinson, A. Meier, J.E. McMahon and S. Boghosian (1991), "The Potential for Electricity Efficiency Improvements in the US Residential Sector", Lawrence Berkeley Laboratory, Berkeley.

Koopmans, C.C. and D.W. te Velde (1996), "NEMO: Netherlands Energy Demand Model", CPB Netherlands Bureau for Economic Research, The Hague.

Kotas, T.J. (1985), *The Exergy Method of Thermal Plant Analysis,* Butterworths, London.

Krause, F., B. Bossel and K.-F. Müller-Reißmann (1980), *Energie-Wende - Wachstum und Wohlstand ohne Erdöl und Uran (Energy Revolution - Growth and Welfare wihout Oil and Uranium),* S. Fischer Verlag, Frankfurt am Main (In German).

Krause, F., E. haites, R, Howarth, J. Koomey (1993) "Energy Policy in the Greenhouse, Vol two, Part 1, Cutting Carbon Emissions: Burden or Benefit?", International Project for Sustainable Paths, El Cerrio, California.

Leach, G., C. Lewis, A. van Buren, F. Romig and G. Foley (1979), *A Low Energy Strategy for the United Kingdom,* The International Institute for Environment and Development, London.

Linhoff, B. (1982), *A User Guide on Process Integration for the Efficient Use of Energy,* The Institution of Chemical Engineers, Rugby, UK.

Lovins, A.B. (1977), *Soft Energy Paths,* Penguin Books, New York.

Lovins, A.B. and L.H. Lovins (1991), "Least-Cost Climatic Stabilization", Ann. Rev. Energy Environ. 16, p433-531.

Manne, A.S. and R.G. Richels (1992), *Buying Greenhouse Insurance - The Economic Costs of Carbon Dioxide Emissions,* The MIT Press, Cambridge.

Marchetti, C. (1979), "Energy Systems - The Broader Context", Technological Forecasting and Social Change 14, p191-203.

Melman, A.G., H. Boot and G. Gerritse (1990), "Energiebesparingspotentiëlen 2015 (Energy Conservation Potentials 2015)", Netherlands Organization for Applied Scientific Research TNO, Apeldoorn.

Mills, E., D. Wilson and T.B. Johansson (1991), "Getting Started: No Regrets Strategies for Reducing Greenhouse Gas Emissions", Energy Policy 19(8-9), p526-542.

Morris, D.R. (1992), "Exergy Analysis and Cumulative Exergy Consumption of Complex Chemical Processes: The Industrial Chlor-Alkali Process", Chemical Engineering Science 46, p459-465.

Nakicenovic, N. (1990), *Dynamics of Change and Long Waves*, in: *Life Cycles and Long Waves*, T. Vasko, R. Ayres, *et al.* (ed.), Springer Verlag, Berlin.

Nakicenovic, N. (Study Director) (1995), "Global Energy Perspectives to 2050 and Beyond", WEC-IIASA, London.

Nelson, R. R. and S.G. Winter (1977), "In Search of Useful Theory of Innovation", Research Policy 6(1), p36-77.

Nieuwlaar, E. (1988), "Developments in Energy Analysis",PhD Thesis, , Utrecht.

Nieuwlaar, E. (1996), "Enerpack 5", Department of Science, Technology and Society, Utrecht University, Utrecht.

Nilsson, L.J. (1993), "Energy Intensity Trends in 31 Industrial and Developing Countries 1950-1980", Energy - The International Journal 18(4), p309-322.

Nilsson, L.J., E.D. Larson, A. Gupta and K. Gilbreath (1996), "Energy Efficiency and the Pulp and Paper Industry", ACEEE, Washington DC.

Novem (1998), "Industriële Databank voor Energie Efficiency *(Industrial Database for Energy Efficiency)*", Ver. 5, Novem, Utrecht (In Dutch).

OECD (1994), *Frascati Manual 1993 - The Measurement of Scientific and Technological Activities,* OECD, Paris.

Olthof, R.J.J. and M. Muradin *(Eds.)* (1996), *Syrene - Kiezen voor een duurzame energiehuishouding. (Syrene - Opting for a Sustainable Energy Consumption Pattern)*, Novem, Utrecht (In Dutch).

Over, J.A. and A.C. Sjoerdsma *(Eds.)* (1974), *Energy Conservation: Ways and Means*, Future Shape of Technology Foundation, The Hague.

PCAST (1997), "Federal Energy Research and Development for the Challenges of the Twenty-First Century", The President's Committee of Advisors on Science and Technology - Energy Research and Development Panel, Washington.

Potma, T. (1979), *Het vergeten scenario (The Forgotten Scenario),* Meulenhof Informatief, Amsterdam (In Dutch).

Potma, T.G. (1977), "Energiebeleid met minder risico (Energy Policy With Less Risk)", Vereniging Milieudefensie/Stroomgroep Stop Kalkar, Amsterdam.

Radgen, P. (1996), "Pinch and Exergy Analysis of a Fertlizer Complex - Part 1", Nitrogen 224, p39-48.

Radgen, P. (1997), "Pinch and Exergy Analysis of a Fertlizer Complex - Part 2", Nitrogen 225, p27-39.

Reddy, A.K.N., R.H. Williams and T.B. Johansson (1997), "Energy After Rio - Prospects and Challenges", United Nations Development Programme, New York.

Riekert, L. (1974), "The Efficiency of Energy-Utilization in Chemical Processes", Chemical Engineering Science 29, p1613-1620.

Rosen, M.A. (1996), "Comparative Assessment of Thermodynamic Efficiencies and Losses for Natural Gas-Based Production Processes for Hydrogen, Ammonia and Methanol", Energy Conversion and Management 37(3), p359-367.

Rosenfeld, A., C. Atkinson, J. Koomey, A. Meier, R.J. Mowris and L. Price (1993), "Conserved Energy Supply Curves for US buildings", Contemporary Policy Issues 11(1), p45-68.

Rossiter, A.P., M.A. Rutkowski and A.S. McMullan (1991), "Pinch Technology Identifies Process Improvements", Hydrocarbon Processing 70(1), p63-66.

Rubin, E.S., R.N. Cooper, R.A. Frosch, T.H. Lee, G. Marland, A.H. Rosenfeld and D.D. Stine (1992), "Realistic Mitigation Options for Global Warming", Science 257(10 July 1992), p148-266.

Sahal, D. (1985), "Technological Guidepost and Innovation Avenues", Research Policy 14, p61-82.

Schipper, L., R. Howarth and E. Carlassare (1992), "Energy Intensity, Sectoral Activity, and Structural Change in the Norwegian Economy", Energy - The International Journal 17, p215-233.

Schipper, L., R.B. Howarth and H. Geller (1990), "United States Energy Use from 1973 to 1987: The Impacts of Improved Energy Efficiency", Annual Review of Energy 15, p455-504.

Schmookler (1962), "Economic Sources of Inventive Activity", Journal of Economic History (March), p1-20.

Schumpeter (1928), "The Instability of Capitalism", Economic Journal, p361-386.

Stoffers, M.J., J.C.G.M. Burghouts and H.J.B.M. Mannaerts (1984), "CENECA - Een model voor het energiegebruik (CENECA - A Model for Energy Use)", Netherlands Bureau for Economic Policy Research (CPB), The Hague (In Dutch).

Sunden, B. (1988), "Analysis of the Heat Recovery in Two Crude Distillation Units", Heat Recovery Systems & CHP 5(8), p483-488.

Szargut, J. and D.R. Morris (1987), "Cumulative Exergy Consumption and Cumulative Degree of Perfection of Chemical Processes", Energy Research 11, p245-261.

Szargut, J., D.R. Morris and F.R. Steward (1988), *Exergy Analysis of Thermal, Chemical. and Metallurgical Processes,* Hemisphere Publishing Corporation, New York.

Wall, G. (1987), "Exergy Conversion in the Swedish Society", Resources and Energy 9, p55-73.

Wall, G. (1988), "Exergy Flows in Industrial Processes", Energy - The International Journal 13(2), p197-208.

WCED, (World Commission on Environment and Development) (1987), *Our Common Future,* Oxford University Press, Oxford, UK.

Webster (1995), *The New International Webster's Dictionary of the English Language,* Trident Press International, Naples, Florida.

Worrell, E. (1994), "Potentials for Improved Use of Industrial Energy and Materials",PhD Thesis, , Utrecht.

Worrell, E., L. Price, N. Martin, J. Farla and R. Schaeffer (1997), "Energy Intensity in the Iron and Steel Industry: A Comparison of Physical and Economic Indicators", Energy Policy 25(7-9), p727-744.

Chapter 3

Albers, R. (1993), "Energie and Emissies van CO_2, NO_x and SO_2 in de scenario's van de derde Nationale Milieu Verkenningen 1990-2015", National Institute of Public Health and Environmental Protection RIVM, Central Planning Bureau CBS, Energy Research Foundation ECN, Bilthoven, The Netherlands.

Beer, J.G. de, M.T. van Wees, E. Worrell and K. Blok (1994), "ICARUS-3 - The Potential of Energy Efficiency Improvement in the Netherlands up to 2000 and 2015", Department of Science, Technology and Society, Utrecht University, Utrecht.

Berkhout, P., J. Jantzen and J.W. Velthuizen (1991), "Energiebesparing in de Nederlandse Produktiehuishouding (Energy Conservation in the Dutch Production Economy)", Stichting voor Economisch Onderzoek, Universiteit van Amsterdam, Amsterdam.

Blok, K. and D. de Jager (1994), "Effectiveness of Non-CO_2 Greenhouse Gas Emission Reduction Technologies", Environmental Monitoring and Assessment **31**, p17-40.

Blok, K., E. Worrell, R.A.W. Albers and R.F.A. Cuelenaere (1990), "Data on Energy Conservation Techniques for the Netherlands", Department of Science, Technology and Society, Utrecht University, Utrecht.

Blok, K., E. Worrell, R.F.A. Cuelenaere and W.C. Turkenburg (1993), "The Cost-Effectiveness of CO_2-Emission Reduction Achieved by Energy Conservation", Energy Policy(June), p656-667.

Burg, T. van den (1992), "Negative Cost Options for Energy-Saving; An Interpretation from the Economic Perpective", Instituut voor Milieuvraagstukken, Amsterdam.

CBS, Netherlands Central Bureau of Statistics (1991), *Nederlandse Energiehuishouding 1990,* SDU/Uitgeverij, 's-Gravenhage.

ECN (1995), "Energie Verslag Nederland (Energy Report the Netherlands)", Netherlands Energy Research Foundation ECN, Petten.

Farla, J.C.M. and K. Blok (1995), "Energy Conservation Investment of Firms: Analysis of Investments in Energy Efficiency in the Netherlands in the 1980's - Paper presented at the ACEEE 1995 Summer Study, Grand Island, NY, USA", Department of Science, Technology and Society, Utrecht University, Utrecht, The Netherlands.

Gillissen, M. and H. Opschoor (1994), "Energy Conservation and Investment Behaviour: an Empirical Analysis of influential Factors and Attitudes", Ecological Economcis Group, Department of Spatial Economics, Free University Amsterdam, Amsterdam.

Gruber, E. and M. Brand (1991), "Promoting Energy Conservation in Small and Medium-Sized Comapnies", Energy Policy(April), p279-287.

Hein, L. and K. Blok (1995), "Transaction Costs of Energy Efficiency Improvements", Presented at Proceedings of the 1995 ECEEE Summer Study: Stustainability and the Reinvention of Government - A Challenge for Energy Efficiency, The European Council for an Energy-Efficient Economy, Stockholm, Sweden.

Houghton, J.T., G.J. Jenkins and J.J. Ephraus *(Eds.)* (1990), *Climate Change, The IPCC Scientific Assessment*, Cambridge University, Cambridge.

Jaffe, A.B. and R.N. Staffe (1994), "The Energy-Efficiency Gap - What Does it Mean?", Energy Policy **22**(10), p804-810.

Koot, L.W., M.H. Brascamp, G. Gerritse, G.J.A.M. Meijer, A.G. Melman and H.J. Munter (1984), "Evaluatie van de werking en effectiviteit van de energietoeslag op de WIR-basispremie en de regeling geldelijke steun beperking energieverbruik in de non-profit sector", Netherlands Organization for Applied Scientific Research TNO, Apeldoorn.

Krause, F., E. Haites, R. Howarth and J. Koomey (1993), "Energy Policy in the Greenhouse, Volume Two, Part 1 - Cutting Carbon Emissions: Burden or Benfit?", International Project for Sustainable Energy Paths, El Cerrito (CA).

Larson, E.D. and L.J. Nilsson (1991), "Electricity Use and Efficiency in Pumping and Air-Handling Systems", ASHRAE Transactions **97**(2), p363-377.

Lövgren, K. (1993), "Economic Instruments for Air Pollution Control in Sweden", Swedish Environmental Protection Agency, Solna, Sweden.

Ministry of Economic Affairs (1993), *Second Memoreandum on Energy Conservation,* Sdu, Uitgeverij Platijnstraat, The Hague.

RIVM (1991), *Nationale Milieuverkenning 2 1990-2010,* Samsom H.D. Tjeenk Willink, Alpen aan den Rijn.

Steering Committe Regulation Energy Taxes (1992), "Regulating Energy Taxes", Ministry of Economic Affairs, The Hague.

Train, K. (1985), "Discount rates and consumers' energy-related decisions: a review of literature", Energy - The International Journal **10**(12), p1243-1253.

Velthuisen, J.W. (1995), "Determinants of Investment in Energy Conservation",PhD Thesis, Groningen University, Groningen, The Netherlands.

Worrell, E., J.G. de Beer, R.F.A. Cuelenaere and K. Blok (1992), "ICARUS - The Potential of Energy Conservation in the Netherlands up to the year 2000", Department of Science, Technology and Society, Utrecht University, Utrecht.

Chapter 4

Anonymous (1994), *CADDET Energy Efficiency Newsletter*, issue 3, 16.

Attwood, B.W. (1996), personal communication, St. Annes Paper and Paperboard Developments Ltd, UK, October 29th 1996.

Back, E.L. (1991), *Tappi Journal,* **74,** 135.

Baum, G.A., Malcolm, E.W., Wahren, D., Swanson, J.W., Easty D.B., Litvay, J.D. and Dugal, H.S. (1981), in *Kirk-Othmer Encyclopedia of Chemical Technology*, 3ed. Wiley, New York, p. 769.

Beer, J.G de, Wees, M.T. van, Worrell, E. and K. Blok (1994), ICARUS-3 - The Potential of Energy Efficiency Improvement in The Netherlands up to 2000 and 2015. report 94013, Department of Science, Technology and Society, Utrecht University, The Netherlands.

Blok, K., Turkenburg, W.C., Eichhammer, W., Farinelli, U. and Johansson, T.B. (1995), Overview of Energy RD&D Options for a Sustainable Future. EUR 16829 EN, European Commission, DG XII, Science, Research and Development, Luxembourg.

Boyd, G. (1986), personal communication, Argonne National Laboratory, April 1996.

Brown, H.L., Hamel, B.H. and Hedman, B.A. (1985), Energy Analysis of 108 Industrial Processes. DoE-contract E(11-1)2862., Department of Energy, Washington DC.

CEC (1983), Energy audit No 3, Pulp, Paper and Board Industry in the European Economic Community (EUR 8792 en), Commission of the European Communities. Directorate-General Information Market and Innovation, Luxembourg.

CEC (1992), 1992 Energy Technology Status Report, California Energy Commission. Sacramento, CA, USA.

Cox, J. (1992), *American Papermaker,* April 1992, 30.

Cutshall, K. and Hutspeth, D. (1987), Hot Pressing. *TAPPI Seminar Notes, 1987 Practical Aspects of Pressing and Drying*, pp. 169-182, TAPPI Press, Atlanta.

Deventer, H. van (1995), personal communication, TNO Institute of Environmental Sciences, Energy Research and Process Innovation, Apeldoorn, The Netherlands, 8 November 1995.

ETSU (1984), Energy Technology Support Unit, UK Department of Energy, *Energy Use and Energy Efficiency in UK Manufacturing Industry up to the Year 2000, volume 2.* Her Majesty's Stationary Office, London.

ETSU (1994) Energy Technology Support Unit, UK Department of Energy, *An Appraisal of UK Energy Research, Development, Demonstration & Dissemination.* Her Majesty's Stationary Office, London.

FAO (1994), Yearbook Forest Products 1993. Food and Agricultural Organization, Rome.

Farla, J., Blok, K. and Schipper, L. (1997), *Energy Policy,* 1997, **25**.

Gavelin, G. (1982), *Drying Technology, Special Handbook X-710* (In Swedish). The Swedish Forest Industries Association, Markaryd, Sweden.

Grubb, M., Edmonds, J., Brink, P. ten, and Morrison, M. (1993), *Annual Review of Energy and Environment,* **18,** 397.

Gunderson, D.E. (1991), An Overview of Press-Drying, Impulse-Drying and Condebelt-Drying Concepts. *Proceedings of The Helsinki Symposium on Alternate Methods of Pulp and Paper Drying.* Helsinki.

Huigens, W.P. (1984), Crown Gelder Heat/Power Installation. Thomassen International, Rheden.

IEA (1994), IEA/OECD Scoping Study: Energy and Environmental Technologies to Respond to Global Climate Change Concerns. OECD/IEA, Paris.

IEA (1996), Comparing Energy Technologies. OECD/IEA, Paris.

Kerttula, R. (1994), New Generation Paper Machines. *Proceedings of Valmet Paper Machine Days 1994,* pp. 101-104, Jyväskylä, Finland.

Komppa, A. (1993a), in *Energy Efficiency in Process Technology,* ed. P.A. Pilavachi. Elsevier Applied Science, London, 1993, p. 101.

Komppa, A. (1993b), personal communication, The Finnish Pulp and Paper Research Institute, 16 April 1993.

Lehtinen, J. (1993), Energy Analysis and Measurements Carried Out During the Year 1992 in Connection with the Condebelt Project of Valmet-Tampella. Valmet-Tampella Company Report 2/93, P.O. Box 267, SF-33101 Tampere, Finland.

Lehtinen, J. (1995a), *Japan Tappi Journal,* June 1995.

Lehtinen, J. (1995b), *Paper Technology,* October 1995, 67.

Lehtinen, J. (1996), personal communication, Valmet-Tampella, Tampere, Finland, April 24 1996.

Lovins, A.B. and Lovins, L.H. (1991), *Annual Review of Energy and Environment,* **16,** 433.

Maier, W. (1986), *Rationelle Energieverwendung durch neue Technologien* (in German). Verlag TÜV Rheinland, Köln.

Melman, A.G., Boot, H. and Gerritse, G. (1990), Energiebesparingspotentiëlen 2015 (in Dutch), TNO-MT, Postbus 342, 7300 AH Apeldoorn, The Netherlands.

Mujumdar, A.S. (1995), in *Handbook of Industrial Drying,* ed. A.S. Mujumdar. Marcel Dekker, New York, p. 1071.

Mulder, B.M.P. and Sinon, A.M.J. (1994), *CADDET Energy Efficiency Newsletter,* 1994, issue 3, 10.

Nederveen, C.J., Hoek, A.L.v.S.-v. and Dijkstra, J.J.F.M. (1991) *Paper Technology*, November, 30.

Nieuwlaar, E. (1998) Developments in Energy Analysis, PhD thesis, Utrecht University.

Nilsson, L.J., Larson, E.D., Gupta, A. and Gilbreath, K. (1996), Energy efficiency and the Pulp and Paper Industry, ACEEE. Washington DC.

Poirier, N.A., Crotogino, R.H., Mujumdar, A.S. and Douglas, W.J.M. (1994), *Journal of Pulp and Paper Science,* 1994, **20,** J97.

Polat, O. and Mujumdar, A.S. (1995), in *Handbook of Industrial Drying*, ed. A.S. Mujumdar, Marcel Dekker, New York, p. 861.

Reese, R.A. (1988), *Pulp & Paper Canada*, **89,** 97.

Schareman, M. (1995), personal communication, Parenco Newspaper, Renkum, 13 November 1995.

Smit, R., Beer, J. de, Worrell, E. and Blok, K. (1994), Long-Term Industrial Energy-Efficiency Improvement: Technology Descriptions. report 94076, Department of Science, Technology and Society, Utrecht University, The Netherlands.

Smith, E. (1993), personal communication, Institute of Paper Science and Technology, Appleton, Wisconsin, USA, 19 March 1993.

Stubbing, T.J. (1990), *Paper Technology*, June, 36.

Stubbing, T.J. (1993a), *Trans IChemE* ,1993, 71 Part A, 488.

Stubbing, T.J. (1993b), Continuous Drying in Superheated Steam. UK Patent Application No. 9317727, 26 August 1993.

Sundqvist, H. (1994), Air Systems for the Future Drying Section. *Proceedings of Valmet Paper Machine Days 1994,* pp. 39-42, Jyvsäkylä, Finland.

Stubbing, T.J. (1996), personal communication, Heat-Win Ltd., UK, October 28th 1996.

Tissari, M. (1994), Peak Productivity on High Speed Machines. *Proceedings of Valmet Paper Machine Days 1994*, pp. 93-100, Jyväskylä, Finland.

Tolonen, T. (1994), Developments in the Dryer Section. *Proceedings of Valmet Paper Machine Days 1994,* pp. 33-37, Jyväskylä, Finland.

Unkila, K., Lehtinen, J. and Juntunen, T. (1991), An Assessment of the Quality Aspects of Condebelt Dried Board and Paper. *Proceedings of Helsinki Symposium on Alternate Methods of Pulp and Paper Drying*. pp. 237-278, Helsinki.

Vincent, J.P. (1993), in *Energy Efficiency in Process Technology*, ed. P.A. Pilavachi. Esevier Applied Science, London, p. 230.

VNP (1995), Annual Review 1994 (in Dutch). Cooperation of the Dutch Paper and Board Manufactures, P.O. Box 3009, 2001 PA Haarlem.

Wahlstrom, B. (1991), *Paper Technology,* February, 18.

WEC (1995) Energy Efficiency Improvement Utilizing High Technology. World Energy Council, London.

Wiedenback, R.D. (1987), Steam Heated Dryers. *TAPPI Seminar Notes, 1987 Practical Aspects of Pressing and Drying*, pp. 241-249, TAPPI Press, Atlanta.

Chapter 5

Anonymous (1990), "Revolution in steel production *(Revolutie in staalproduktie)*", Intermediair **26**(41 (In Dutch)), p41.

Anonymous (1996a), "DIOS Pilot Plant Operation Ends, Process Awaits Commercialization", Iron & Steelmaker **23**(3), p9.

Anonymous (1996b), "Today's Steel Technology", Advanced Materials & Processes **149**(1), p32-35.

Anonymous (1997a), "Hoogovens Considers Combined Casting-Milling (Hoogovens denkt aan gietwalsen)", Trouw, p(In Dutch).

Anonymous (1997b), "Thyssen Makes Considerable Investments in Thin Slab Casting (Thyssen investeert fors in dunne-plakgieten)", Technisch Weekblad **11**(28 (In Dutch)).

Badra, C. (1995), "Industry, Trade and Technology Review", Office of Industries - United States International Trade Commission, Washington DC.

Beer, J. de, E. Worrell and K. Blok (1997), "Long-Term Energy-Efficiency Improvements in the Paper and Board Industry", Energy - The International Journal **23**(1), p21-42.

Bisio, G. (1993), "Exergy Method for Efficient Energy Resource Use in the Steel Industry", Energy - The International Journal **18**(9), p971-985.

Bisio, G. (1997), "Energy Recovery from Molten Slag and Exploitation of the Recovered Energy", Energy - The International Journal **22**(5), p501-509.

Bisio, G. and S. Poggi (1990), "Efficient Energy Resource Use in the Steel Industry with Particular Reference to Italy", 25th Intersociety Energy Conversion Engineering Conference, New York.

Bisio, G. and S. Poggi (1991), "The Possible Uitlization of the Thermal Energy Discharged from Sintering Plants of Steel Works", 26th Intersociety Energy Conversion Engineering Conference, New York.

Bock, M., G.A. Boyd, S.H. Karlson and M. Ross (1994), "Best Practice Electricity Use in Steel Minimills", Iron & Steelmaker **21**(5), p63-66.

Bosley, J.J., J.P. Clark, T.E. Dancy, R.J. Fruehan and E.H. McIntyre (1987), "Tcchnoeconomic Assessment of Electric Steelmaking Through the Year 2000", Center for Metals Production, Pittsburgh.

Boustead, I. and G.F. Hancock (1979), *Handbook of Industrial Analysis,* John Wiley & Sons, New York.

Chatterjee, A. (1996), *Beyond the Blast Furnace,* CRC Press, Boca Raton.

Cheeley, R.B., E. Eichberger and A. Bolkart (1996), "COREX/MIDREX On the Way to Commercialization", Direct from Midrex(4), p8-9.

Considine, I. and D. Maxwell *(Eds.)* (1974), *Chemical and Process Technology Encyclopedia*, McGraw-Hill, New York.

COREX (1996), "COREX - Revolution in Ironmaking", Voest-Alpine Industrieanlagenbau GmbH, Linz.

Cusack, B.L., G.S. Wingrove and G.J. Hardie (1995), "Initial Operation of the HISmelt Research and Development Facility", Iron and Steelmaker **2**(22), p13-20.

Dijk, K.M. van, R. van Dijk, V.J.L van Eekhout, H. van Hulst, W. Schipper and J.H. Stam (1994), "Methanol from Natural Gas", Delft University of Technology, Delft.

Dungs, H and U. Tschirner (1994), "Energy and Material Conversion in Coke Dry Quenching Plants as Found in Existing Facilities", Coke Making International **6**(1), p19-29.

Eberle, A., W. Schiffer and S. Siuka (1996), "Start-Up and First Operational Results of the COREX Plant C-2000 at POSCO's Pohang Works", VOEST-ALPINE Industrieanlagenbau GmbH, Linz.

EZ (1994), "From World Market to End User (Van Wereldmarkt tot Eindverbruiker)", Ministry of Economic Affairs, The Hague.

Farley, J.M. and P.J. Koros (1992), "Progress Towards Direct Steelmaking in the USA", Steel Times International(March), p22-23.

Faure, H. (1993a), "The New Technologies for Iron and Steelmaking", 27th IISI Annual Meeting and Conference, Paris, IISI.

Faure, H. A. (1993b), "Development, State of the Art and Future Aspects of Steelmaking", Metallurgical Plant and Technology International **16**(3), p32-41.

Flemming, G. (1995), Personal Communication, SMS Schloemann-Siemag, 30 May 1995.

Flemming, G., P. Kappes, W. Rohde and L. Vogtmann (1988), "Rolling of Continuously Cast Strips and the Technical Consequences With Respect to the Design of Hot Strip Production Plants", MPT **11**(1), p16-35.

Fruehan, R.J. (1994), "Effect of Emerging Technologies on the Competitiveness in the Steel Industry", Iron & Steelmaker **21**(2), p17-21.

Furukawa, T. (1994), "5,000 Daily Tons of Direct Ion-Ore Smelting by 2000", New Steel **10**(11), p36-38.

Ghamarin, A. and A.B. Cambel (1982), "Exergy Analysis of Illinois No. 6 Coal", Energy The International Journal **7**(6), p483-488.

Gielen, D.J. and A.W.N. van Dril (1997), "The Basic Metal Industry and its Energy Use", ECN, Petten, The Netherlands.

Gomez, M., J.G. Gayle and A.R. taylor (1965), "Heat Content on Specific Heat of Coals and Related Products", US Bureau of Mines Report of Investigation 66076, Washington.

Gool, W. van *(Ed.)* (1986), *Poly-energy Handbook (Poly-energyie zakboekje)*, PBNA, The Hague.

Gretz, J., W. Korf and R. Lyons (1991), "Hydrogen in the Steel Industry", Int. J. Hydrogen Energy **16**(10), p691-693.

Hammersley, G. (1973), "The Charcoal Iron Industry and its Fuel", Econ. Hist. Rev., p593-613.

Heal, D.W. (1975), "Modern Perspectives on the History of Fuel Economy in the Iron and Steel Industry", Ironmaking and Steelmaking (Quarterly) **4**, p222-227.

Hendricks, C. (1995), "Strip Casting - A Revolution in the Steel Industry?", Metallurgical Plant Technology **18**(3), p42-49.

Hendriks, C. (1994), "Carbon Dioxide Removal from Coal-Fired Power Plants", Thesis, Utrecht University,

Hofer, L. (1996), "Steel Production Technologies - Fuchs and VAI Have Joined Forces", Steel Standard - VAI Steelmaking Technology **7**(6), p2-3.

Hofer, L. (1997a), Personal Communication, Voest Alpine Industrieanlagenbau GmbH, Linz, 23 May 1997.
Hofer, L. (1997b), "Electric Steelmaking with FUCHS Shaft Furnace Technology", VAI, Linz, Austria.
Hoffman, J.P. (1992), "Presidential Address: Oxygen-coal In-bath Smelting Reduction - a Future Process for the Production of Iron and Stainless Steel?", Journal of the South African Institute of Mining and Metallurgy **92**(8/9), p253-273.
Hoogovens (1995), "Cyclone Converter Furnace - leaflets", Hoogovens Staal, IJmuiden, Netherlands.
Hoogovens (1996), "Environmental Report 1995 (Milieujaarverslag 1995)", Hoogovens Staal (In Dutch), IJmuiden.
Hyde, C.K. (1977), *Technological Change and the British Iron Industry, 1700-1870,* Princeton University Press, Princeton.
IEA (1993), "Energy Prices and Taxes", IEA, Paris.
IISI (1982), "Energy and the Steel Industry", International Iron and Steel Institute, Brussels.
IISI (1990), "Statistics on Energy in the Steel Industry (1990 Update)", International Iron and Steel Institute, Brussels.
IISI (1992), "Steel Statistics Yearbook 1992", International Iron and Steel Institute, Brussels.
IISI (1996a), "Statistics on Energy in the Steel Industry (1996 Update)", International Iron and Steel Institute, Brussels.
IISI (1996b), "Steel Statistics Yearbook 1995", International Iron and Steel Institute, Brussels.
Innes, J.A. (1995), "From Direct Reduction to Direct Smelting - Options for an Expanding China", Pacific Economic Co-operation Council Minerals and Energy Forum, Beijing.
Jeans, J.S. (1882), "On the Consumption and Economy of Fuel in the Iron and Steel Manufacture", Journal of the Iron and Steel Institute **3**, p128-179.
Juleff, G. (1996), "An Ancient Wind-Powered Iron Smelting Technology in Sri Lanka", Nature **379**(4 January), p60-63.
Kirk-Othmer (1981), , John Wiley & Sons, New York.
Kreulitsch, H., W. Egger, H. Wiesinger and A. Eberle (1993), "Iron and Steelmaking of the Future", Steel Times **221**(5), p217-220.
Kruger, D. (1995), "CPR- A combined Casting/Rolling Process for Producing Steel Strip", Iron and Steel Engineer **72**(45), p31-36.
Kudrin, V.A. (1985), *Steel Making,* Mir Publishers, Moscow.
Langen, J. van, K. Meijer, M. Corbett and G. Malgarini (1992), "The Cyclone Convertor Furnace", Journees Siderurgiques ATS 1992, Paris.
Lassat de Pressigny, Y. de (1993), "Jupiter - A Virgin Iron Production Process for Scarp Based Steel Making", Steel Times **221**(12), p513.
Maier, W. and G. Angerer (1986), *Rational Energy Use by New Technologies (Rationelle Energieverwendung durch neue Technologien),* Verlag TüV Rheinland GmbH (in German), Köln.

McIntyre, E.H. and E.R. Landry (1992), "Electric Arc Furnace Efficiency", The EPRI Center for Materials Production, Pittsburgh.

Meijer (1995), Personal Communication, Hoogovens Corporate Research Laboratorium, IJmuiden, 17 July 1995.

Meijer, H.K.A., C.P. Teerhuis, G.A. Flierman and R. Boom (1995), "The Cyclone Converter Furnace - Recent Developments", La revue de Metallurgie - ATS-JS, p42-43.

Meijer, K. (1996), Personal Communication, Hoogovens Corporate Research Laboratorium, IJmuiden, 13 June 1996.

Meijer, K. (1997), Personal Communication, Hoogovens Corporate Research Laboratorium, IJmuiden, 17 March 1997.

Midrex (1991), "Midrex - Economical Ironmaking Alternatives for Modern Steelmaking", Midrex Direct Reduction Corporation, Charlotte, NC, USA.

Midrex (1996), *1995 World Direct Reduction Statistics,* Midrex Direct Reduction Corporation, Charlotte, NC, USA.

Millbank, P. (1995), "Direct Route to Iron Gathers Momentum", Metal Bulletin Monthly (Metals Technology Supplement)(April), p21, 24-25.

Mott, R., D. Chase, F. Hofmann and F.J. Küper (1994), "CSP plant Nucor Steel, Hickman: The Outstanding Performance and the Expansion to Two Strands", 2nd European Conference on Continuous Casting, Düsseldorf, June 20-22, The German Iron and Steel Institute.

Nieuwlaar, E. (1996), "Enerpack 5", Utrecht University, Dept. of Science, Technology and Society, Utrecht.

Norris, P (1996), Internet Publication on Page made on August, 30, 1996 1996, IISI, Brussel.

Parodi, G.G. (1993), "Near Net Shape Casting of Flat Products", 27th IISI Annual Meeting and Conference, Paris, IISI.

Patuzzi, A., K. Antlinger, H. Grabner and W. Krieger (1990), "Comparative Study of Modern Scrap Melting Processes", The Sixth International Iron and Steel Congress, Nagoya, Japan, Iron and Steel Institute Japan.

Porterfield, W.W. (1984), *Inorganic Chemistry,* Addison-Wesley Publishing Company, Reading, UK.

Prideaux, R.N. (1996), "The HIsmelt Process; Premium Grade Metallicss for Asia", SEAISI 25th Anniversary Conference, Bangkok.

Puehringer, O., H. Wiesinger, B.H.P. Havenga, R. Hauk, W.L. Kepplinger and F. Wallner (1991), "Practical Experience with the Corex-process and its Potential for Development (Betriebserfahrungen mit dem Corex-Verfahren und dessen Entwicklungspotential)", Eisen und Stahl **111**(9), p37-44 (In German).

Ree, R. van, A.B.J. Oudhuis, A. Faaij and A.P.W.M. Curvers (1995), "Energy from Biomass: An Assessment of Two Promising Systems for Energy Production - Report 3.3: Modelling of a BIG/CC System With Aspen+.", Netherlands Energy Research Foundation, Petten, The Netherlands.

Rohde, W. and G. Flemming (1995), "Stand, Leistungsvermögen und Weiterentwicklung der CSP-Technologie", Stahl und Eisen **115**(9), p89-100.

Rong, Shen Wen, Liu Jian and Gerhard Fuchs (1996), "Compact Mill Concepts in Electric Furnace Steel Making", Steel Standard - VAI Steelmaking Technology **7**(6), p9-14.

Rosenberg, N. (1982), *Inside the Black Box,* Cambridge University Press, Cambridge.

Rossillo-Calle, F., M.A.A. de Rezende, P. Furtado and D.O. Hall (1996), *The Charcoal Dilemma,* Intermediate Technology Publications, London.

Salak, D. (1995), *Ferrous Powder Metallurgy,* Cambridge International Science Publishing, Cambridge.

Siemens, C.W. (1873), "On the Manufacture of Iron and Steel by Direct Process", Journal of the Iron and Steel Institute **5**, p37-91.

Smith, T. (1995), "European Electric Steel Congress Attracts 1000 Delegates", Steel Times **223**(7), p272-273, 278.

SMS (1995), "CSP Technology Revolutionizing Hot Strip Production -Leaflet", SMS Schloemann-Siemag, Dusseldorf.

Soldaat, A. (1997), "Twelve Projects Against CO_2-emission Receive 280 million Dutch Guilders (Twaalf projecten tegen CO_2-uitstoot krijgen 280 miljoen gulden)", Energie- en Milieuspectrum(5), p28-29 (In Dutch).

Spielmann, P.E. (1924), *The Constituents of Coal Tar,* Longmans Green, London.

Stam, B. (1997), "Hoogovens Considers Thin Slab Casting (Hoogovens overweegt dunne-plakgieten)", Technisch Weekblad **11**(28), p1.

Stelco (1993), "Present and Future Use of Energy in the Canadian Steel Industry", Efficiency and Alternative Energy Technology Branch, CANMET, Energy, Mines and Resources Canada, Ottawa.

Stepanov, V.S. (1993), *Analysis of Energy Efficiency of Industrial Processes,* Springer-Verlag, Berlin.

Szargut, J. and D.R. Morris (1987), "Cumulative Exergy Consumption and Cumulative Degree of Perfection of Chemical Processes", Energy Research **11**, p245-261.

Szargut, J., D.R. Morris and F.R. Steward (1988), *Exergy Analysis of Thermal, Chemical, and Metallurgical Processes,* Hemisphere Publishing Coporation, New York.

Szekely, J. and G. Trapaga (1994), "Some Perspectives on New Steelmaking Technologies", Metallurgical Plant and Technology International **17**(4), p30-47.

Takano, H., Y. Kitauchi and H. Hiura (1988), "Design for the 145MW Blast Furnace Gas Firing Gas Turbine Combined Cycle Plant", Gas Turbine and Aeroengine Congress and Exposition, Amsterdam, June 5-9, ASME.

Teoh, L.L. (1989), "Electric Arc Furnace Technology: Recent Developments and Future Trends", Ironmaking and Steelmaking **16**(5), p303-313.

Thümmler and R. Oberacker (1993), *Introduction to Powder Metallurgy,* Institute of Materials, London.

Tierney, B. and P. Linehan *(Eds.)* (1994), *Ullman's Encyclopedia of Industrial Chemistry,* VHC, Weinheim, Germany.

Tilton, J.E. *(Ed.)* (1990), *World Metal Demand*, Resources of the Future, Washington, D.C.

Tomassetti, G. (1995), *The Steel Industry in Italy: Thin Slab Casting in Cremona*, in: *Energy Efficiency Utilising High Techology*(ed.), World Energy Council, London.

Torlet, J.P. (1997), Personal Communication, Research & Development, Cockerill Sambre, Vivegnis, Belgium, 3 November.

VAI (1994), "Beam Blank Minimill for Stahlwerk Thueringen", Strand News **8**, p6-7.

Verweij (1997), Personal Communication, Hoogovens, IJmuiden, 1 April.

Weast, R.C. *(Ed.)* (1983), *CRC Handbook of Chemistry and Physics (64 ed.)*, CRC, Boca Raton.

WEC (1995), "Energy Efficiency Improvement Utilizing High Technology - An Assessment of Energy Use in Industry and Buildings", World Energy Council, London.

Weston, T.R. and M.W. Thompson (1996a), "Bright Future seen for the Romelt Process", MBM Metals Technology Supplement **308**(1 August), p6-9.

Weston, T.R. and M.W. Thompson (1996b), "The Romelt Process - Applications in the 21st Century Steel Industry", ICF Kaiser, Fairfax, Virginia, USA.

Worrell, E. (1994), "Potentials for Improved Use of Industrial Energy and Materials", Thesis, Utrecht University,

Worrell, E. (1995), "Advanced Technologies and Energy Efficiency in the Iron and Steel Industry in China", Energy for Sustainable Development **11**(4), p27-40.

Worrell, E. and J. de Beer (1991), "Gross Energy Requirements - Partial Report on Steel (Energiekentallen - Deelrapport Staal)", Castro Consulting Engineer (In Dutch), Amersfoort.

Worrell, E., J. de Beer and K. Blok (1993), *Energy Conservation in the Iron and Steel Industry*, in: *Energy Efficiency in Process Technology*, P.A. Pilavachi (ed.), Elsevier, Amsterdam.

Worrell, E., R.J.J. van Heijningen, J.F.M. de Castro, J.H.O. Hazewinkel, J.G. de Beer, A.P.C. Faaij and K. Vringer (1994), "New gross Energy-requirement figures for materials production", Energy - The International Journal **19**(6), p627-640.

Yemin, L. and X. Zhihong (1996), "Smelting Reduction Process for Iron and Chemicals", Laboratory of Computer Chemistry, Institute of Chemical Metallurgy, Chinese Academy for Sciences, Peking.

Yoshida, K., T. Kobayashi, M. Tanaka and T. Watanabe (1982), "Energy Savings in Continuous Casting", 4th International Iron and Steel Congress, London, May.

Zervas, T., J.T. McMullam and B.C. Williams (1996), "Gas-Based Direct Reduction Processes for Iron and Steel Production", International Journal of Energy Research **20**(2), p157-185.

Chapter 6

Adris, A.M., S.S.E.H. Elnashaie and R. Hughes (1991), "A Fluidized Bed Membrane Reactor for the Steam Reforming of Methane", The Canadian Journal of Chemical Engineering **69**, p1061-1069.

Aika, K.-i. and K. Tamaru (1995), *Ammonia Synthesis over Non-Iron Catalysts and Related Phenomena*, in: *Ammonia - Catalysis and Manufacture*, Nielsen (ed.), Springer Verlag, Berlin Heidelberg.

Alqahtany, H., D. Eng and M. Stoukides (1993), "Methane Steam Reforming over Fe Electrodes in a Solid Electrolyte Cell", Energy & Fuels **7**, p495-504.

Anonymous (1988), "Nitric Acid Process Technology", Fertilizer Focus(March), p13, 15.

Anonymous (1990a), "Nitric Acid Process Technology", Fertilizer Focus(March), p38-44.

Anonymous (1990b), "Reforming the Front End", Nitrogen(195), p22-36.

Anonymous (1993), "GT26 Promises 240 MWe of Record Efficiency", Modern Power Systems(October), ppp. 17-19.

Anonymous (1994), "Catalysts for Ammonia Manufacture", Fertilizer International(329), p32, 34-36, 38.

Anonymous (1997), "Petrochemical Processes '97", Hydrocarbon Processing(March), p108.

Appl, M. (1992a), "Modern Ammonia Technology: Where have we got to, where are we going?", Nitrogen(199), p46-75.

Appl, M. (1992b), "Modern Ammonia Technology: Where have we got to, where are we going? (part 2)", Nitrogen(200), p27-38.

Appleby, A.J. (1996), "Fuel Cell Technology: Status and Future Prospects", Energy - The International Journal **21**(7/8), p521-653.

Appleby, A.J. and F.R. Foulkes (1989), *Fuel Cell Handbook,* Van Nostrand Reinhold, New York.

Armor, J.N. (1995), "Membrane Catalysis: Where is it Now, What Needs to be Done?", Catalysis Today **25**, p199-207.

Atkins, P.W. (1983), *Physical Chemistry,* Oxford University Press, Oxford.

Ayres, R.U., W.H. Schlesinger. R.H. Socolow (1994), *Human Impacts on the Carbon and Nitrogen Cycles*, in: *Industrial Ecology and Social Change*, R.H. Socolow (ed.), Cambridge University Press, Cambridge.

Barbieri, G. and F.P. Di Maio (1997), "Simulation of Methane Steam Reforming Process in a Catalytic Pd-Membrane Reactor", Industrial & Engineering Chemical Research **36**(6), p2121-2127.

Barbieri, G., V. Violante, F.P. Di Maio, A. Criscuoli and E. Drioli (1997), "Methane Steam Reforming Analysis in a Palladium-Based Catalytic Membrane Reactor", Industrial & Engineering Chemical Research **36**(8), p3369-3374.

Beer, J. de, E. Worrell, K.Blok (1998), "Future Technologies for Energy-Efficient Iron and Steel Making", *Annual Review of Energy and Environment,* **23**, p. 123-205. .

Beer, J. de, E. Worrell and K. Blok (1998), "Long-Term Energy-Efficient Improvements in the Paper and Board Industry", Energy - The International Journal **23**, 1, p.21-42.

Beringer, J.E., N. Brewin and A.W.B. Johnston (1982), *Symbitioc Nitrogen Fixation in Plants*, in: *Bacteria and Plants*, M.E. Rhodeș-Roberts (ed.), Academic Press, London.

Blok, K. (1993), "The Development of Industrial CHP in the Netherlands", Energy Policy **21**(February), p158-175.

Blok, K. and W.C. Turkenburg (1994), "CO_2 Emission Reduction by Means of Industrial CHP in the Netherlands", Energy Conservation and Management **35**, p317-340.

Bodewes, H. and M.C. Lugten (1992), "A Cogeneration Unit at the Shell Refinery in Frederica", Comprimo Engineers and Contractors, Amsterdam.

Bryan, P.F. (1996), "Composite-Metal Membranes For High-Temperature Hydrogen Separations", Internet Publication on www.bendresearch.co, Page made on 30 September 1996, Bend Research, Bend OR.

Chauvel, A. and G. Lefebre (1989), *Petrochemical Processes,* Insitut Francais du Petrole Publications, Paris.

Clarke, S.J. and W.J. Mazzafro (1993), *Nitric Acid*, in: *Kirk-Othmer Encyclopedia of Chemical Technology*, J.I. Kroschwitz (ed.), vol. 17, John Wiley & Sons, New York.

Cremer, H. (1980), *Thermodynamic Balance and Analysis of a Synthesis Gas and Ammonia Plant*, in: *Thermodynamics: Second Law Analysis*, R.A. Gaggioli (ed.), ACS Symposium Series, vol. 122, American Chemical Society, Washington D.C.

Czuppon, T.A. (1994), "Advanced Ammonia Technologies", Fertiliser Industry - Annual Review **17**, p63-68.

Czuppon, T.A., S.A. Knez and J.M. Rovner (1993), *Ammonia*, in: *Kirk-Othmer Encyclopedia of Chemical Technology*, J.I. Kroschwitz (ed.), vol. 2, John Wiley & Sons, New York.

Czuppon, T.A., S.A. Knez and R.B. Strait (1996), "Commercial Experience Review of KAAP and KRES", Presented at AIChE Safety Symposium, Boston (MA).

Deshmukh, D. (1993), "Energy Savings and Capacity Utilization of Advanced Control Systems", Ammonia Plant Safety **33**, p278-288.

Dickman, S. (1987), "West German Release of Altered Bacteria Causes Furore", Nature **328**, p568.

Drioli, E. and L. Giorno (1996), "Catalytic Membrane Reactors", Chemistry and Industry(Jan 1), p19-22.

Dudnikov, Y.S., M.Y. Kuprikov and V.I. Vakhrushev (1988), "Production of Nitric Acid by the Plasma Chemical Method at the Point of Use", Soviet Chemical Industry **20**(10), p56-58.

Dybkjaer, I. (1995a), *Ammonia Production Processes*, in: *Ammonia - Catalysis and Manufacture*, Nielsen (ed.), Springer-Verlag, Berlin Heidelberg.

Dybkjaer, I. (1995b), "Tubular Reforming and Autothermal Reforming of Natural Gas - An Overview of Available Processes", Fuel Processing Technology **42**, p85-107.

Dybkjaer, I. and E.A. Gram (1984), "Energy Savings in Ammonia Synthesis - Design of Converters and High Activity Catalysts", Chemical Economy and Engineering Review **16**(9), p29-35.

Edlund, D.J. and J.R. Peterson (1995), "Hydrogen Separation and Purification using a Practical Metal Membrane", Presented at 6th Annual US Hydrogen Conference, Alexandria, Virginia.

Ermenc, E.D. (1956a), "Wisconsin Process Pebble Furnace Fixes Atmospheric Nitrogen", Chemical Engineering Progress **52**(4), p149-153.

Ermenc, E.D. (1956b), "Wisconsin Process System for Recovery of Dilute Oxides of Nitrogen", Chemical Engineering Progress **52**(11), p488-492.

FAO (1996), "FAO Fertilizer Yearbook", .

Foster, E.G. and F. Daniels (1951), "Recovery of Nitrogen Oxides by Silica Gel", Industrial and Engineering Chemistry **43**(4), p986-992.

Fryzuk, M., J.B. Love, S.J. Rettig and V.G. Young (1997), "Transformation of Coordinated Dinitrogen by Reaction with Dihydrogen and Primary Silanes", Science **275**(5305), p1445-1447.

Fujimoto, N., A. Saikusa, K. Hada and Y. Sudo (1992), *Safety Analysis and Considerations for HTTR Steam Reforming Hydrogen/Methanol Co-Production System*, in: *High Temperature Applications of Nuclear Energy*(ed.), IEAE, Oarai, Japan.

Gilbert, N. and F. Daniels (1948), "Fixation of Atmospheric Nitrogen in a Gas Heated Furnace", Industrial and Engineering Chemistry **40**(9), p1719-1723.

Gosnell, J. (1997), Personal Communication, Brown & Root, August 28, 1997.

Grotz, B.J. and L. Grisolia (1992), "Fifth Anniversary of Ammonia Unit E", Nitrogen(199), p39-45.

Haber, L.F. (1971), *The Chemical Industry 1900-1930 - Internationl Growth and Technological Change,* Clarendon Press, Oxford.

Hada, K., N. Fujimoto and Y. Sudo (1992), *Design of Steam Reforming Hydrogen and Methanol Co-Production System to be Connected to the HTTR*, in: *High Temperature Applications of Nuclear Energy*(ed.), IEAE, Oarai, Japan.

Heath, R., J. Mulckhuyse and S. Venkataram (1985), "The Potential for Energy Efficiency in the Fertilizer Industry", The World Bank, Washington, D.C.

Hendriks, C. (1994), "Carbon Dioxide Removal from Coal-Fired Power Plants",PhD Thesis, , Utrecht.

Hinderink, A.P., F.P.J.M. Kerkhof, A.B.K. Lie, J. De Swaan Arons and H.J. van der Kooii (1996), "Exergy Analysis With a Flowsheeting Simulator - II. Application; Synthesis Gas Production From Natural Gas", Chemical Engineering Science **51**(20), p4701-4715.

Hoffmeister, G. (1993), *Fertilizers*, in: *Kirk-Othmer Encyclopedia of Chemical Technology*, J.I. Kroschwitz (ed.), vol. 10, John Wiley & Sons, New York.

Honti, G.D. *(Ed.)* (1976), *The Nitrogen Industry*, Akademiaa Kiado, Budapest.

Ihde, A.J. (1964), *The Development of Modern Chemistry,* Harper & Row, New York.

Kinzig, A.P. and R.H. Socolow (1994), "Human Impacts on the Nitrogen Cycle", Physics Today(November), p24-31.

Koros, W.J. and G.K. Flemming (1993), "Membrane-Based Gas Separation", Journal of Membrane Science **93**, p1-80.

Krachtwerktuigen (1986), "Survey of Gas Turbines Available in the Netherlands", Krachtwerktuigen, Amersfoort, the Netherlands.

Kreith, F. and R.E. West *(Eds.)* (1997), *CRC Handbook of Energy Efficiency*, CRC Press, Boca Raton.

Kropff, M.J. and C.T.J. Spitters (1990), "Grondslagen Plantaardige Produktie", Vakgroep Theoretische Plantenkunde, Landbouwuniversiteit Wageningen, Wageningen.

Kugeler, K. and R. Schulten (1989), *Hochtemperaturreaktortechnik,* Springer-Verlag, Berlin.

Laciak, D.V. and G.P. Pez (1988), Ammonia Separation Using Ion Exchange Polymeric Membranes and Sorbents, US, Air Products and Chemicals, Inc.,

Lægsgaard Jørgensen, S., P.E. Højlund Nielsen and P. Lehrmann (1995), "Steam Reforming of Methane in a Membrane Reactor", Catalysis Today **25**(3/4), p303-307.

Laplaza, C.E. and C.C. Cummins (1995), "Dinitrogen Cleavage by a Three-Coordinate Molybdenum(III) Complex", Science **268**(12 May), p861-863.

LeBlanc, J.R. (1984), "Make Ammonia with less Energy", Hydrocarbon Processing(July).

LeBlanc, J.R. (1996), "Ammonia 2000 - Kellogg Technology for the Future", Presented at Asia Nitrogen '96, Singapore.

Leigh, G.J. (1991), *What Does the Future Hold? - A Survey of Possible Technological Developments*, in: *Catalytic Ammonia Synthesis*, J.R. Jennings (ed.), Fundamental and Applied Catalysis, M.V. Twigg, Plenum Press, New York.

Leigh, G.J. (1997), "Biological Nitrogen Fixation and Model Chemistry", Science **275**(5305), p1442.

Livingstone, J.G. and A. Pinto (1983), "The AMV process provides a simplified plant with reduced capital costs; optimum use is made of natural gas as feedstock source", Chemical Engineering Progress(May), p62-66.

Mittasch, A. (1951), *Geschichte der Ammoniaksynthese,* Verlag Chemie, Weinheim.

Moore, T. (1997), "Market Potential High for Fuel Cells", EPRI Journal(May/June), p6-17.

MEM-kast, Mulder, A. (1993), "Development of Metal-based Silica Membranes for Gas Separation",PhD Thesis, ,

Mulder, M. (1996), *Basic Principles of Membrane Technology,* Kluwer Academic Publishers, Dordrecht.

Nielsen, S.E. (1997), "Key Parameters Affecting Process Flexibility and Operating Costs", Haldor Topsoe, Copenhagen.

Nieuwlaar, E. (1988), "Developments in Energy Analysis",PhD Thesis, , Utrecht.

Nieuwlaar, E. (1996), "ENERPACK 5 - A Package for the Thermodynamic Analysis of Energy Systems", Utrecht Univeristy, Department of Science, Technology and Society, Utrecht.

OECD (1997), *CO_2 Emissions From Fuel Combustion,* OECD, Paris.

Pez, G.P. and D.V. Laciak (1988), Ammonia Separation Using Semipermeable Membranes, US, Air Products and Chemicals, Inc.,

Postgate, J. (1987), *Nitrogen Fixation,* London, Edward Arnold.

Radgen, P. (1996), "Pinch and Exergy Analysis of a Fertlizer Complex - Part 1", Nitrogen **224**, p39-48.

Radgen, P. (1997), "Pinch and Exergy Analysis of a Fertlizer Complex - Part 2", Nitrogen **225**, p27-39.

Riekert, L. (1974), "The Efficiency of Energy-Utilization in Chemical Processes", Chemical Engineering Science **29**, p1613-1620.

RIVM (1996), *Achtergronden bij de Milieubalans 96 (Background Information on the Environmental Balance 1996),* Samsom H.D. Tjeenk Willink, Alphen aan den Rijn (The Netherlands).

Rostrup-Nielsen, J.R. (1994), "Catalysis and Large-Scale Conversion of Natural Gas", Catalysis Today **21**, p257-267.

Saracco, G., G.F. Versteeg and W.P.M. van Swaaij (1994), "Current Hurdles to the Succes of High-Temperature Membrane Reactors", Journal of Membrane Science **95**, p105-123.

Shah, R.K. (1997), *Recuperators, Regenerators and Storage: Recuperators, Regenerators and Compact Heat Exchangers*, in: *CRC Handbook of Energy Efficiency*, F. Kreith and R.E. West (ed.), CRC, Boca Raton.

Shan, H., Y. Yang, A.J. James and P.R. Sharp (1997), "Dinitrogen Bridged Gold Clusers", Science **275**(5303), p1460-1462.

Smil, V. (1997), "Global Population and the Nitrogen Cycle", Scientific American(July 1997), p58-63.

Smit, R. and J. de Beer (1995), "Applicability of the High Temperature Nuclear Reactor in the Production of Methanol and Ammonia *(Inpasbaarheid van de hoge-temperatuurreactor in de produktie van methanol en ammoniak)*", Department of Science, Technology and Society, Utrecht.

Smith, M.D. (1996), *Plasma Technology*, in: *Kirk-Othmer Encyclopedia of Chemical Technology*, J. Kroschwitz (ed.), vol. 19, John Wiley & Sons, New York.

Sogge, J. and T. Strøm (1997), *Membrane Reactors - A New Technology for Production of Synthesis Gas by Steam Reforming*, in: *Natural Gas Conversion IV*, M. de Pontes, R.I. Espinoza, *et al.* (ed.), Studies in Surface Science and Catalysis, vol. 107, Elsevier Science, Amsterdam.

Soria, R. (1995), "Overview on Industrial Membranes", Catalysis Today **25**(3/4), p285-290.

Strøm, T., T. Pettersen, T. Sundset and J. Sogge (1997), "Use of Membrane Reactor in Production of Synthesis Gas - Process Options", Presented at ECCE-1, May 5-7, Florence, Italy.

Stryer, L. (1981), *Biochemistry,* W.H. Freeman and Company, San Francisco.

Szargut, J., D.R. Morris and F.R. Steward (1988), *Exergy Analysis of Thermal. Chemical and Metallurgical Processes,* Hemispere, New York.

Tamaru, K. (1991), *The History of the Development of Ammonia Synthesis*, in: *Catalytic Ammonia Synthesis - Fundamentals and Practice*, J.R. Jennings (ed.), Fundamentals and Applied Catalysis, M.V. Twigg and M.S. Spencer, Plenum Press, New York.

Tennison, S.R. (1997), "An Overview of the Joule Project - "Microporous Ceramic Membranes for Gas Separation Processes" - JOE-CT95-0018", Materials and Science Technology Int. Ltd., Henley Park, Guildford GU32AF, UK.

Timmins, R.S. and P.R. Ammann (1967), *Nitrogen Fixation*, in: *The Application of Plasma to Chemical Processing*, R.B. Baddour and R.S. Timmins (ed.), Pergamon Press, Oxford.

Topham, S.A. (1985), *The History of the Catalytic Synthesis of Ammonia*, in: *Catalysis, Science and Technology*, J.R. Anderson and M. Boudart (ed.), vol. 7, Springer-Verlag, Berlin.

Topsoe (1997), "Highlights of the Topsoe Low Energy Ammonia Process", Haldor Topsoe, Lyngby (Denmark).

Vayenas, C.G. and R.D. Farr (1980), "Cogeneration of Electric Energy and Nitric Oxide", Science **208**, p593-594.

VDI (1990), "VDI-Guidelines: Emission Control Nitric Fertilizers", Verein Deutsher Ingenieure, Düsseldorf.

Veen, H. van (1997), Personal Communication, Project Manager Inorganic Membranes, Netherlands Energy Research Foundation ECN - Unit Energy Efficiency, Petten, 23 October 1997.

Venugopolan, M. and S. Veprek (1983), *Kinetics and Catalysis in Plasma Chemistry*, in: *Plasma Chemistry IV*, F. Boschke (ed.), Topics in Current Chemistry, vol. 107, Springer-Verlag, Berlin.

Voet, E. van de, Klein R and H.A. Udo de Haes (1996), "Nitrogen Pollution in the European Uinio - Orgins and Proposes Soluitons", Environmental Conservation **23**(2), p120-132.

Weast, R.C. *(Ed.)* (1983), *CRC Handbook of Chemistry and Physics*, CRC Press, Boca Raton, Florida.

Weatherly (1997), "Nitric Acid Plant Reference List", Weatherly Inc., Chemateur Engineering Group, Atlanta.

WEC (1995), "Energy Efficiency Improvement Utilizing High Technology", World Energy Council, London.

Westerterp, K.R. (1993), *New Methanol Process*, in: *Energy Efficiency in Process Technology*, P.A. Plivachi (ed.), Elsevier Applied Science, London.

Westerterp, K.R., T.N. Bodewes, M.S.A. Vrijland and M. Kuczynski (1988), "Two New Methanol Converters", Hydrocarbon Processing(November), p6973.

Westerterp, K.R. and M. Kuczynski (1987), "A Model for a Countercurrent Gas-Solid-Solid Trickle Flow Reactor for Equilibrium Reactors. The Methanol Syntheis", Chemical Engineering Science **42**(8), p1871-1885.

Westerterp, K.R., M. M. Kuczymski and c.H.M. Kamphuis (1989), "Synthesis of Methanol in a Reactor with Interstage Product Removal", Ind. Eng. Chem. Res. **28**, p763-771.

Weston, C.W. (1993), *Ammonia Compounds*, in: *Kirk-Othmer Encyclopedia of Chemical Technology*, J.I. Kroschwitz (ed.), vol. 2, John Wiley & Sons, New York.

Williams, R.H., E.D. Larson, R.E. Katofsky and J. Chen (1995), "Methanol and Hydrogen from Biomass for Transportation, with Comparison to Methanol adn Hydrogen from Natural Gas and Coal", The Center for Energy and Environmental Studies, Princeton University, Princeton (NJ).

Wordragen, M. van (1996), "Bacteriën in plaats van kunstmest", Intermediair **32**(50), p43.

Worrell, E. and J.G. de Beer (1995), "Energy Savings in the Nitrogen Fertilizer Industry", Fertilizer Interantional **18**, p123-132.

Worrell, E. and K. Blok (1994), "Energy Savings in the Nitrogen Fertilizer Industry in the Netherlands", Energy **19**(2), p195-209.

Worrell, E., B. Meuleman and K. Blok (1995), "Energy Savings by Efficient Application of Fertilizer", Energy (accepted for publication).

Zaman, J. and A. Chakma (1994), "Inorganic Membrane Reactors", Journal of Membrane Science **92**, p1-28.

MIX
Papier aus verantwortungsvollen Quellen
Paper from responsible sources
FSC® C105338
FSC
www.fsc.org

If you have any concerns about our products,
you can contact us on
ProductSafety@springernature.com

In case Publisher is established outside the EU,
the EU authorized representative is:
Springer Nature Customer Service Center GmbH
Europaplatz 3, 69115 Heidelberg, Germany

Printed by Libri Plureos GmbH
in Hamburg, Germany